Control and Filtering of Fuzzy Systems Under Communication Channels

Xiao-Heng Chang · Jun Xiong · Zhi-Min Li ·
Bo Wu

Control and Filtering of Fuzzy Systems Under Communication Channels

Springer

Xiao-Heng Chang
School of Information Science and Engineering
Wuhan University of Science and Technology
Wuhan, Hubei, China

Jun Xiong
School of Information Science and Engineering
Wuhan University of Science and Technology
Wuhan, Hubei, China

Zhi-Min Li
School of Electronic and Control Engineering
North China Institute of Aerospace Engineering
Langfang, Hebei, China

Bo Wu
School of Information Science and Engineering
Wuhan University of Science and Technology
Wuhan, Hubei, China

ISBN 978-981-99-4345-6 ISBN 978-981-99-4346-3 (eBook)
https://doi.org/10.1007/978-981-99-4346-3

This Springer imprint is published by the registered company Springer Nature Singapore Pte Ltd.
The registered company address is: 152 Beach Road, #21-01/04 Gateway East, Singapore 189721, Singapore

Contents

List of Figures

Chapter 1
Introduction and Preview

1.1 Introduction

In the past two decades, increasing attention has been paid to deal with the analysis and design problems for nonlinear networked systems based on Takagi–Sugeno (T–S) fuzzy model and some interesting results have been published in the open literature. In this monograph, some novel advances on the study of nonlinear networked systems based on T–S fuzzy model will be given. Firstly, the various robust control strategies for nonlinear networked systems with dynamic quantization are proposed. Secondly, the various filtering problems are addressed for nonlinear networked systems subject to multiple communication constraints. Thirdly, the various output feedback controller design methods are presented for nonlinear networked systems with various communication constraints. Fourthly, in the presence of quantization and event-triggered communication scheme, the tracking control and security filtering schemes are developed for nonlinear networked systems, respectively. Finally, the quantized fault detection and fault-tolerant control problems are considered for nonlinear networked systems.

1.1.1 T–S Fuzzy Systems

It is generally known that nonlinearities are unavoidable in almost all practical control systems and it is difficult to deal with the analysis and design problems for nonlinear systems in both fields of control theory and control engineering. Therefore, considerable efforts have been made to investigate the nonlinear systems and different approaches have been proposed over the past several decades. Among the existing approaches, fuzzy logic control is regarded as one of the most effective and powerful strategies to investigate nonlinear systems [1]. As a result, the study of fuzzy logic control has received a great deal of attention from both scientists and engineers and the corresponding results have been reported in the existing literature, for

X. Chang et al., *Control and Filtering of Fuzzy Systems Under Communication Channels*,
https://doi.org/10.1007/978-981-99-4346-3_1

instance, T–S fuzzy model based control, adaptive fuzzy control, fuzzy proportional-integral-derivative (PID) control, and so on. In addition, the fuzzy logic control has been successfully found a great variety of practical applications, e.g., artificial intelligence and robotics, management and intelligent decisionmaking, motor industry and driver assistance system, etc.

Particularly, in the existing fuzzy logic control methods, the T–S fuzzy model-based control approach has been widely utilized to study nonlinear systems [2]. The main reason is that the considered nonlinear systems can be approximated as some local linear time-invariant system models connected by IF–THEN rules. In this way, the conventional linear system theory can be utilized to investigate the considered nonlinear systems. In fact, the study of the T–S fuzzy model can be traced back to the publication of pioneering work developed by Takagi and Sugeno in 1985 [3]. Subsequently, it has been proven that the T–S fuzzy model can approximate any smooth nonlinear function to any specified accuracy within any compact set. In recent years, more and more researchers have shown great interest in investigating nonlinear systems based on T–S fuzzy model and a number of significant results on the stability analysis, filter and controller designs have been reported. The stability analysis and stabilization problems were addressed for T–S fuzzy systems based on the parallel distributed compensation (PDC) strategy in [4]. By representing the interactions among the fuzzy subsystems in a single matrix and solving it by linear matrix inequality (LMI), new approaches to relaxed quadratic stability and stabilizability conditions for T–S fuzzy systems were proposed in [5]. Based on the fuzzy Lyapunov function approach, the stability conditions for open-loop fuzzy systems and stabilization conditions for closed-loop fuzzy systems were developed in [6]. Based on fuzzy Lyapunov functions and non-PDC control laws, new stabilization conditions were represented for continuous-time T–S fuzzy systems in [7]. By using new control laws and new nonquadratic Lyapunov functions, LMI-based relaxed non-quadratic stabilization conditions for discrete-time T–S fuzzy systems were given in [8]. The robust stabilization and $\mathscr{H}_\infty$ control problems for uncertain discrete-time T–S fuzzy systems were addressed in [9]. In [10, 11], the problem of robust static output feedback $\mathscr{H}_\infty$ control for discrete-time T–S fuzzy systems with time-varying norm-bounded uncertainties was considered. By considering the properties of system output matrices, an $\mathscr{H}_\infty$ static output feedback control strategy was developed for discrete-time T–S fuzzy systems in [12]. The observer-based output feedback $\mathscr{H}_\infty$ controller design problem was studied for continuous-time T–S fuzzy systems in [13]. In [14], the observer-based output feedback $\mathscr{H}_\infty$ tracking controller was designed for continuous-time T–S fuzzy systems. The problem of designing an output feedback controller with pole placement constraints for singular perturbed T–S fuzzy systems was researched in [15]. In [16], the authors discussed the problem of robust dynamic output feedback $\mathscr{H}_\infty$ control for uncertain discrete-time T–S fuzzy systems with time delays. Based on the piecewise Lyapunov function approach, the state feedback $\mathscr{H}_\infty$ control problem for continuous-time T–S fuzzy systems was addressed in [17]. By utilizing the descriptor representation approach, the observer-based output feedback $\mathscr{H}_\infty$ control problem was investigated for discrete-time T–S fuzzy systems in [18]. In [19], the nonfragile $\mathscr{H}_\infty$ filtering problem was studied for continuous-time T–S fuzzy systems with additive gain variations. By introducing more slack matrices, less-conservative results on $\mathcal{H}_\infty$ filtering for discrete-time T–S fuzzy systems were given in [20]. In [21], the $\mathcal{H}_\infty$ filter was designed for discrete-time T–S fuzzy systems in finite-frequency domain. The nonfragile $\mathcal{H}_\infty$ filter design approach for uncertain continuous-time T–S fuzzy systems with linear fractional parametric gain variations was developed in [22]. The nonfragile $\mathcal{H}_\infty$ filtering problem for continuous-time T–S fuzzy systems with multiplicative gain variations was considered in [23]. In [24], the problem of $\mathcal{H}_\infty$ filtering for continuous-time T–S fuzzy systems with D stability constraints was studied. The problem of robust $\mathcal{H}_\infty$ filtering was investigated for continuous-time T–S fuzzy systems in [25]. In [26], the $\mathcal{H}_\infty$ filtering problem was addressed for continuous-time T–S fuzzy systems with partly immeasurable premise variables. By using basis-dependent Lyapunov functions, the $\mathcal{H}_\infty$ filter design problem was solved for discrete-time T–S fuzzy systems in [27]. The resilient energy-to-peak filtering strategy was developed for continuous-time T–S fuzzy systems based on the two-step approach in [28]. In [29], the authors investigated the $\mathcal{H}_-/\mathcal{H}_\infty$ fault detection observer design problem for discrete-time T–S fuzzy systems with sensor faults and unknown bounded disturbances. The fault detection problem in the finite frequency domain was studied for continuous-time T–S fuzzy systems with sensor faults in [30]. The problem of robust $\mathcal{H}_\infty$ fault-detection filter design for uncertain continuous-time T–S fuzzy systems was discussed in [31]. In [32], the problem of state feedback fault-tolerant control for continuous-time T–S fuzzy systems with actuator faults was considered. The problem of robust fault estimation and observer-based output feedback fault tolerant control for continuous-time T–S fuzzy systems was investigated in [33].

In this monograph, we use the following T–S fuzzy model to represent the investigated nonlinear networked systems, in which the ith rule is described as follows:

Plant Rule i_{th}: **IF** $\varepsilon_1(t)$ is M_{1i}, $\varepsilon_2(t)$ is M_{2i}, and, ..., and $\varepsilon_d(t)$ is M_{di}, **THEN**

$$\begin{aligned} \mathcal{F}\,(x(t)) &= A_i x(t) + B_i u(t) + E_i w(t), \\ z(t) &= C_i x(t) + D_i u(t) + F_i w(t), \\ y(t) &= L_i x(t) + H_i w(t), \end{aligned} \tag{1.1}$$

where $\varepsilon(t) = [\varepsilon_1(t), \varepsilon_2(t), \ldots, \varepsilon_d(t)]$ is used to represent the premise variable, M_{pi} $(i = 1, 2, \ldots, r,\ p = 1, 2, \ldots, d)$ is used to denote the fuzzy sets, and r refers to the number of fuzzy rules. $\mathcal{F}\,(\cdot)$ stands for the time-derivative for continuous-time systems and forward operator for discrete-time systems. $x(t) \in \mathbb{R}^{n_x}$ stands for the state variable, $u(t) \in \mathbb{R}^{n_u}$ denotes the control input, $y(t) \in \mathbb{R}^{n_y}$ represents the measured output, $w(t) \in \mathbb{R}^{n_w}$ is the noise input which belongs to $\mathcal{L}_2[0, \infty)$ for continuous-time systems and $\ell_2[0, \infty)$ for discrete-time systems, and $z(t) \in \mathbb{R}^{n_z}$ is utilized to denote the controlled output. $A_i \in \mathbb{R}^{n_x \times n_x}$, $B_i \in \mathbb{R}^{n_x \times n_u}$, $E_i \in \mathbb{R}^{n_x \times n_w}$, $C_i \in \mathbb{R}^{n_z \times n_x}$, $D_i \in \mathbb{R}^{n_z \times n_u}$, $F_i \in \mathbb{R}^{n_z \times n_w}$, $L_i \in \mathbb{R}^{n_y \times n_x}$, and $H_i \in \mathbb{R}^{n_y \times n_w}$ for $i = 1, 2, \ldots, r$ are the given system matrices.

Denote

$$m_i(\varepsilon(t)) = \prod_{p=1}^{d} M_{pi}(\varepsilon_p(t)), i = 1, 2, \ldots, r, \tag{1.2}$$

where $M_{pi}(\varepsilon_p(t))$ stands for the grade of membership of $\varepsilon_p(t)$ in M_{pi}.

In this monograph, it is assumed that

$$m_i(\varepsilon(t)) > 0, \sum_{i=1}^{r} m_i(\varepsilon(t)) > 0, i = 1, 2, \ldots, r. \tag{1.3}$$

By formulating the fuzzy basis functions as

$$\rho_i(\varepsilon(t)) = \frac{m_i(\varepsilon(t))}{\sum_{j=1}^{r} m_j(\varepsilon(t))}, i = 1, 2, \ldots, r. \tag{1.4}$$

Then, it can be obtained that

$$\rho_i(\varepsilon(t)) \geq 0, \sum_{i=1}^{r} \rho_i(\varepsilon(t)) = 1, i = 1, 2, \ldots, r. \tag{1.5}$$

Moreover, the T–S fuzzy model in (1.1) can be inferred as follows:

$$\begin{aligned} \mathscr{F}(x(t)) &= A(\rho)x(t) + B(\rho)u(t) + E(\rho)w(t), \\ z(t) &= C(\rho)x(t) + D(\rho)u(t) + F(\rho)w(t), \\ y(t) &= L(\rho)x(t) + H(\rho)w(t), \end{aligned} \tag{1.6}$$

where

$$A(\rho) = \sum_{i=1}^{r} \rho_i(\varepsilon(t))A_i, \quad B(\rho) = \sum_{i=1}^{r} \rho_i(\varepsilon(t))B_i,$$

$$C(\rho) = \sum_{i=1}^{r} \rho_i(\varepsilon(t))C_i, \quad D(\rho) = \sum_{i=1}^{r} \rho_i(\varepsilon(t))D_i,$$

$$E(\rho) = \sum_{i=1}^{r} \rho_i(\varepsilon(t))E_i, \quad F(\rho) = \sum_{i=1}^{r} \rho_i(\varepsilon(t))F_i,$$

$$L(\rho) = \sum_{i=1}^{r} \rho_i(\varepsilon(t))L_i, \quad H(\rho) = \sum_{i=1}^{r} \rho_i(\varepsilon(t))H_i.$$

1.1.2 Networked Systems

Networked systems are a class of control systems in which the communication network is utilized to exchange information between different components. In the last two decades, as one of the main research topics in both academic and industrial fields, networked systems have been widely investigated by scholars at home and abroad. One reason for this trend is that networked systems have obvious advantages over well-known point-to-point control systems, such as low cost, reduced weight and power requirements, simple installation and maintenance, and high reliability [34]. Another reason is that networked systems have been successfully applied in a great deal of modern complicated industry processes, such as intelligent manufacturing, industrial automation, unmanned vehicles, and teleoperation systems [35]. In networked systems, the most important feature is that the communication network employed for the exchange of information is unreliable and bandwidth-limited. Therefore, a significant problem in networked systems is how to achieve the reasonable and efficacious utilization of limited communication resources. In addition, how to analyze and design networked systems with network-induced constraints caused by the unreliable and bandwidth-limited communication network is also a difficult problem to be investigated. At last, how to develop the security control and filtering strategies for networked systems against cyber attacks is another challenging problem to be addressed.

In the existing results, communication protocol, quantization strategy, and event-triggered strategy are regarded as the three most effective measures to realize the reasonable and efficacious utilization of the limited communication resources in networked systems. The communication protocol is able to decide which nodes have the privileges to transmit information at the prescribed transmission instant, which implies that the data collisions can be relieved to a large extent. Recently, a great number of achievements have been reported for networked systems with communication protocols. Among these results, there are mainly three types of communication protocols, i.e., weighted try-once-discard protocol [36], round-robin protocol [36, 37], and stochastic communication protocol [38]. For the quantization strategy, the amount of data in the transmission of information can be greatly reduced by introducing the quantizer, which implies that the transmission burden of the communication network can be relieved in a great measure. Up to now, there are mainly two quite different approaches to investigating networked systems with quantization in the literature, i.e., static quantization methodology and dynamic quantization methodology. In the static quantization methodology, the quantizer is considered to be static and time-invariant. While in dynamic quantization methodology, the quantizer is considered to be dynamic and time-varying. Based on static quantization methodology, the control and filtering problems have been addressed for networked systems in [39–44] and references therein. The study of networked systems based on the dynamic quantization methodology has also attracted considerable attention in the past few decades, see, for instance, [37, 45–55]. For the event-triggered strategy, the main idea is to design a suitable event-triggered condition to determine whether the signal should be transmitted to the next node or not. In contrast to the conventional time-triggered strategy where the signal is transmitted in a periodic manner, in the event-triggered strategy, the signal will be transmitted only when the presupposed event-triggered condition is satisfied. Therefore, the event-triggered strategy can effectively decrease the unnecessary waste of communication resources, which illustrates that the rational and efficacious utilization of limited communication resources can be realized. Recent years have witnessed a growing interest in studying networked systems based on event-triggered strategy. To mention a few, the event-triggered control problem was considered in [53–58] and the event-triggered filtering problem was studied in [59–61].

On the other hand, due to introducing unreliable and bandwidth-limited communication networks into feedback control loops, some network-induced constraints, such as channel fading, packet dropouts, and communication delays, are unavoidable in networked systems. Among all the network-induced constraints that emerged, packet dropouts and communication delays are considered to be two of the main causes for the performance deterioration or even the instability of the networked systems. For the study of networked systems subject to packet dropouts, there mainly exist two approaches to model the phenomenon of packet dropouts in the existing literature, i.e., Bernoulli stochastic processes and Markov chains. Over the past few decades, the study of networked systems with packet dropouts has attracted growing interest from numerous scholars and a great number of important results have been reported. To mention a few, the filtering problem has been considered in [44, 62–64] and the control problem has been investigated in [42, 43, 50, 51, 54, 55, 65–67]. In the past few decades, the investigation of networked systems with communication delays has also attracted considerable attention and a great deal of results have been published, for instance, [37, 43, 44, 51, 52, 58, 61, 67–69] and references therein.

In addition, the unreliable and bandwidth-limited communication networks used in most networked systems are open, which indicates that networked systems are vulnerable to cyber attacks. In the past several years, there have been increasing research interests in developing security control and filtering strategies for networked systems under cyber attacks, and some interesting results have been reported in the open literature. Among these results, there mainly exist two kinds of cyber attacks. One kind is called denial-of-service attacks and another kind is called deception attacks. For networked systems with denial-of-service attacks, security filtering and security control problems have been addressed in [70–72]. For networked systems with deception attacks, security filtering and security control problems have been investigated in [73–75].

1.1.3 Fuzzy-Model-Based Nonlinear Networked Systems

In recent years, the study of nonlinear networked systems based on the T–S fuzzy model has attracted a growing interest and a number of significant results have been proposed. Noticeable works include [76–116]. In the presence of weighted try-once-discard protocol, the sliding mode control problem was considered for T–S fuzzy networked systems and T–S fuzzy networked singularly perturbed systems with static quantization in [76, 77], respectively. For T–S fuzzy networked systems with round-robin communication protocol, the $\mathscr{H}_\infty$ PID control problem was addressed in [78] and the robust $\mathscr{H}_\infty$ fault detection problem was discussed in [79]. By considering the effect of stochastic communication protocol, the efficient model-predictive control strategy was given for T–S fuzzy networked systems in [80] and the robust dynamic output feedback control strategy was developed for T–S fuzzy networked systems subject to dynamic quantization in [81]. The peak-to-peak filtering problem was considered for T–S fuzzy networked systems with static quantization in [82] and the $\mathscr{H}_\infty$ filtering problem was addressed for T–S fuzzy networked systems with static quantization and packet dropouts in [83]. The output feedback control problem was investigated for T–S fuzzy networked systems with static quantization in [84, 85]. In [86], the nonfragile $\mathscr{H}_\infty$ state feedback control problem was considered for T–S fuzzy networked systems with static quantization. In the presence of input and output static quantization, the output feedback tracking control strategy was proposed in [87] and the event-triggered output feedback tracking control strategy was presented in [88]. For T–S fuzzy networked systems with static quantization and packet dropouts, the controller design problem was discussed in [89]. By taking static quantization, network induced delays, and packet dropouts into simultaneous consideration, the robust $\mathscr{H}_\infty$ control problem and guaranteed cost control problem were studied for T–S fuzzy networked systems in [90, 91], respectively. The $\mathscr{H}_\infty$ static output feedback control problem was researched for T–S fuzzy networked systems with input and output dynamic quantization in [92], where a novel adjusting scheme was proposed for the parameters of the dynamic quantizers. In [93], a guaranteed cost static output feedback control strategy was proposed for T–S fuzzy networked systems with input and output dynamic quantization. Both the full- and reduced-order energy-to-peak filtering problems were addressed for T–S fuzzy networked systems with dynamic quantization in [94]. For T–S fuzzy networked systems with input and output dynamic quantization, the observer-based output feedback control problem was studied in [95]. In the presence of input dynamic quantization, the robust state feedback control problem was investigated in [96]. In [97], the event-triggered output feedback tracking control strategy was presented for T–S fuzzy networked systems subject to input and output dynamic quantization. In [98], the problem of robust $\mathscr{H}_\infty$ filtering was investigated for T–S fuzzy networked systems with dynamic quantization and data dropouts. For T–S fuzzy networked systems with dynamic quantization, communication delays, and impulsive effects, the problem of output feedback control was discussed in [99]. In [100], the $\mathscr{H}_\infty$ state feedback control problem was solved for T–S fuzzy networked systems with dynamic quantization, communication delays, and packet dropouts. In the presence of dynamic quantization and stochastic cyber attacks, both the full- and reduced-order event-triggered $\mathscr{H}_\infty$ filters were designed for T–S fuzzy networked systems in [101]. In [102], the problem of event-triggered observer-based output feedback non-parallel distribution compensation control was researched for T–S fuzzy networked systems. An observer-based event-triggered sliding mode control strategy was proposed for T–S fuzzy networked systems in [103]. The reduced-order event-triggered energy-to-peak filtering problem was investigated for T–S fuzzy networked systems in [104]. For T–S fuzzy networked systems with packet dropouts, the $\mathscr{H}_\infty$ filter design problem was solved in [105] and the peak-to-peak filter design problem was considered in [106]. In the presence of multiple packet dropouts, the asynchronous output feedback $\mathscr{H}_\infty$ control strategy was developed for T–S fuzzy networked systems in [107]. For T–S fuzzy networked systems with packet dropouts, the state feedback $\mathscr{H}_\infty$ control problem was discussed in [108] and the static output feedback $\mathscr{H}_\infty$ control was studied in [109]. In [110], the dynamic output feedback $\mathscr{H}_\infty$ controller design problem was addressed for T–S fuzzy networked systems subject to multiple probabilistic communication delays and multiple packet dropouts. The robust stabilization problem was investigated for T–S fuzzy networked systems with communication delays in [111]. In [112], the fault detection problem was addressed for T–S fuzzy networked systems with communication delays. For T–S fuzzy networked systems with communication delays, both state feedback and output feedback fuzzy delay compensation control strategies were proposed in [113]. In the presence of stochastic deception attacks, the state feedback stabilization problem was studied for T–S fuzzy networked systems with quantization and hybrid-triggered scheme in [114]. For T–S fuzzy networked systems with denial-of-service attacks, a security event-triggered $\mathscr{H}_\infty$ filtering strategy was proposed in [115]. In [116], the security finite-time filter design problem was solved for T–S fuzzy networked systems subject to deception attacks, packet dropouts, and static quantization.

On the basis of the above results, the authors make a further study on controller and filter design problems for nonlinear networked systems based on the T–S fuzzy model in this monograph, which contains some novel technologies and new results. These include the fuzzy robust $\mathscr{H}_\infty$ control problem with dynamic quantization, which is addressed in Chap. 2, the fuzzy filtering problem with multiple communication constraints, which is considered in Chap. 3, the fuzzy output feedback control problem with communication constraints, which is discussed in Chap. 4, the event-triggered fuzzy control and filtering problems with communication constraints, which are investigated in Chap. 5, and the fuzzy fault detection and fault-tolerant control problems with quantization, which are researched in Chap. 6.

1.2 Preliminaries

The following lemmas will be needed for deriving the main results in this monograph.

Lemma 1.1 *[110] (Schur complement) For matrices Λ_1, Λ_2, and Λ_3 with $\Lambda_1 = \Lambda_1^T$ and $\Lambda_2 = \Lambda_2^T > 0$. Then, it can be obtained that $\Lambda_1 + \Lambda_3^T \Lambda_2^{-1} \Lambda_3 < 0$ if and only if*

$$\begin{bmatrix} \Lambda_1 & * \\ \Lambda_3 & -\Lambda_2 \end{bmatrix} < 0. \tag{1.7}$$

Lemma 1.2 *[82] Let $X_1 = X_1^T$, X_2, X_3, and Δ be real matrices with appropriate dimensions and $\Delta^T \Delta \leq I$. Then, it can be obtained that $X_1 + X_2 \Delta X_3 + X_3^T \Delta^T X_2^T < 0$, if and only if there exists a scalar $\lambda > 0$ such that*

$$X_1 + \lambda^{-1} X_2 X_2^T + \lambda X_3^T X_3 < 0. \tag{1.8}$$

Lemma 1.3 *[45] (S-procedure) Given quadratic functions for $\eta \in \mathbb{R}^m$, $\Pi_0(\eta) = \eta^T \Sigma_0 \eta$, $\Pi_1(\eta) = \eta^T \Sigma_1 \eta, \ldots, \Pi_\theta(\eta) = \eta^T \Sigma_\theta \eta$, $\Sigma_i = \Sigma_i^T, i = 1, 2, \ldots, \theta$. Then, it can be obtained that $\Pi_0(\eta) < 0$ with $\Pi_1(\eta) \geq 0$, $\Pi_2(\eta) \geq 0, \ldots, \Pi_\theta(\eta) \geq 0$, if there exist scalars $\varepsilon_1 > 0$, $\varepsilon_2 > 0, \ldots, \varepsilon_\theta > 0$ satisfying*

$$\Sigma_0 + \varepsilon_1 \Sigma_1 + \varepsilon_2 \Sigma_2 + \cdots + \varepsilon_\theta \Sigma_\theta < 0. \tag{1.9}$$

Lemma 1.4 *[10] For matrices M_1, M_2, M_3, and M_4 with appropriate dimensions. Then, it can be obtained that $M_1 + M_2 M_3 + M_3^T M_2^T < 0$, if there exists a scalar λ satisfying*

$$\begin{bmatrix} M_1 & * \\ \lambda M_2^T + M_4 M_3 & -\lambda M_4 - \lambda M_4^T \end{bmatrix} < 0. \tag{1.10}$$

Lemma 1.5 *[51] In terms of any positive definite matrix W, the matrices (or scalars) R and S with appropriate dimensions, the following inequality holds*

$$-2R^T S \leq R^T W^{-1} R + S^T W S. \tag{1.11}$$

Lemma 1.6 *[48] For matrices $X > 0$ and $Y > 0$ with same dimension, the following statements are equivalent:*

$$-Y^{-1} + X < 0, \tag{1.12}$$

$$Y - X^{-1} < 0. \tag{1.13}$$

1.3 Outline of the Monograph

This monograph is organized into 6 chapters, as follows.

Chapter 2 is devoted to studying the $\mathscr{H}_\infty$ control problem for T–S fuzzy systems with dynamic quantization. More specifically, a two-step design strategy is proposed to deal with the robust $\mathscr{H}_\infty$ state feedback control problem for uncertain discrete-time T–S fuzzy systems with control input quantization based on the fuzzy Lyapunov functional method in Sect. 2.1, a novel two-step design scheme is developed to investigate the non-fragile $\mathscr{H}_\infty$ state feedback control problem for discrete-time T–S fuzzy systems with the controller gain perturbation and state quantization in Sect. 2.2, the descriptor representation approach is employed to design the resilient $\mathscr{H}_\infty$ static output feedback controller for continuous-time T–S fuzzy systems with control input and measurement output quantization, parameter uncertainties, and controller gain perturbation in Sect. 2.3.

Chapter 3 is concerned with the filtering problem for T–S fuzzy systems with multiple communication constraints. More specifically, a novel ℓ_2–ℓ_∞ filtering strategy is presented for discrete-time T–S fuzzy systems with the effects of static quantization on the measurement output and the performance output in Sect. 3.1, the induced ℓ_∞ filter design problem is considered for discrete-time T–S fuzzy systems with the effects of data packet dropouts on the measurement output and the performance output in Sect. 3.2, the $\mathscr{H}_\infty$ filtering problem is discussed for discrete-time T–S fuzzy systems with the effects of dynamic quantization on the measurement output and the performance output in Sect. 3.3.

Chapter 4 is focused on the output feedback control problem for T–S fuzzy systems with communication constraints. More specifically, the static output feedback stabilization and $\mathscr{H}_\infty$ control problems are addressed for discrete-time T–S fuzzy systems with input and output dynamic quantization in Sect. 4.1, the static output feedback $\mathscr{H}_\infty$ controller design approach is developed for continuous-time T–S fuzzy systems with the effects of transmission delay and data dropouts in sensor-to-controller and controller-to-actuator communication channels in Sect. 4.2, and the dynamic output feedback $\mathscr{H}_\infty$ control problem is solved for discrete-time T–S fuzzy systems with the effect of stochastic communication protocol in the communication channel from the plant to the controller based on the fuzzy Markov jump model approach in Sect. 4.3.

Chapter 5 studies the event-triggered output feedback tracking control and event-triggered filtering problems for T–S fuzzy systems with different communication constraints. More specifically, the event-triggered static output feedback tracking control problem is investigated for discrete-time T–S fuzzy systems with input and output static quantization in Sect. 5.1, the event-triggered $\mathscr{L}_2$–$\mathscr{L}_\infty$ filtering problem is discussed for continuous-time T–S fuzzy systems with time-delay and DoS attacks based on the piecewise Lyapunov–Krasovskii functional approach in Sect. 5.2, and the event-triggered full- and reduced-order $\mathscr{H}_\infty$ filters design approach is proposed for discrete-time T–S fuzzy systems with dynamic quantization and stochastic deception attacks in Sect. 5.3.

Chapter 6 investigates the problems of fault detection and fault-tolerant control for T–S fuzzy systems with quantization. More specifically, the resilient $\mathscr{H}_\infty$ fault detection filter design approach is presented for uncertain discrete-time T–S fuzzy systems with dynamic quantization and filter gain perturbations in Sect. 6.1, the $\mathscr{H}_\infty$ fault-tolerant static output feedback control problem is studied for discrete-time T–S fuzzy Markov jump systems with actuator fault and dynamic quantization in Sect. 6.2, and the non-fragile guaranteed cost fault-tolerant state feedback control problem is considered for discrete-time T–S fuzzy systems with input dynamic quantization and actuator fault based on the two-step design approach in Sect. 6.3.

References

1. Feng G (2006) A survey on analysis and design of model-based fuzzy control systems. IEEE Trans Fuzzy Syst 14:676–697
2. Tanaka K, Wang HO (2001) Fuzzy control systems design and analysis: A linear matrix inequality approach. Wiley, New York
3. Takagi T, Sugeno M (1985) Fuzzy identification of systems and its applications to modeling and control. IEEE Trans Syst Man Cybern 15:116–132
4. Wang HO, Tanaka K, Griffin MF (1996) An approach to fuzzy control of nonlinear systems: Stability and design issues. IEEE Trans Fuzzy Syst 4:14–23
5. Kim E, Lee H (2000) New approaches to relaxed quadratic stability condition of fuzzy control systems. IEEE Trans Fuzzy Syst 8:523–534
6. Tanaka K, Hori T, Wang HO (2003) A multiple Lyapunov function approach to stabilization of fuzzy control systems. IEEE Trans Fuzzy Syst 11:582–589
7. Chang XH, Yang GH (2010) Relaxed stabilization conditions for continuous-time Takagi–Sugeno fuzzy control systems. Inf Sci 180:3273–3287
8. Guerra TM, Vermeiren L (2004) LMI-based relaxed nonquadratic stabilization conditions for nonlinear systems in the Takagi–Sugeno's form. Automatica 40:823–829
9. Cao YY, Frank PM (2000) Robust $\mathscr{H}_\infty$ disturbance attenuation for a class of uncertain discrete-time fuzzy systems. IEEE Trans Fuzzy Syst 8:406–415
10. Chang XH, Zhang L, Park JH (2015) Robust static output feedback $\mathscr{H}_\infty$ control for uncertain fuzzy systems. Fuzzy Sets Syst 273:87–104
11. Kau SW, Lee HJ, Yang CM, Lee CH, Hong L, Fang CH (2007) Robust $\mathscr{H}_\infty$ fuzzy static output feedback control of T–S fuzzy systems with parametric uncertainties. Fuzzy Sets Syst 158:135–146
12. Dong J, Yang GH (2009) Static output feedback $\mathscr{H}_\infty$ control of a class of nonlinear discrete-time systems. Fuzzy Sets Syst 160:2844–2859
13. Liu X, Zhang Q (2003) New approaches to $\mathscr{H}_\infty$ controller designs based on fuzzy observers for T–S fuzzy systems via LMI. Automatica 39:1571–1582
14. Tseng CS, Chen BS, Uang HJ (2001) Fuzzy tracking control design for nonlinear dynamic systems via T–S fuzzy model. IEEE Trans Fuzzy Syst 9:381–392
15. Assawinchaichote W, Nguang SK (2006) Fuzzy $\mathscr{H}_\infty$ output feedback control design for singularly perturbed systems with pole placement constraints: An LMI approach. IEEE Trans Fuzzy Syst 14:361–371
16. Xu S, Lam J (2005) Robust $\mathscr{H}_\infty$ control for uncertain discrete-time-delay fuzzy systems via output feedback controllers. IEEE Trans Fuzzy Syst 13:82–93
17. Feng G (2003) Controller synthesis of fuzzy dynamic systems based on piecewise Lyapunov functions. IEEE Trans Fuzzy Syst 11:605–612
18. Chang XH, Yang GH (2011) A descriptor representation approach to observer-based $\mathscr{H}_\infty$ control synthesis for discrete-time fuzzy systems. Fuzzy Sets Syst 185:38–51
19. Chang XH, Yang GH (2011) Nonfragile $\mathscr{H}_\infty$ filtering of continuous-time fuzzy systems. IEEE Trans Signal Process 59:1528–1538
20. Zhang H, Shi Y, Mehr AS (2012) On $\mathscr{H}_\infty$ filtering for discrete-time Takagi–Sugeno fuzzy systems. IEEE Trans Fuzzy Syst 20:396–401
21. Ding DW, Yang GH (2010) Fuzzy filter design for nonlinear systems in finite-frequency domain. IEEE Trans Fuzzy Syst 18:935–945
22. Chang XH (2012) Robust nonfragile $\mathscr{H}_\infty$ filtering of fuzzy systems with linear fractional parametric uncertainties. IEEE Trans Fuzzy Syst 20:1001–1011
23. Chang XH, Yang GH (2014) Nonfragile $\mathscr{H}_\infty$ filter design for T–S fuzzy systems in standard form. IEEE Trans Ind Electron 61:3448–3458
24. Nguang SK, Assawinchaichote W (2003) $\mathscr{H}_\infty$ filtering for fuzzy dynamical systems with D stability constraints. IEEE Trans Circuits Syst I Fundam Theory Appl 50:1503–1508
25. Feng G (2005) Robust $\mathscr{H}_\infty$ filtering of fuzzy dynamic systems. IEEE Trans Aerosp Electron Syst 41:658–670

26. Dong J, Yang GH (2017) $\mathscr{H}_\infty$ filtering for continuous-time T–S fuzzy systems with partly immeasurable premise variables. IEEE Trans Syst, Man, Cybern: Syst 47:1931–1940
27. Zhou S, Lam J, Xue A (2007) $\mathscr{H}_\infty$ filtering of discrete-time fuzzy systems via basis-dependent Lyapunov function approach. Fuzzy Sets Syst 158:180–193
28. Chang XH, Park JH, Shi P (2017) Fuzzy resilient energy-to-peak filtering for continuous-time nonlinear systems. IEEE Trans Fuzzy Syst 25:1576–1588
29. Chadli M, Abdo A, Ding SX (2013) $\mathscr{H}_-/\mathscr{H}_\infty$ fault detection filter design for discrete-time Takagi–Sugeno fuzzy system. Automatica 49:1996–2005
30. Li XJ, Yang GH (2014) Fault detection in finite frequency domain for Takagi–Sugeno fuzzy systems with sensor faults. IEEE Trans Cybern 44:1446–1458
31. Nguang SK, Shi P, Ding S (2007) Fault detection for uncertain fuzzy systems: An LMI approach. IEEE Trans Fuzzy Syst 15:1251–1262
32. Shen Q, Jiang B, Cocquempot V (2012) Fault-tolerant control for T–S fuzzy systems with application to near-space hypersonic vehicle with actuator faults. IEEE Trans Fuzzy Syst 20:652–665
33. Zhang K, Jiang B, Staroswiecki M (2010) Dynamic output feedback-fault tolerant controller design for Takagi–Sugeno fuzzy systems with actuator faults. IEEE Trans Fuzzy Syst 18:194–201
34. Zhang XM, Han QL, Ge X, Ding D, Ding L, Yue D, Peng C (2020) Networked control systems: A survey of trends and techniques. IEEE/CAA J Autom Sinica 7:1–17
35. Zhang L, Gao H, Kaynak O (2013) Network-induced constraints in networked control systems–A survey. IEEE Trans Industr Inf 9:403–416
36. Zou L, Wang Z, Gao H (2016) Set-membership filtering for time-varying systems with mixed time-delays under round-robin and weighted try-once-discard protocols. Automatica 74:341–348
37. Liu K, Fridman E, Johansson KH, Xia Y (2016) Quantized control under round-robin communication protocol. IEEE Trans Ind Electron 63:4461–4471
38. Guo G, Wang L (2015) Control over medium-constrained vehicular networks with fading channels and random access protocol: A networked systems approach. IEEE Trans Veh Technol 64:3347–3358
39. Elia N, Mitter SK (2001) Stabilization of linear systems with limited information. IEEE Trans Autom Control 46:1384–1400
40. Fu M, Xie L (2005) The sector bound approach to quantized feedback control. IEEE Trans Autom Control 50:1698–1711
41. Gao H, Chen T (2008) A new approach to quantized feedback control systems. Automatica 44:534–542
42. Su L, Chesi G (2018) Robust stability of uncertain linear systems with input and output quantization and packet loss. Automatica 87:267–273
43. Gao H, Chen T, Lam J (2008) A new delay system approach to network-based control. Automatica 44:39–52
44. Gao H, Chen T (2007) $\mathscr{H}_\infty$ estimation for uncertain systems with limited communication capacity. IEEE Trans Autom Control 52:2070–2084
45. Chang XH, Xiong J, Li ZM, Park JH (2018) Quantized static output feedback control for discrete-time systems. IEEE Trans Industr Inf 14:3426–3435
46. Chang XH, Huang R, Wang H, Liu L (2020) Robust design strategy of quantized feedback control. IEEE Trans Circuits Syst II Exp Briefs 67:730–734
47. Chen N, Zhai G, Gui W, Yang C, Liu W (2010) Decentralised $\mathscr{H}_\infty$ quantisers design for uncertain interconnected networked systems. IET Control Theory Appl 4:177–185
48. Chang XH, Huang R, Park JH (2020) Robust guaranteed cost control under digital communication channels. IEEE Trans Industr Inf 16:319–327
49. Che WW, Yang GH (2009) Quantised $\mathscr{H}_\infty$ filter design for discrete-time systems. Int J Control 82:195–206
50. Niu Y, Ho DWC (2014) Control strategy with adaptive quantizer's parameters under digital communication channels. Automatica 50:2665–2671

51. Chang XH, Liu Y (2022) Quantized output feedback control of AFS for electric vehicles with transmission delay and data dropouts. IEEE Trans Intell Transp Syst 23:16026–16037
52. Liu K, Fridman E, Johansson KH (2015) Dynamic quantization of uncertain linear networked control systems. Automatica 59:248–255
53. Zheng BC, Yu X, Xue Y (2018) Quantized feedback sliding-mode control: An event-triggered approach. Automatica 91:126–135
54. Wu C, Zhao X, Xia W, Liu J, Başar T (2021) $\mathscr{L}_2$-gain analysis for dynamic event-triggered networked control systems with packet losses and quantization. Automatica 129:109587
55. Yang H, Xu Y, Zhang J (2017) Event-driven control for networked control systems with quantization and Markov packet losses. IEEE Trans Cybern 47:2235–2243
56. Lunze J, Lehmann D (2010) A state-feedback approach to event-based control. Automatica 46:211–215
57. Wang X, Lemmon MD (2009) Self-triggered feedback control systems with finite-gain $\mathscr{L}_2$ stability. IEEE Trans Autom Control 54:452–467
58. Yue D, Tian E, Han QL (2013) A delay system method for designing event-triggered controllers of networked control systems. IEEE Trans Autom Control 58:475–481
59. Zhang XM, Han QL (2015) Event-based $\mathscr{H}_\infty$ filtering for sampled-data systems. Automatica 51:55–69
60. Meng X, Chen T (2014) Event triggered robust filter design for discrete-time systems. IET Control Theory Appl 8:104–113
61. Hu S, Yue D (2012) Event-based $\mathscr{H}_\infty$ filtering for networked system with communication delay. Signal Process 92:2029–2039
62. Zhang WA, Yu L, Song H (2009) $\mathscr{H}_\infty$ filtering of networked discrete-time systems with random packet losses. Inf Sci 179:3944–3955
63. Wang Z, Yang F, Ho DWC, Liu X (2006) Robust $\mathscr{H}_\infty$ filtering for stochastic time-delay systems with missing measurements. IEEE Trans Signal Process 54:2579–2587
64. Shen B, Wang Z, Hung YS (2010) Distributed $\mathscr{H}_\infty$-consensus filtering in sensor networks with multiple missing measurements: The finite-horizon case. Automatica 46:1682–1688
65. Wang Z, Yang F, Ho DWC, Liu X (2007) Robust $\mathscr{H}_\infty$ control for networked systems with random packet losses. IEEE Trans Syst Man Cybern B Cybern 37:916–924
66. Wu J, Chen T (2007) Design of networked control systems with packet dropouts. IEEE Trans Autom Control 52:1314–1319
67. Yue D, Han QL, Lam J (2005) Network-based robust $\mathscr{H}_\infty$ control of systems with uncertainty. Automatica 41:999–1007
68. Yang F, Wang Z, Hung YS, Gani M (2006) $\mathscr{H}_\infty$ control for networked systems with random communication delays. IEEE Trans Autom Control 51:511–518
69. Zhang L, Shi Y, Chen T, Huang B (2005) A new method for stabilization of networked control systems with random delays. IEEE Trans Autom Control 50:1177–1181
70. Chen W, Ding D, Dong H, Wei G (2019) Distributed resilient filtering for power systems subject to denial-of-service attacks. IEEE Trans Syst, Man, Cybern: Syst 49:1688–1697
71. Zhou Y, Chang XH (2022) Event-triggered quantized $\mathscr{L}_2$–$\mathscr{L}_\infty$ filtering for neural networks under denial-of-service attacks. Int J Robust Nonlinear Control 32:5897–5918
72. Yuan Y, Yuan H, Guo L, Yang H, Sun S (2016) Resilient control of networked control system under DoS attacks: A unified game approach. IEEE Trans Industr Inf 12:1786–1794
73. Ding D, Wang Z, Ho DWC, Wei G (2017) Distributed recursive filtering for stochastic systems under uniform quantizations and deception attacks through sensor networks. Automatica 78:231–240
74. Pang ZH, Liu GP (2012) Design and implementation of secure networked predictive control systems under deception attacks. IEEE Trans Control Syst Technol 20:1334–1342
75. Tian E, Peng C (2020) Memory-based event-triggering $\mathscr{H}_\infty$ load frequency control for power systems under deception attacks. IEEE Trans Cybern 50:4610–4618
76. Zhang Z, Niu Y, Lam HK (2020) Sliding-mode control of T–S fuzzy systems under weighted try-once-discard protocol. IEEE Trans Cybern 50:4972–4982

77. Wang J, Yang C, Xia J, Wu ZG, Shen H (2022) Observer-based sliding mode control for networked fuzzy singularly perturbed systems under weighted try-once-discard protocol. IEEE Trans Fuzzy Syst 30:1889–1899
78. Wang Y, Wang Z, Zou L, Dong H (2022) $\mathscr{H}_\infty$ PID control for discrete-time fuzzy systems with infinite-distributed delays under round-robin communication protocol. IEEE Trans Fuzzy Syst 30:1875–1888
79. Luo Y, Wang Z, Wei G, Alsaadi FE (2017) $\mathscr{H}_\infty$ fuzzy fault detection for uncertain 2-D systems under round-robin scheduling protocol. IEEE Trans Syst, Man, Cybern: Syst 47:2172–2184
80. Dong Y, Song Y, Wei G (2021) Efficient model-predictive control for networked interval type-2 T–S fuzzy system with stochastic communication protocol. IEEE Trans Fuzzy Syst 29:286–297
81. Wu B, Chang XH, Zhao X (2021) Fuzzy $\mathscr{H}_\infty$ output feedback control for nonlinear NCSs with quantization and stochastic communication protocol. IEEE Trans Fuzzy Syst 29:2623–2634
82. Chang XH, Wang YM (2018) Peak-to-peak filtering for networked nonlinear DC motor systems with quantization. IEEE Trans Industr Inf 14:5378–5388
83. Zhang C, Feng G, Gao H, Qiu J (2011) $\mathscr{H}_\infty$ filtering for nonlinear discrete-time systems subject to quantization and packet dropouts. IEEE Trans Fuzzy Syst 19:353–365
84. Qiu J, Feng G, Gao H (2012) Observer-based piecewise affine output feedback controller synthesis of continuous-time T–S fuzzy affine dynamic systems using quantized measurements. IEEE Trans Fuzzy Syst 20:1046–1062
85. Wang H, Yang GH (2016) Dynamic output feedback controller design for affine T–S fuzzy systems with quantized measurements. ISA Trans 64:202–215
86. Zheng Q, Xu S, Zhang Z (2022) Nonfragile $\mathscr{H}_\infty$ control for uncertain Takagi–Sugeno fuzzy systems under digital communication channels and its application. IEEE Trans Syst, Man, Cybern: Syst 52:3638–3647
87. Li ZM, Park JH (2020) Dissipative fuzzy tracking control for nonlinear networked systems with quantization. IEEE Trans Syst, Man, Cybern: Syst 50:5130–5141
88. Li ZM, Chang XH, Park JH (2021) Quantized static output feedback fuzzy tracking control for discrete-time nonlinear networked systems with asynchronous event-triggered constraints. IEEE Trans Syst, Man, Cybern: Syst 51:3820–3831
89. Shen H, Men Y, Wu ZG, Cao J, Lu G (2019) Network-based quantized control for fuzzy singularly perturbed semi-Markov jump systems and its application. IEEE Trans Circuits Syst I Reg Papers 66:1130–1140
90. Mahmoud MS, Saif AWA (2012) Robust quantized approach to fuzzy networked control systems. IEEE J Emerging Sel Topics Circuits Syst 2:71–81
91. Lu R, Cheng H, Bai J (2015) Fuzzy-model-based quantized guaranteed cost control of nonlinear networked systems. IEEE Trans Fuzzy Syst 23:567–575
92. Chang XH, Yang C, Xiong J (2019) Quantized fuzzy output feedback $\mathscr{H}_\infty$ control for nonlinear systems with adjustment of dynamic parameters. IEEE Trans Syst, Man, Cybern: Syst 49:2005–2015
93. Zheng Q, Xu S, Du B (2022) Quantized guaranteed cost output feedback control for nonlinear networked control systems and its applications. IEEE Trans Fuzzy Syst 30:2402–2411
94. Chang XH, Li ZM, Park JH (2018) Fuzzy generalized $\mathscr{H}_2$ filtering for nonlinear discrete-time systems with measurement quantization. IEEE Trans Syst, Man, Cybern: Syst 48:2419–2430
95. Chang XH, Jin X (2022) Observer-based fuzzy feedback control for nonlinear systems subject to transmission signal quantization. Appl Math Comput 414:126657
96. Pan TT, Chang XH, Liu Y (2022) Robust fuzzy feedback control for nonlinear systems with input quantization. IEEE Trans Fuzzy Syst 30:4905–4914
97. Li ZM, Chang XH, Xiong J (2022) Event-based fuzzy tracking control for nonlinear networked systems subject to dynamic quantization. IEEE Trans Fuzzy Syst 31:941–954
98. Chang XH, Liu Y (2020) Robust $\mathscr{H}_\infty$ filtering for vehicle sideslip angle with quantization and data dropouts. IEEE Trans Veh Technol 69:10435–10445
99. Zhang H, Yan H, Yang F, Chen Q (2011) Quantized control design for impulsive fuzzy networked systems. IEEE Trans Fuzzy Syst 19:1153–1162

100. Chu H, Fei S, Yue D, Peng C, Sun J (2011) $\mathscr{H}_\infty$ quantized control for nonlinear networked control systems. Fuzzy Sets Syst 174:99–113
101. Li ZM, Xiong J (2022) Event-triggered fuzzy filtering for nonlinear networked systems with dynamic quantization and stochastic cyber attacks. ISA Trans 121:53–62
102. Peng C, Ma S, Xie X (2017) Observer-based non-PDC control for networked T–S fuzzy systems with an event-triggered communication. IEEE Trans Cybern 47:2279–2287
103. Liu X, Su X, Shi P, Shen C (2019) Observer-based sliding mode control for uncertain fuzzy systems via event-triggered strategy. IEEE Trans Fuzzy Syst 27:2190–2201
104. Su X, Wen Y, Shi P, Lam HK (2019) Event-triggered fuzzy filtering for nonlinear dynamic systems via reduced-order approach. IEEE Trans Fuzzy Syst 27:1215–1225
105. Gao H, Zhao Y, Lam J, Chen K (2009) $\mathscr{H}_\infty$ fuzzy filtering of nonlinear systems with intermittent measurements. IEEE Trans Fuzzy Syst 17:291–300
106. Chang XH, Liu Q, Wang YM, Xiong J (2019) Fuzzy peak-to-peak filtering for networked nonlinear systems with multipath data packet dropouts. IEEE Trans Fuzzy Syst 27:436–446
107. Qiu J, Feng G, Gao H (2011) Asynchronous output-feedback control of networked nonlinear systems with multiple packet dropouts: T–S fuzzy affine model-based approach. IEEE Trans Fuzzy Syst 19:1014–1030
108. Gao H, Zhao Y, Chen T (2009) $\mathscr{H}_\infty$ fuzzy control of nonlinear systems under unreliable communication links. IEEE Trans Fuzzy Syst 17:265–278
109. Qiu J, Feng G, Gao H (2010) Fuzzy-model-based piecewise $\mathscr{H}_\infty$ static-output-feedback controller design for networked nonlinear systems. IEEE Trans Fuzzy Syst 18:919–934
110. Dong H, Wang Z, Ho DWC, Gao H (2010) Robust $\mathscr{H}_\infty$ fuzzy output-feedback control with multiple probabilistic delays and multiple missing measurements. IEEE Trans Fuzzy Syst 18:712–725
111. Peng C, Yang TC (2010) Communication-delay-distribution-dependent networked control for a class of T–S fuzzy systems. IEEE Trans Fuzzy Syst 18:326–335
112. Zheng Y, Fang H, Wang HO (2006) Takagi–Sugeno fuzzy-model-based fault detection for networked control systems with Markov delays. IEEE Trans Syst Man Cybern B Cybern 36:924–929
113. Zhang J, Shi P, Xia Y (2013) Fuzzy delay compensation control for T–S fuzzy systems over network. IEEE Trans Cybern 43:259–268
114. Liu J, Wei L, Xie X, Tian E, Fei S (2018) Quantized stabilization for T–S fuzzy systems with hybrid-triggered mechanism and stochastic cyber-attacks. IEEE Trans Fuzzy Syst 26:3820–3834
115. Gu Z, Ahn CK, Yue D, Xie X (2022) Event-triggered $\mathscr{H}_\infty$ filtering for T–S fuzzy-model-based nonlinear networked systems with multisensors against DoS attacks. IEEE Trans Cybern 52:5311–5321
116. Cheng J, Huang W, Park JH, Cao J (2022) A hierarchical structure approach to finite-time filter design for fuzzy Markov switching systems with deception attacks. IEEE Trans Cybern 52:7254–7264

Chapter 2
Fuzzy Robust $\mathscr{H}_\infty$ Control with Dynamic Quantization

Abstract This chapter investigates the quantized $\mathscr{H}_\infty$ control problem for T–S fuzzy systems, where the dynamic quantizer is adopted for signal quantization before transmission to reduce the communication burden. Firstly, the robust $\mathscr{H}_\infty$ control is addressed for uncertain discrete-time T–S fuzzy systems with the control input quantization via a two-step design approach and the fuzzy Lyapunov functional method. Secondly, a new two-step strategy is proposed to onlinely update the dynamic quantizer's parameter and design the non-fragile controller for discrete-time T–S fuzzy systems with the controller gain perturbation and state quantization, so that the quantized closed-loop system is asymptotically stable and maintains the same specified $\mathscr{H}_\infty$ performance as the one without quantization. Thirdly, by fully considering the effect of system parameter uncertainties, controller gain perturbation, control input and measurement output quantization on continuous-time T–S fuzzy systems, the resilient $\mathscr{H}_\infty$ control problem is solved based on two descriptor representation approaches, and a novel adjusting rule is given for dynamic quantizers' parameters. Finally, some examples are provided to show the feasibility and effectiveness of the proposed design methods, respectively.

Keywords T–S fuzzy systems · Norm-bounded uncertainties · Dynamic quantization · Two-step design method · Descriptor representation approach

2.1 Robust Control with Input Quantization

In this section, a two-step method is given to deal with the state feedback control problem of uncertain T–S fuzzy systems with control input quantization. The fuzzy Lyapunov function is used to judge that the T–S fuzzy closed-loop system is asymptotically stable when there are norm-bounded uncertain parameters, and has the same $\mathscr{H}_\infty$ performance as the system without control input quantization.

X. Chang et al., *Control and Filtering of Fuzzy Systems Under Communication Channels*,
https://doi.org/10.1007/978-981-99-4346-3_2

2.1.1 Problem Formulation

Consider a discrete-time T–S fuzzy model with norm-bounded uncertainties as follows:

Plant Rule i_{th}: **IF** $\varepsilon_1(k)$ is M_{1i}, $\varepsilon_2(k)$ is M_{2i}, and, ..., and $\varepsilon_d(k)$ is M_{di}, **THEN**

$$\begin{aligned} x(k+1) &= (A_i + \Delta_{Ai})x(k) + (B_i + \Delta_{Bi})\mathbb{Q}(u(k)) + (E_i + \Delta_{Ei})w(k), \\ z(k) &= (C_i + \Delta_{Ci})x(k) + (D_i + \Delta_{Di})\mathbb{Q}(u(k)) + (F_i + \Delta_{Fi})w(k), \end{aligned} \tag{2.1}$$

where $\varepsilon_1(k), \varepsilon_2(k), \ldots, \varepsilon_d(k)$ represent the premise variables; M_{pi} with $i = 1, 2, \ldots, r$, $p = 1, 2, \ldots, d$ represents the fuzzy sets, where r is the number of fuzzy rules; $x(k) \in \mathbb{R}^{n_x}$ and $u(k) \in \mathbb{R}^{n_u}$ represent the state variable, control input, respectively; $w(k) \in \mathbb{R}^{n_w}$ denotes the noise signal in $\ell_2[0, \infty)$; $z(k) \in \mathbb{R}^{n_z}$ is the controlled output; $A_i \in \mathbb{R}^{n_x \times n_x}$, $B_i \in \mathbb{R}^{n_x \times n_u}$, $C_i \in \mathbb{R}^{n_z \times n_x}$, $D_i \in \mathbb{R}^{n_z \times n_u}$, $E_i \in \mathbb{R}^{n_x \times n_w}$, and $F_i \in \mathbb{R}^{n_z \times n_w}$, with $i = 1, 2, \ldots, r$ denote the system matrices; Δ_{Ai}, Δ_{Bi}, Δ_{Ci}, Δ_{Di}, Δ_{Ei}, and Δ_{Fi}, $i = 1, \ldots, r$ are the uncertain parameter matrices, which can be described as:

$$\begin{pmatrix} \Delta_{Ai} & \Delta_{Bi} & \Delta_{Ei} \\ \Delta_{Ci} & \Delta_{Di} & \Delta_{Fi} \end{pmatrix} = \begin{pmatrix} X_1 \\ X_2 \end{pmatrix} \Delta \begin{pmatrix} Y_{1i} & Y_{2i} & Y_{3i} \end{pmatrix}, \tag{2.2}$$

where $\Delta^T \Delta \leq I$, Δ represents the norm-bounded uncertainty, Y_{1i}, Y_{2i}, Y_{3i}, X_1 and X_2 are known constant matrices.

By introducing the following fuzzy basis functions:

$$\rho_i(\varepsilon(k)) = \frac{\prod_{p=1}^{d} M_{pi}(\varepsilon_p(k))}{\sum_{i=1}^{r} \prod_{p=1}^{d} M_{pi}(\varepsilon_p(k))} \geq 0, \tag{2.3}$$

one has

$$\sum_{i=1}^{r} \rho_i(\varepsilon(k)) = 1, \quad i = 1, 2, \ldots, r, \tag{2.4}$$

where $M_{pi}(\varepsilon_p(k))$ is the grade of membership function of $\varepsilon_p(k)$ in M_{pi}.

The uncertain T–S fuzzy model (2.1) can be deduced as follows:

$$\begin{aligned} x(k+1) &= (A(\rho) + \Delta_A(\rho))\, x(k) + (B(\rho) + \Delta_B(\rho))\, \mathbb{Q}(u(k)) \\ &\quad + (E(\rho) + \Delta_E(\rho))\, w(k), \\ z(k) &= (C(\rho) + \Delta_C(\rho))\, x(k) + (D(\rho) + \Delta_D(\rho))\, \mathbb{Q}(u(k)) \\ &\quad + (F(\rho) + \Delta_F(\rho))\, w(k), \end{aligned} \tag{2.5}$$

where

$$A(\rho) = \sum_{i=1}^{r} \rho_i(\varepsilon(k))A_i, \quad B(\rho) = \sum_{i=1}^{r} \rho_i(\varepsilon(k))B_i,$$

$$C(\rho) = \sum_{i=1}^{r} \rho_i(\varepsilon(k))C_i, \quad D(\rho) = \sum_{i=1}^{r} \rho_i(\varepsilon(k))D_i,$$

$$E(\rho) = \sum_{i=1}^{r} \rho_i(\varepsilon(k))E_i, \quad F(\rho) = \sum_{i=1}^{r} \rho_i(\varepsilon(k))F_i,$$

$$\Delta_A(\rho) = X_1 \Delta Y_1(\rho) = \sum_{i=1}^{r} \rho_i(\varepsilon(k))X_1 \Delta Y_{1i},$$

$$\Delta_B(\rho) = X_1 \Delta Y_2(\rho) = \sum_{i=1}^{r} \rho_i(\varepsilon(k))X_1 \Delta Y_{2i},$$

$$\Delta_C(\rho) = X_2 \Delta Y_1(\rho) = \sum_{i=1}^{r} \rho_i(\varepsilon(k))X_2 \Delta Y_{1i},$$

$$\Delta_D(\rho) = X_2 \Delta Y_2(\rho) = \sum_{i=1}^{r} \rho_i(\varepsilon(k))X_2 \Delta Y_{2i},$$

$$\Delta_E(\rho) = X_1 \Delta Y_3(\rho) = \sum_{i=1}^{r} \rho_i(\varepsilon(k))X_1 \Delta Y_{3i},$$

$$\Delta_F(\rho) = X_2 \Delta Y_3(\rho) = \sum_{i=1}^{r} \rho_i(\varepsilon(k))X_2 \Delta Y_{3i}.$$

Considering the control input signal needs to be quantized by the quantizer before being sent to the data channel, the dynamic quantizer which is the same as the one in [1] is adopted to quantize the signal variable $s(k) \in \mathbb{R}^{n_s}$. Then we define the $\Delta_s > 0$ and $M_s > 0$ to denote the error bound and the range of quantizer, respectively, which satisfy the following condition:

$$\|q(s(k)) - s(k)\| \le \Delta_s, \quad if \ \|s(k)\| \le M_s, \tag{2.6a}$$

$$\|q(s(k))\| > M_s - \Delta_s, \quad if \ \|s(k)\| > M_s. \tag{2.6b}$$

Next, we will give the following dynamic quantizer to the networked system:

$$\mathbb{Q}(s(k)) = \mu_s(k) q\left(\frac{s(k)}{\mu_s(k)}\right), \tag{2.7}$$

where $\mu_s(k) > 0$ denotes the dynamic quantizer's parameter. That is, $\mu_s(k)\Delta_s$ and $\mu_s(k)M_s$ mean that the error bound and the range of quantizer can be updated with the adjustment of dynamic quantizer's parameter $\mu_s(k)$. Then, the quantized control input signal can be described as

$$\mathbb{Q}(u(k)) = \mu_u(k) q\left(\frac{u(k)}{\mu_u(k)}\right), \tag{2.8}$$

with

$$u(k) = Kx(k), \tag{2.9}$$

where K is controller gain matrix.

Substituting (2.8) and (2.9) into (2.5), the quantized fuzzy closed-loop system can be given as follows:

$$\begin{aligned} x(k+1) &= \left(\tilde{A}(\rho) + \tilde{B}(\rho)K\right)x(k) + \tilde{B}(\rho)e_u(k) + \tilde{E}(\rho)w(k), \\ z(k) &= \left(\tilde{C}(\rho) + \tilde{D}(\rho)K\right)x(k) + \tilde{D}(\rho)e_u(k) + \tilde{F}(\rho)w(k), \end{aligned} \tag{2.10}$$

with

$$\begin{aligned} e_u(k) &= \mathbb{Q}(u(k)) - u(k) = \mu_u(k)\left(q\left(\frac{u(k)}{\mu_u(k)}\right) - \frac{u(k)}{\mu_u(k)}\right), \\ \tilde{A}(\rho) &= A(\rho) + \Delta_A(\rho), \quad \tilde{B}(\rho) = B(\rho) + \Delta_B(\rho), \\ \tilde{C}(\rho) &= C(\rho) + \Delta_C(\rho), \quad \tilde{D}(\rho) = D(\rho) + \Delta_D(\rho), \\ \tilde{E}(\rho) &= E(\rho) + \Delta_E(\rho), \quad \tilde{F}(\rho) = F(\rho) + \Delta_F(\rho). \end{aligned}$$

The quantized robust state feedback $\mathscr{H}_\infty$ control problem considered in this section is to design the robust state feedback controller in (2.9) and the dynamic quantizer in (2.8) such that

(1) The quantized fuzzy closed-loop system in (2.10) is asymptotically stable when $w(k) = 0$.
(2) The quantized fuzzy closed-loop system in (2.10) has a prescribed $\mathscr{H}_\infty$ performance γ, i.e., under the initial condition $x(0) = 0$, $\sum_{k=0}^{\infty} z^T(k)z(k) < \gamma^2 \sum_{k=0}^{\infty} w^T(k)w(k)$ is satisfied for any nonzero $w(k) \in \ell_2[0, \infty)$.

2.1.2 *Main Results*

As mentioned above, this section will give an improved two-step algorithm to design the controller and dynamic quantizer. In Step 1, a robust controller is designed, and in Step 2, the dynamic quantizer's parameter $\mu_u(k)$ is determined.

Theorem 2.1 *For the given positive scalar γ, selected quantization range M_u and quantization error bound Δ_u, if there exist scalars $\sigma_u(\rho) > 0$, $\tau_u > 0$, and matrix $P(\rho) > 0$, such that the following matrix inequalities hold:*

$$\frac{1}{M_u} - \sqrt{\sigma_u(\rho)} < 0, \tag{2.11}$$

$$\begin{aligned} &\chi_1^T(\rho)P(\rho^+)\chi_1(\rho) + \chi_2^T(\rho)\chi_2(\rho) - diag\{P(\rho), 0, \gamma^2 I\} \\ &+ \tau_u\left(\chi_3^T\chi_3 - diag\{0, I, 0\}\right) < 0, \end{aligned} \tag{2.12}$$

where

$$\begin{aligned} \chi_1(\rho) &= \left[\tilde{A}(\rho) + \tilde{B}(\rho)K \quad \tilde{B}(\rho) \quad \tilde{E}(\rho)\right], \\ \chi_2(\rho) &= \left[\tilde{C}(\rho) + \tilde{D}(\rho)K \quad \tilde{D}(\rho) \quad \tilde{F}(\rho)\right], \\ \chi_3 &= \left[2\sqrt{\sigma_u(\rho)}\Delta_u K \quad 0 \quad 0\right], \end{aligned}$$

and the online adjustment strategy of dynamic quantizer's parameter $\mu_u(k)$ is given as:

$$\sqrt{\sigma_u(\rho)}\,\|u(k)\| \le \mu_u(k) \le 2\sqrt{\sigma_u(\rho)}\,\|u(k)\|. \tag{2.13}$$

Then, the quantized fuzzy closed-loop system (2.10) is asymptotically stable under the specified $\mathcal{H}_\infty$ performance index γ.

Proof If the condition (2.11) holds and the online adjustment strategy (2.13) can be guaranteed, then the control input signal does not exceed the quantization range. According to (2.6) and (2.13), the following inequality can be obtained:

$$\begin{aligned} \|e_u(k)\| &= \mu_u(k)\left\|q\left(\frac{u(k)}{\mu_u(k)}\right) - \frac{u(k)}{\mu_u(k)}\right\| \\ &\le \mu_u(k)\Delta_u \\ &\le 2\sqrt{\sigma_u(\rho)}\Delta_u\,\|u(k)\| \\ &= 2\sqrt{\sigma_u(\rho)}\Delta_u\,\|Kx(k)\|, \end{aligned} \tag{2.14}$$

i.e.,

$$e_u^T(k)e_u(k) \le 4\sigma_u(\rho)\Delta_u^2 x^T(k)K^T K x(k). \tag{2.15}$$

In order to obtain the scalar variable $\sqrt{\sigma_u(\rho)}$ related to the dynamic quantizer's parameter $\mu_u(k)$, the following inequality is obtained from (2.15):

$$\varphi^T(k)\left(Ha\left\{\left[2\sqrt{\sigma_u(\rho)}\Delta_u K \;\; 0 \;\; 0\right]\right\} - diag\left\{0,\, I,\, 0\right\}\right)\varphi(k) \ge 0, \tag{2.16}$$

where $\varphi(k) = \left[x^T(k)\ \ e_u^T(k)\ \ w^T(k)\right]^T$. Next, the Lyapunov function will be constructed as $V(x(k)) = x^T(k)P(\rho)x(k)$ with $P(\rho) = \sum_{i=1}^r \rho_i(\varepsilon(k))P_i > 0$, then one can get:

$$\begin{aligned}
&V(x(k+1)) - V(x(k)) + z^T(k)z(k) - \gamma^2 w^T(k)w(k)\\
&\quad = \Delta V(x(k)) + z^T(k)z(k) - \gamma^2 w^T(k)w(k)\\
&\quad = \varphi^T(k)\Big(\chi_1^T(\rho)P(\rho^+)\chi_1(\rho) + \chi_2^T(\rho)\chi_2(\rho) - diag\left\{P(\rho),\ 0,\ \gamma^2 I\right\}\Big)\varphi(k).
\end{aligned}$$

If the matrix inequality (2.12) is satisfied, it is easy to get that when the variable $\varphi(k) \neq 0$, one has:

$$\begin{aligned}
&\varphi^T(k)\Big\{\chi_1^T(\rho)P(\rho^+)\chi_1(\rho) + \chi_2^T(\rho)\chi_2(\rho) - diag\left\{P(\rho),\ 0,\ \gamma^2 I\right\}\\
&\quad + \tau_u\Big(\chi_3^T\chi_3 - diag\{0,\ I,\ 0\}\Big)\Big\}\varphi(k) < 0,
\end{aligned} \tag{2.17}$$

then based on $\tau_u > 0$ and (2.16),

$$\varphi^T(k)\Big(\chi_1^T(\rho)P(\rho^+)\chi_1(\rho) + \chi_2^T(\rho)\chi_2(\rho) - diag\{P(\rho),\ 0,\ \gamma^2 I\}\Big)\varphi(k) < 0,$$

can be obtained according to the S-procedure (Lemma 1.3).

It shows that the $\mathscr{H}_\infty$ performance γ can be guaranteed with the quantized fuzzy closed-loop system (2.10). The proof is completed.

It is worth noting that Theorem 2.1 gives a useful $\mathscr{H}_\infty$ performance analysis criterion based on matrix inequality, but matrix inequality (2.12) contains some nonlinearity and uncertainties, which leads to matrix inequality (2.12) being nonconvex (non-strict LMI), it means that matrix inequality conditions cannot be directly used in controller design. The following theorem proposes a new $\mathscr{H}_\infty$ performance analysis criterion for quantized fuzzy system (2.10) by defining new variables and using matrix inequality transformation technology to solve this nonlinear problem.

Theorem 2.2 *For the given quantization error bound Δ_u and quantization range M_u, the quantized fuzzy closed-loop system (2.10) is asymptotically stable with the prescribed $\mathscr{H}_\infty$ performance index $\gamma > 0$, if there exist matrices $Q(\rho) > 0$, G, and N, scalars $\tau_u > 0$ and $\nu_u(\rho) > 0$, such that the following matrix inequalities hold:*

$$M_u^2 \nu_u(\rho) > \tau_u, \tag{2.18}$$

$$\tilde{\Lambda}(\rho) = \begin{bmatrix}
-G - G^T + Q(\rho) & * & * & * & * & * \\
0 & -\gamma^2 I & * & * & * & * \\
\tilde{A}(\rho)G + \tilde{B}(\rho)N & \tilde{E}(\rho) & -Q(\rho^+) & * & * & * \\
\tilde{C}(\rho)G + \tilde{D}(\rho)N & \tilde{F}(\rho) & 0 & -I & * & * \\
0 & 0 & \tilde{B}^T(\rho) & \tilde{D}^T(\rho) & -\tau_u I & * \\
2\nu_u(\rho)\Delta_u N & 0 & 0 & 0 & 0 & -\nu_u(\rho)I
\end{bmatrix} < 0, \tag{2.19}$$

and the online adjusting strategy of dynamic quantizer's parameter $\mu_u(k)$ is given the same as (2.13) with $\sigma_u(\rho) = \nu_u(\rho)/\tau_u$.

Proof By using Schur complement (Lemma 1.1) for the matrix inequality (2.12) in Theorem 2.1, the following inequality can be obtained

$$\begin{bmatrix} -P(\rho) & * & * & * & * & * \\ 0 & -\tau_u I & * & * & * & * \\ 0 & 0 & -\gamma^2 I & * & * & * \\ \tilde{A}(\rho)+\tilde{B}(\rho)K & \tilde{B}(\rho) & \tilde{E}(\rho) & -P^{-1}(\rho^+) & * & * \\ \tilde{C}(\rho)+\tilde{D}(\rho)K & \tilde{D}(\rho) & \tilde{F}(\rho) & 0 & -I & * \\ 2\tau_u\sigma_u(\rho)\Delta_u K & 0 & 0 & 0 & 0 & -\tau_u\sigma_u(\rho)I \end{bmatrix} < 0. \tag{2.20}$$

Defining $Q(\rho) = P^{-1}(\rho)$, $Q(\rho^+) = P^{-1}(\rho^+)$ and $\nu_u(\rho) = \tau_u\sigma_u(\rho)$, the above inequality (2.20) can be rewritten as:

$$\begin{bmatrix} -Q^{-1}(\rho) & * & * & * & * & * \\ 0 & -\tau_u I & * & * & * & * \\ 0 & 0 & -\gamma^2 I & * & * & * \\ \tilde{A}(\rho)+\tilde{B}(\rho)K & \tilde{B}(\rho) & \tilde{E}(\rho) & -Q(\rho^+) & * & * \\ \tilde{C}(\rho)+\tilde{D}(\rho)K & \tilde{D}(\rho) & \tilde{F}(\rho) & 0 & -I & * \\ 2\nu_u(\rho)\Delta_u K & 0 & 0 & 0 & 0 & -\nu_u(\rho)I \end{bmatrix} < 0. \tag{2.21}$$

By introducing a transformation matrix Υ as

$$\Upsilon = \begin{bmatrix} G^T & 0 & 0 & 0 & 0 & 0 \\ 0 & 0 & I & 0 & 0 & 0 \\ 0 & 0 & 0 & I & 0 & 0 \\ 0 & 0 & 0 & 0 & I & 0 \\ 0 & I & 0 & 0 & 0 & 0 \\ 0 & 0 & 0 & 0 & 0 & I \end{bmatrix},$$

and performing congruence transformations with Υ to (2.21), the inequality $-(G - Q(\rho))^T Q^{-1}(\rho)(G - Q(\rho)) \le 0$ can be rewritten as:

$$-G^T Q^{-1}(\rho) G \le -G^T - G + Q(\rho),$$

where the symmetric matrix $Q(\rho) = \sum_{i=1}^{r} \rho_i(\varepsilon(k)) Q_i > 0$.

By defining a matrix variable $N = KG$, the inequality (2.18) and (2.19) in Theorem 2.2 can be guaranteed. The proof is completed.

Theorem 2.2 proposes a new $\mathscr{H}_\infty$ performance analysis criterion based on the quantized fuzzy system (2.10), in which matrix inequalities (2.18) and (2.19) effectively avoid product terms (nonlinear coupling terms). However, it is obvious that matrix inequalities (2.18) and (2.19) are still nonconvex, including uncertainty Δ

and nonlinear coupling term $\nu_u(\rho)\Delta_u N$, which makes it difficult to design the robust controller and the dynamic quantizer for the quantized fuzzy closed-loop system in (2.10) through the $\mathcal{H}_\infty$ performance analysis criterion in Theorem 2.2. Next, this chapter presents an improved two-step design method in the following algorithm to solve the above problem. Firstly, in Step 1, a robust controller is designed so that the quantized fuzzy closed-loop system (2.10) with norm-bounded uncertainties has the same performance as the system without control input quantization, and then in Step 2, the scalar variable $\sigma(\rho)$ related to the dynamic quantizer's parameter $\mu_u(k)$ is calculated.

Algorithm 2.1 For the quantized fuzzy closed-loop system (2.10) with norm-bounded uncertainties, it is asymptotically stable under the given $\mathcal{H}_\infty$ performance index γ:

Step 1. Solving the following LMIs to obtain the values of matrices N, G, $Q_i > 0$, and the controller gain $K = NG^{-1}$:

$$\begin{bmatrix} -G-G^T+Q_i & * & * & * & * \\ 0 & -\gamma^2 I & * & * & * \\ A_iG+B_iN & E_i & -Q_j+\alpha X_1X_1^T & * & * \\ C_iG+D_iN & F_i & \alpha X_2X_1^T & -I+\alpha X_2X_2^T & * \\ Y_{1i}G+Y_{2i}N & Y_{3i} & 0 & 0 & -\alpha I \end{bmatrix} < 0, \tag{2.22}$$

for $i, j = 1, 2, \ldots, r$.

Step 2. Defining $\nu_u(\rho) = \sum_{i=1}^r \rho_i(\varepsilon(k))\nu_{ui}$, with $\nu_{ui} > 0$ and solving the following LMIs with known Q_i, G, N, and K obtained by Step 1, then the values of scalar variables τ_u and ν_{ui} can be determined:

$$M_u^2\nu_{ui} > \tau_u, \tag{2.23}$$

$$\Lambda_{ij} = \begin{bmatrix} -G-G^T+Q_i & * & * & * & * & * & * \\ 0 & -\gamma^2 I & * & * & * & * & * \\ A_iG+B_iN & E_i & -Q_j+\alpha X_1X_1^T & * & * & * & * \\ C_iG+D_iN & F_i & \alpha X_2X_1^T & -I+\alpha X_2X_2^T & * & * & * \\ Y_{1i}G+Y_{2i}N & Y_{3i} & 0 & 0 & -\alpha I & * & * \\ 0 & 0 & B_i^T & D_i^T & Y_{2i}^T & -\tau_u I & * \\ 2\nu_{ui}\Delta_u N & 0 & 0 & 0 & 0 & 0 & -\nu_{ui} I \end{bmatrix} < 0, \tag{2.24}$$

for $i, j = 1, 2, \ldots, r$.

Then, the scalar variable $\sigma_u(\rho) = \nu_u(\rho)/\tau_u$. Through the online adjustment strategy in Theorem 2.2, the dynamic quantizer's parameter $\mu_u(k)$ can be adjusted by the scalar variable $\sigma_u(\rho)$.

Discussion 2.1 (1) According to the membership function given in (2.4), if the inequalities in (2.22) are satisfied, then one can obtain

$$\begin{bmatrix} -G-G^T+Q(\rho) & * & * & * & * \\ 0 & -\gamma^2 I & * & * & * \\ A(\rho)G+B(\rho)N & E(\rho) & -Q(\rho^+)+\alpha X_1 X_1^T & * & * \\ C(\rho)G+D(\rho)N & F(\rho) & \alpha X_2 X_1^T & -I+\alpha X_2 X_2^T & * \\ Y_1(\rho)G+Y_2(\rho)N & Y_3(\rho) & 0 & 0 & -\alpha I \end{bmatrix} < 0. \tag{2.25}$$

By using the Schur complement (Lemma 1.1), the inequality (2.25) can be rewritten as:

$$\begin{aligned} &\begin{bmatrix} -G-G^T+Q(\rho) & * & * & * \\ 0 & -\gamma^2 I & * & * \\ A(\rho)G+B(\rho)N & E(\rho) & -Q(\rho^+) & * \\ C(\rho)G+D(\rho)N & F(\rho) & 0 & -I \end{bmatrix} + \alpha \begin{bmatrix} 0 \\ 0 \\ X_1 \\ X_2 \end{bmatrix} \begin{bmatrix} 0 & 0 & X_1^T & X_2^T \end{bmatrix} \\ &+ \frac{1}{\alpha} \begin{bmatrix} (Y_1(\rho)G+Y_2(\rho)N)^T \\ Y_3^T(\rho) \\ 0 \\ 0 \end{bmatrix} \begin{bmatrix} Y_1(\rho)G+Y_2(\rho)N & Y_3(\rho) & 0 & 0 \end{bmatrix} < 0. \end{aligned} \tag{2.26}$$

Then, based on Lemma 1.2 for $\alpha > 0$ and $\Delta^T \Delta \le I$, (2.26) means that

$$\begin{aligned} &\begin{bmatrix} -G-G^T+Q(\rho) & * & * & * \\ 0 & -\gamma^2 I & * & * \\ A(\rho)G+B(\rho)N & E(\rho) & -Q(\rho^+) & * \\ C(\rho)G+D(\rho)N & F(\rho) & 0 & -I \end{bmatrix} \\ &+ He\left\{ \begin{bmatrix} 0 \\ 0 \\ X_1 \\ X_2 \end{bmatrix} \Delta \begin{bmatrix} Y_1(\rho)G+Y_2(\rho)N & Y_3(\rho) & 0 & 0 \end{bmatrix} \right\} < 0. \end{aligned} \tag{2.27}$$

(2) Similar to (1), if the condition $\Lambda_{ij} < 0$ in (2.24) holds, then one can obtain

$$\Lambda(\rho) = \begin{bmatrix} -G-G^T+Q(\rho) & * & * & * & * & * & * \\ 0 & -\gamma^2 I & * & * & * & * & * \\ A(\rho)G+B(\rho)N & E(\rho) & -Q(\rho^+)+\alpha X_1 X_1^T & * & * & * & * \\ C(\rho)G+D(\rho)N & F(\rho) & \alpha X_2 X_1^T & -I+\alpha X_2 X_2^T & * & * & * \\ Y_1(\rho)G+Y_2(\rho)N & Y_3(\rho) & 0 & 0 & -\alpha I & * & * \\ 0 & 0 & B^T(\rho) & D^T(\rho) & Y_2^T(\rho) & -\tau_u I & * \\ 2\nu_u(\rho)\Delta_u N & 0 & 0 & 0 & 0 & 0 & -\nu_u(\rho) I \end{bmatrix} < 0. \tag{2.28}$$

By using Schur complement (Lemma 1.1) to (2.28), one has:

$$\tilde{\Lambda}(\rho) = \begin{bmatrix} -G - G^T + Q(\rho) & * & * & * & * & * \\ 0 & -\gamma^2 I & * & * & * & * \\ A(\rho)G + B(\rho)N & E(\rho) & -Q(\rho^+) & * & * & * \\ C(\rho)G + D(\rho)N & F(\rho) & 0 & -I & * & * \\ 0 & 0 & B^T(\rho) & D^T(\rho) & -\tau_u I & * \\ 2\nu_u(\rho)\Delta_u N & 0 & 0 & 0 & 0 & -\nu_u(\rho)I \end{bmatrix}$$

$$+ \frac{1}{\alpha}\begin{bmatrix} (Y_1(\rho)G + Y_2(\rho)N)^T \\ Y_3^T(\rho) \\ 0 \\ 0 \\ Y_2^T(\rho) \\ 0 \end{bmatrix} \begin{bmatrix} Y_1(\rho)G + Y_2(\rho)N & Y_3(\rho) & 0 & 0 & Y_2(\rho) & 0 \end{bmatrix} \quad (2.29)$$

$$+ \alpha \begin{bmatrix} 0 \\ 0 \\ X_1 \\ X_2 \\ 0 \\ 0 \end{bmatrix} \begin{bmatrix} 0 & 0 & X_1^T & X_2^T & 0 & 0 \end{bmatrix} < 0.$$

That is

$$\bar{\Lambda}(\rho) = \begin{bmatrix} -G - G^T + Q(\rho) & * & * & * & * & * \\ 0 & -\gamma^2 I & * & * & * & * \\ A(\rho)G + B(\rho)N & E(\rho) & -Q(\rho^+) & * & * & * \\ C(\rho)G + D(\rho)N & F(\rho) & 0 & -I & * & * \\ 0 & 0 & B^T(\rho) & D^T(\rho) & -\tau_u I & * \\ 2\nu_u(\rho)\Delta_u N & 0 & 0 & 0 & 0 & -\nu_u(\rho)I \end{bmatrix}$$

$$+ He\left\{ \begin{bmatrix} 0 \\ 0 \\ X_1 \\ X_2 \\ 0 \\ 0 \end{bmatrix} \Delta \begin{bmatrix} Y_1(\rho)G + Y_2(\rho)N & Y_3(\rho) & 0 & 0 & Y_2(\rho) & 0 \end{bmatrix} \right\} < 0. \quad (2.30)$$

Remark 2.1 According to Discussion 2.1, Algorithm 2.1 finally guarantees that the condition $M_u^2 \nu_u(\rho) > \tau_u$ holds by the definition of τ_u and $\nu_u(\rho)$ with $\bar{\Lambda}(\rho) < 0$, that is, (2.18) and (2.19) are established. Therefore, the two-step design strategy proposed in Algorithm 2.1 is effective for the robust control design of the quantized fuzzy system (2.10) with norm-bounded uncertainties.

Remark 2.2 The quantized control input signal can be obtained by determining the dynamic quantizer's parameter $\mu_u(k)$ as a unique value and transforming it into a communication channel. The adjustment rule of the dynamic quantizer's parameter $\mu_u(k)$ considered in this chapter is as [2]:

The adjusting rule:

$$\mu_u(k) = \begin{cases} \text{floor}\left(2\vartheta_u \|u(k)\| \times 10^{\beta}\right) \times 10^{-\beta}, & 0 \leq 2\vartheta_u \|u(k)\| < 1, \\ \text{floor}\left(2\vartheta_u \|u(k)\|\right), & 1 \leq 2\vartheta_u \|u(k)\|, \end{cases}$$

where $\beta = \min\left\{\beta \in \mathcal{N}^{+} | \left(2\vartheta_u \|u(k)\| \times 10^{\beta}\right) > 1\right\}$ and floor(θ) is the integer function and gives the maximum integer that does not exceed θ. It is worth noting that, in Algorithm 2.1, the $\mu_u(k)$ of the dynamic quantizer (2.7) is updated by using $\vartheta_u = \sqrt{\frac{\nu_u(\rho)}{\tau_u}}$ through the above adjusting rule.

2.1.3 Simulation Example

This section uses the truck-trailer system shown in Fig. 2.1 as an example to verify the effectiveness and feasibility of the proposed method. Set $L = 6.0$ m, $l = 3.5$ m, $V_m = -15$ m/s, and $T_m = 0.1$ s, where V_m represents the speed of backing up, and T_m means the sampling time [3].

Then the state equation of the truck trailer system in [4] can be presented as follows:

$$\begin{aligned} x_1(k+1) &= \left(1 - \frac{V_m T_m}{L}\right) x_1(k) + \frac{V_m T_m}{l} u(k), \\ x_2(k+1) &= \frac{V_m T_m}{L} x_1(k) + x_2(k) + 0.2w(k), \\ x_3(k+1) &= x_3(k) + V_m T_m \sin(\delta(k)) + 0.1w(k), \\ z(k) &= \frac{V_m T_m}{l} u(k), \end{aligned} \tag{2.31}$$

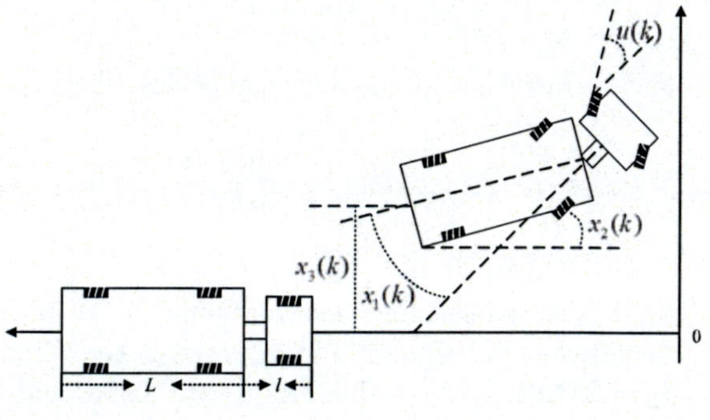

Fig. 2.1 Truck-trailer system

the nonlinear truck-trailer system can be approximately expressed by two local linear systems as:

$$\begin{aligned} R^1 : x(k+1) &= (A_1 + \Delta_{A1})x(k) + B_1u(k) + E_1w(k), \\ z(k) &= D_1u(k), \\ R^2 : x(k+1) &= (A_2 + \Delta_{A2})x(k) + B_2u(k) + E_2w(k), \\ z(k) &= D_2u(k), \end{aligned} \tag{2.32}$$

where $x(k) = \begin{bmatrix} x_1(k) & x_2(k) & x_3(k) \end{bmatrix}^T$, the R^1 denotes a region around $x_2(k) + \frac{V_mT_m}{2L}x_1(k) = 0$, and R^2 denotes a region around $x_2(k) + \frac{V_mT_m}{2L}x_1(k) = \pm\frac{179.997\pi}{180}$, and $d = \frac{0.01}{LT_m}$ [4]:

$$\begin{aligned} A_1 &= \begin{bmatrix} 1 - \frac{V_mT_m}{L} & 0 & 0 \\ \frac{V_mT_m}{L} & 1 & 0 \\ \frac{V_m^2T_m^2}{2L} & V_mT_m & 1 \end{bmatrix}, \quad A_2 = \begin{bmatrix} 1 - \frac{V_mT_m}{L} & 0 & 0 \\ \frac{V_mT_m}{L} & 1 & 0 \\ \frac{dV_m^2T_m^2}{2L} & dV_mT_m & 1 \end{bmatrix}, \\ E_1 &= E_2 = \begin{bmatrix} 0 & 0.2 & 0.1 \end{bmatrix}^T, \quad B_1 = B_2 = \begin{bmatrix} \frac{V_mT_m}{l} & 0 & 0 \end{bmatrix}^T, \\ X_1 &= \begin{bmatrix} 0 & 0 & 0.0018 \end{bmatrix}^T, \quad Y_{11} = Y_{12} = \begin{bmatrix} \frac{V_mT_m}{L} & 0 & 0 \end{bmatrix}, \\ D_1 &= D_2 = \frac{V_mT_m}{l}. \end{aligned}$$

By setting the system parameters in this example to be the same as [4], and substituting these parameters into the above known parameter matrices, one can get:

$$\begin{aligned} A_1 &= \begin{bmatrix} 1.25 & 0 & 0 \\ -0.25 & 1 & 0 \\ 0.1875 & -1.5 & 1 \end{bmatrix}, \quad A_2 = \begin{bmatrix} 1.25 & 0 & 0 \\ -0.25 & 1 & 0 \\ 0.0031 & -0.025 & 1 \end{bmatrix}, \\ E_1 &= E_2 = \begin{bmatrix} 0 & 0.2 & 0.1 \end{bmatrix}^T, \quad B_1 = B_2 = \begin{bmatrix} -0.4286 & 0 & 0 \end{bmatrix}^T, \\ X_x &= \begin{bmatrix} 0 & 0 & 0.0018 \end{bmatrix}^T, \quad Y_{11} = Y_{12} = \begin{bmatrix} -0.25 & 0 & 0 \end{bmatrix}, \\ D_1 &= D_2 = -0.4286, \end{aligned}$$

and the membership functions can be described as:

$$\rho_1(k) = \frac{\sin(\delta(k))}{\delta(k)}, \quad \rho_2(k) = 1 - \rho_1(\delta(k)). \tag{2.33}$$

In this simulation example, set the performance index $\gamma = 1$, quantization range $M_u = 100$ and quantization error bound $\Delta_u = 0.1$ of dynamic quantizer. According to the two-step design method proposed in Algorithm 2.1, one gets the scalar variables $\tau_u = 232.9847$, $\nu_{u1} = 0.0468$, $\nu_{u2} = 0.0468$, and the robust controller gain $K = \begin{bmatrix} 2.4138 & -1.2604 & 0.0966 \end{bmatrix}$.

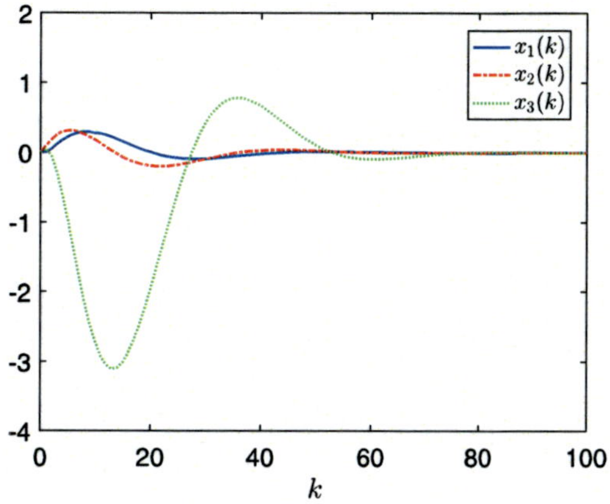

Fig. 2.2 Response of the system state $x(k)$

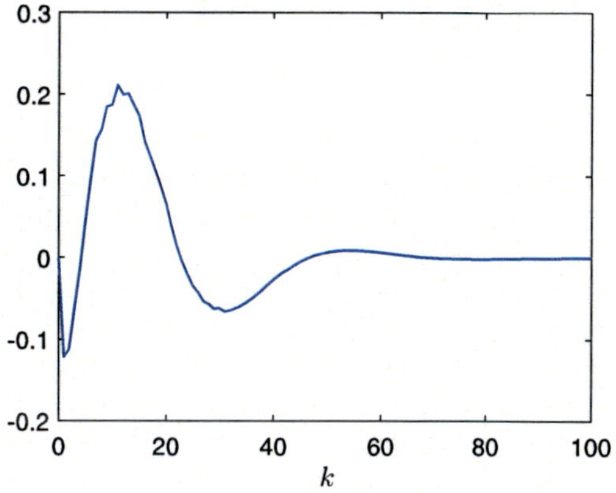

Fig. 2.3 Response of the control input $u(k)$

Define the initial condition $x(0) = \begin{bmatrix} 0 & 0 & 0 \end{bmatrix}^T$ and the disturbance noise signal $w(k) = 0.5e^{-0.1k}\cos(0.1k)$. Then, Fig. 2.2 presents the dynamic response of system state $x(k)$. Figure 2.3 shows the response of the control input signal $u(k)$. Figure 2.7 presents the dynamic quantizer's parameter $\mu_u(k)$. Through the adjusting rule in Remark 2.2, it shows that the adaptive adjustment of dynamic quantizer's parameters is related to the control input $u(k)$. Figure 2.4 shows the response of the quantized input signal $\mathbb{Q}(u(k))$. Figure 2.5 presents the response of the quantization error $\mathbb{Q}(u(k)) - u(k)$. Figure 2.6 presents the real-time values of the ratio of $\sqrt{\sum_{\zeta=0}^{k} z^T(\zeta)z(\zeta)/\sum_{\zeta=0}^{k} w^T(\zeta)w(\zeta)}$. Given the disturbance noise signal $w(k)$, this ratio eventually tends to stabilize, and the convergence value is lower than the performance index $\gamma = 1$. In conclusion, these simulation results show that the proposed two-step design method is robust and can ensure that the quantized fuzzy closed-loop system (2.10) satisfies the prescribed $\mathscr{H}_\infty$ performance.

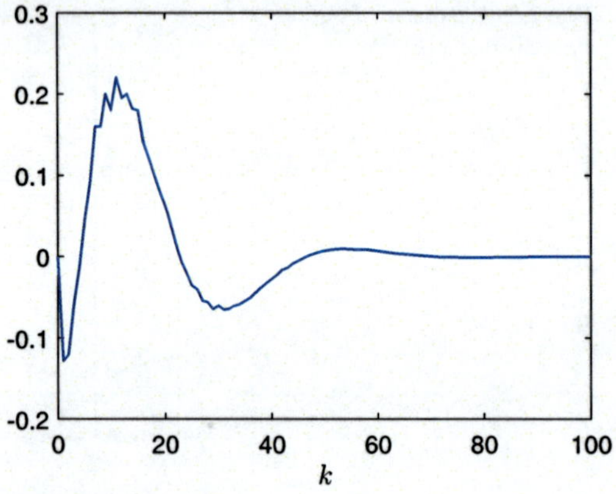

Fig. 2.4 Response of the quantized input signal $\mathbb{Q}(u(k))$

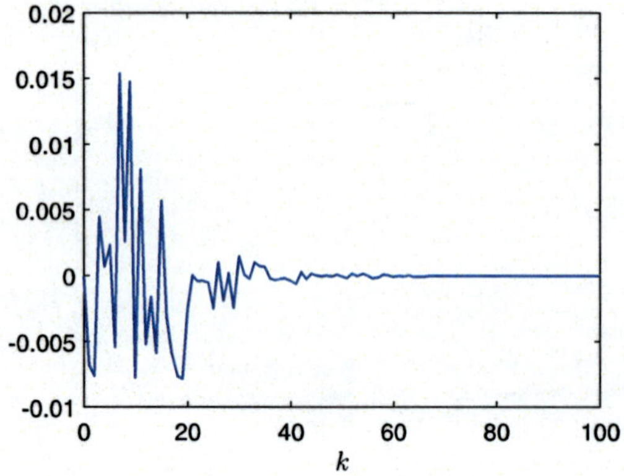

Fig. 2.5 Response of the quantization error $\mathbb{Q}(u(k)) - u(k)$

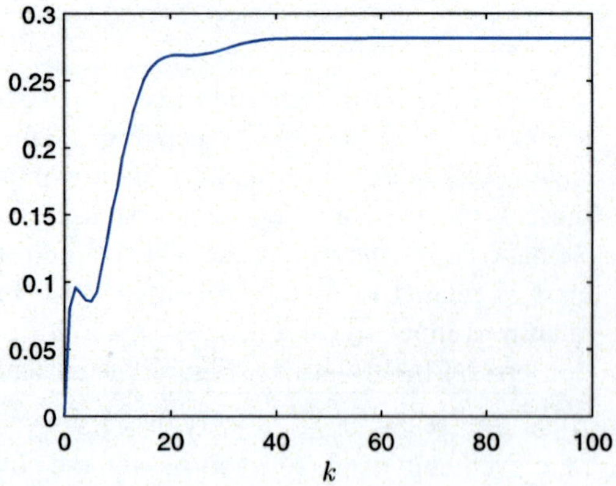

Fig. 2.6 History of $\sqrt{\sum_{\zeta=0}^{k} z^T(\zeta)z(\zeta) / \sum_{\zeta=0}^{k} w^T(\zeta)w(\zeta)}$

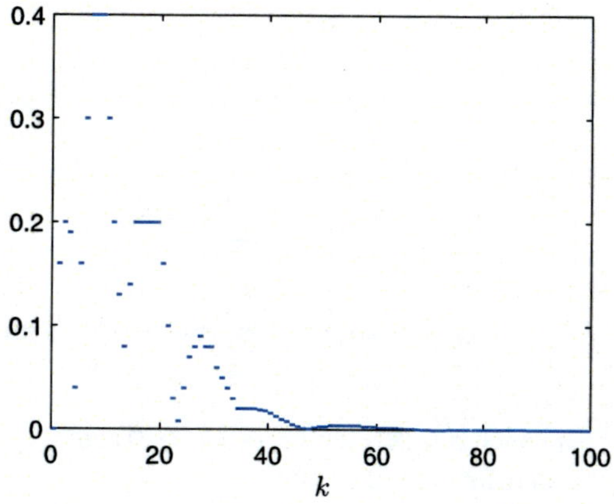

Fig. 2.7 Dynamic quantizer's parameter $\mu_u(k)$

2.2 Non-fragile Control with State Quantization

In this section, a new two-step strategy is proposed to study the non-fragile $\mathscr{H}_\infty$ control problem of discrete-time fuzzy systems, which can directly provide a simpler and more intuitive design for the non-fragile controller and dynamic quantizer based on the proposed conditions, and make the closed-loop system meet the same specified $\mathscr{H}_\infty$ performance index as the one without quantization.

2.2.1 *Problem Formulation*

Consider a class of T–S fuzzy models as follows:

Plant Rule i_{th}: **IF** $\varepsilon_1(k)$ is M_{1i}, $\varepsilon_2(k)$ is M_{2i}, and, …, and $\varepsilon_d(k)$ is M_{di}, **THEN**

$$\begin{aligned} x(k+1) &= A_i x(k) + B_i u(k) + E_i w(k), \\ z(k) &= C_i x(k) + D_i u(k) + F_i w(k), \end{aligned} \tag{2.34}$$

where $x(k) \in \mathbb{R}^{n_x}$ and $u(k) \in \mathbb{R}^{n_u}$ represent the state variable, control input, respectively; $w(k) \in \mathbb{R}^{n_w}$ denotes the noise signal in $\ell_2[0, \infty)$; $z(k) \in \mathbb{R}^{n_z}$ denotes the controlled output. Matrices $A_i \in \mathbb{R}^{n_x \times n_x}$, $B_i \in \mathbb{R}^{n_x \times n_u}$, $C_i \in \mathbb{R}^{n_z \times n_x}$, $D_i \in \mathbb{R}^{n_z \times n_u}$, $E_i \in \mathbb{R}^{n_x \times n_w}$, and $F_i \in \mathbb{R}^{n_z \times n_w}$, $i = 1, 2, \ldots, r$ represent the constant matrices of appropriate dimensions; $\varepsilon_1(k), \varepsilon_2(k), \ldots, \varepsilon_d(k)$ represent the premise variables; M_{pi}, $i = 1, 2, \ldots, r$, $p = 1, 2, \ldots, d$ represents the fuzzy sets; r is the number of fuzzy rules.

As in Sect. 2.1, the T–S fuzzy model (2.34) can be deduced as follows:

$$\begin{aligned} x(k+1) &= A(\rho)x(k) + B(\rho)u(k) + E(\rho)w(k), \\ z(k) &= C(\rho)x(k) + D(\rho)u(k) + F(\rho)w(k), \end{aligned} \tag{2.35}$$

where

$$A(\rho) = \sum_{i=1}^{r} \rho_i(\varepsilon(k))A_i, \quad B(\rho) = \sum_{i=1}^{r} \rho_i(\varepsilon(k))B_i,$$
$$C(\rho) = \sum_{i=1}^{r} \rho_i(\varepsilon(k))C_i, \quad D(\rho) = \sum_{i=1}^{r} \rho_i(\varepsilon(k))D_i,$$
$$E(\rho) = \sum_{i=1}^{r} \rho_i(\varepsilon(k))E_i, \quad F(\rho) = \sum_{i=1}^{r} \rho_i(\varepsilon(k))F_i,$$

in which $\rho_i(\varepsilon(k))$ with $\rho_i(\varepsilon(k)) \geq 0$ and $\sum_{i=1}^{r} \rho_i(\varepsilon(k)) = 1$ is the membership function defined in (2.3) and (2.4).

In this section, it is assumed that the system state is available and needs to be quantized by the quantizer before being sent to the data channel, then the dynamic quantizer which is the same as the one in (2.7) is adopted to quantize the system state $x(k)$.

According to the definition in Sect. 2.1, the quantized state signal can be described as:

$$\mathbb{Q}(x(k)) = \mu_x(k) q\left(\frac{x(k)}{\mu_x(k)}\right). \tag{2.36}$$

Due to the inaccuracy of actuator implementation process, environment change, and other factors, the controller gain may have some uncertainties to a certain extent. Therefore, the non-fragile controller based on state feedback can be described as follows:

$$u(k) = (K + \Delta_K)\mathbb{Q}(x(k)), \tag{2.37}$$

where K represents the controller gain, $\Delta_K = K_1 \Delta_k(k) K_2$ denotes gain uncertainty, K_1 and K_2 are known matrices, and $\Delta_k^T(k)\Delta_k(k) \leq I$ is an uncertain term.

Substituting (2.36) and (2.37) into (2.35), the quantized fuzzy closed-loop system can be given as follows:

$$\begin{aligned} x(k+1) &= \Big(A(\rho) + B(\rho)K + B(\rho)\Delta_K\Big)x(k) + \Big(B(\rho)K \\ &\quad + B(\rho)\Delta_K\Big)e_x(k) + E(\rho)w(k), \\ z(k) &= \Big(C(\rho) + D(\rho)K + D(\rho)\Delta_K\Big)x(k) + \Big(D(\rho)K \\ &\quad + D(\rho)\Delta_K\Big)e_x(k) + F(\rho)w(k), \end{aligned} \tag{2.38}$$

where $e_x(k) = \mu_x(k)\left(q\left(\frac{x(k)}{\mu_x(k)}\right) - \frac{x(k)}{\mu_x(k)}\right)$.

The quantized non-fragile state feedback $\mathscr{H}_\infty$ control problem considered in this section is to design the non-fragile state feedback controller in (2.37) and the dynamic quantizer in (2.36) such that

(1) The quantized fuzzy closed-loop system in (2.38) is asymptotically stable when $w(k) = 0$.
(2) The quantized fuzzy closed-loop system in (2.38) has a prescribed $\mathscr{H}_\infty$ performance γ, i.e., under the initial condition $x(0) = 0$, $\sum_{k=0}^{\infty} z^T(k)z(k) < \gamma^2 \sum_{k=0}^{\infty} w^T(k)w(k)$ is satisfied for any nonzero $w(k) \in \ell_2[0, \infty)$.

Remark 2.3 For the quantized closed-loop system without uncertainty, [5–8] proposed some effective two-step methods based on system structure constraints to ensure $\mathscr{H}_\infty$ performance, that is, the system matrices, $F = 0$ [5], or $D = 0$ [6, 7], or $D = 0$ and $F = 0$ [8]. Once $D \neq 0$ and $F \neq 0$, it will generate a product term between $e_x(k)$ and $w(k)$ [5–8], that is, the design of the quantizer parameter $\mu_x(k)$ may depend on $w(k)$, which will increase the design difficulty. It means that these results are not applicable to system (2.38) synchronously with the control input signal and the external disturbance in $z(k)$. Therefore, compared with the existing results [5–8], the design strategy proposed in this section does not have these constraints and is more general.

2.2.2 Main Results

Theorem 2.3 *Based on the quantized fuzzy closed-loop system (2.38), given a scalar $\gamma > 0$, for known quantization range bound M_x and quantization error Δ_x, if there exist matrix $P > 0$, scalars $\sigma_x > 0$ and $\tau_x > 0$, such that the following matrix inequality holds:*

$$\chi_1^T P \chi_1 + \chi_2^T \chi_2 + \tau_x \left(\chi_3^T \chi_3 - diag\{0, I, 0\} \right) - diag\{P, 0, \gamma^2 I\} < 0, \quad (2.39)$$

where

$$\chi_1 = \begin{bmatrix} A(\rho) + B(\rho)(K + \Delta_K) & B(\rho)(K + \Delta_K) & E(\rho) \end{bmatrix},$$
$$\chi_2 = \begin{bmatrix} C(\rho) + D(\rho)(K + \Delta_K) & D(\rho)(K + \Delta_K) & F(\rho) \end{bmatrix},$$
$$\chi_3 = \begin{bmatrix} 2\sqrt{\sigma_x}\Delta_x & 0 & 0 \end{bmatrix},$$

and the online adjustment strategy of dynamic quantizer's parameter $\mu_x(k)$ is given as:

$$\sqrt{\sigma_x}\,\|x(k)\| \leq \mu_x(k) \leq 2\sqrt{\sigma_x}\,\|x(k)\|\,. \quad (2.40)$$

Then, through the online adjustment strategy (2.40), the given quantized feedback controller (2.37) can guarantee that the quantized fuzzy closed-loop system (2.38) is asymptotically stable with the prescribed $\mathscr{H}_\infty$ performance γ. And the quantizer range M_x is set to be large enough, that is, $M_x > 1/\sqrt{\sigma_x}$.

Proof Substituting (2.36) into (2.6), we can obtain that:

$$\left\| q\left(\frac{x(k)}{\mu_x(k)}\right) - \frac{x(k)}{\mu_x(k)} \right\| \leq \Delta_x, \tag{2.41}$$

where $\|x(k)/\mu_x(k)\| \leq M_x$. If the dynamic parameter $\mu_x(k)$ of the quantizer (2.36) is designed as (2.40), one has that:

$$\frac{1}{2\sqrt{\sigma_x}}\mu_x(k) \leq \|x(k)\| \leq \frac{1}{\sqrt{\sigma_x}}\mu_x(k) < M_x\mu_x(k). \tag{2.42}$$

Consider the homogeneity property of Euclidean norm and combine the equation $e_x(k) = \mu_x(k)\left(q\left(\frac{x(k)}{\mu_x(k)}\right) - \frac{x(k)}{\mu_x(k)}\right)$. Then the following condition can be obtained,

$$\|e_x(k)\| \leq 2\sqrt{\sigma_x}\Delta_x \|x(k)\|. \tag{2.43}$$

Defining $\zeta(k) = \left[x^T(k)\ e_x^T(k)\ w^T(k)\right]^T$, the following conclusion can be obtained:

$$\zeta^T(k)\left(\chi_3^T\chi_3 - diag\{0,\ I,\ 0\}\right)\zeta(k) \geq 0. \tag{2.44}$$

If the matrix inequality (2.39) is satisfied, then considering the variable $\zeta(k) \neq 0$, the following inequality can be guaranteed

$$\zeta^T(k)\left(\Xi_1 + \Xi_2\right)\zeta(k) < 0, \tag{2.45}$$

with

$$\Xi_1 = \chi_1^T P\chi_1 + \chi_2^T\chi_2 - diag\{P,\ 0,\ \gamma^2 I\},$$
$$\Xi_2 = \tau_x\left(\chi_3^T\chi_3 - diag\{0,\ I,\ 0\}\right),$$

where the scalar $\tau_x > 0$. Combining (2.44) with scalar $\sigma_x > 0$ can give $\zeta^T(k)\Xi_1 \zeta(k) < 0$. Next, constructing the Lyapunov function as $V(x(k)) = x^T(k)Px(k),\ P > 0$, one can get

$$V(x(k+1)) - V(x(k)) + z^T(k)z(k) - \gamma^2 w^T(k)w(k) = \zeta^T(k)\Xi_1\zeta(k) < 0, \tag{2.46}$$

which shows that the $\mathcal{H}_\infty$ performance γ can be guaranteed for the quantized fuzzy closed-loop system (2.38) [9, 10]. The proof is completed.

It should be noted that the performance condition (2.39) in Theorem 2.3 adds some additional terms to the standard $\mathcal{H}_\infty$ performance analysis criterion, such as Ξ_2, $B(\rho)K$, and $D(\rho)K$, which reflects the impact of the quantization error. In [1, 11], based on the one-step strategy, the controllers (filters) are designed by considering the quantization error (those additional terms), but compared with the controllers (filters) without quantization error, the system performance may be degraded. In Theorem 2.3, the controller and quantizer parameters are designed by a two-step method, and the $\mathcal{H}_\infty$ performance analysis conditions in (2.39) are applied to the quantized feedback $\mathcal{H}_\infty$ control design. This method can ensure that the quantized closed-loop system (2.38) has the same performance as the one without quantization.

Theorem 2.4 *Based on the quantized fuzzy closed-loop system (2.38), given a scalar $\gamma > 0$, for known quantization range bound M_x and quantization error Δ_x, if there exist matrices K, N, G, and $Q > 0$, scalar $\eta_x > 0$ such that the following matrix inequalities hold:*

$$M_x > \eta_x, \tag{2.47}$$

$$\Lambda(\rho) = \begin{bmatrix} Q - G - G^T & * & * & * & * & * \\ 0 & -\gamma^2 I & * & * & * & * \\ A(\rho)G + B(\rho)N + B(\rho)\Delta_K G & E(\rho) & -Q & * & * & * \\ C(\rho)G + D(\rho)N + D(\rho)\Delta_K G & F(\rho) & 0 & -I & * & * \\ 0 & 0 & (B(\rho)(K + \Delta_K))^T & (D(\rho)(K + \Delta_K))^T & -\eta_x I & * \\ 2\Delta_x G & 0 & 0 & 0 & 0 & -\eta_x I \end{bmatrix} < 0, \tag{2.48}$$

and the online adjustment strategy of dynamic quantizer's parameter $\mu_x(k)$ is given as:

$$\frac{1}{\eta_x} \|x(k)\| \leq \mu_x(k) \leq \frac{2}{\eta_x} \|x(k)\| . \tag{2.49}$$

Then, through the online adjustment strategy (2.49), the given quantized feedback controller (2.37) can guarantee that the quantized fuzzy closed-loop system (2.38) is asymptotically stable with the prescribed $\mathcal{H}_\infty$ performance γ.

Proof By using Schur complement (Lemma 1.1) for the matrix inequality (2.39) in Theorem 2.3, the following inequality can be obtained

$$\begin{bmatrix} -P & * & * & * & * & * \\ 0 & -\tau_x I & * & * & * & * \\ 0 & 0 & -\gamma^2 I & * & * & * \\ A(\rho) + B(\rho)(K + \Delta_K) & B(\rho)(K + \Delta_K) & E(\rho) & -P^{-1} & * & * \\ C(\rho) + D(\rho)(K + \Delta_K) & D(\rho)(K + \Delta_K) & F(\rho) & 0 & -I & * \\ 2\Delta_x & 0 & 0 & 0 & 0 & -\frac{1}{\tau_x \sigma_x} I \end{bmatrix} < 0. \tag{2.50}$$

By defining $\eta_x = \tau_x = \frac{1}{\sqrt{\sigma_x}}$ and $P = Q^{-1}$, the inequality (2.50) can be rewritten as

$$\begin{bmatrix} -Q^{-1} & * & * & * & * & * \\ 0 & -\eta_x I & * & * & * & * \\ 0 & 0 & -\gamma^2 I & * & * & * \\ A(\rho)+B(\rho)(K+\Delta_K) & B(\rho)(K+\Delta_K) & E(\rho) & -Q & * & * \\ C(\rho)+D(\rho)(K+\Delta_K) & D(\rho)(K+\Delta_K) & F(\rho) & 0 & -I & * \\ 2\Delta_x & 0 & 0 & 0 & 0 & -\eta_x I \end{bmatrix} < 0. \quad (2.51)$$

Set a nonsingular matrix Υ which is expressed as

$$\Upsilon = \begin{bmatrix} G^T & 0 & 0 & 0 & 0 & 0 \\ 0 & 0 & I & 0 & 0 & 0 \\ 0 & 0 & 0 & I & 0 & 0 \\ 0 & 0 & 0 & 0 & I & 0 \\ 0 & I & 0 & 0 & 0 & 0 \\ 0 & 0 & 0 & 0 & 0 & I \end{bmatrix},$$

and perform the congruence transformation with Υ to (2.51). Further, based on the following fact

$$-(G-Q)^T Q^{-1}(G-Q) \le 0,$$

where Q is a symmetric matrix and $Q > 0$, which means that $-G^T Q^{-1} G \le -G^T - G + Q$, the inequality (2.48) is obtained by defining a new matrix variable $N = KG$. The proof is completed.

It should be pointed out that, unlike the standard LMIs that cannot be solved directly, because $N = KG$, (2.48) of Theorem 2.4 is a bilinear matrix inequality (BLMI) on matrix variables G and K. Next, the following algorithm based on Theorem 2.4 and Lemma 1.2 is given, which proposes a new two-step strategy to realize the design of quantized non-fragile feedback controller (2.37) with adjusting strategy (2.49).

Algorithm 2.2 For the quantized fuzzy closed-loop system (2.38), it is asymptotically stable under the given $\mathcal{H}_\infty$ performance index γ:

Step 1. Finding the matrix variables $Q > 0$, G, and N satisfying $\Lambda_i < 0$ for $i = 1, \ldots, r$, where

$$\Lambda_i = \begin{bmatrix} Q-G-G^T & * & * & * & * \\ 0 & -\gamma^2 I & * & * & * \\ A_i G + B_i N & E_i & -Q+\alpha B_i K_1 K_1^T B_i^T & * & * \\ C_i G + D_i N & F_i & \alpha D_i K_1 K_1^T B_i^T & -I+\alpha D_i K_1 K_1^T D_i^T & * \\ K_2 G & 0 & 0 & 0 & -\alpha I \end{bmatrix}. \quad (2.52)$$

Step 2. Solving the following LMI (2.53) with Q, G, N obtained by Step 1, γ and $K = NG^{-1}$ ($\Lambda_i < 0$, with $i = 1, \ldots, r$ means that the matrix G is nonsingular), and (2.47) can determine the value of scalar variable η_x:

$$\begin{bmatrix} Q-G-G^T & * & * & * & * & * & * \\ 0 & -\gamma^2 I & * & * & * & * & * \\ A_iG+B_iN & E_i & -Q+\alpha B_iK_1K_1^TB_i^T & * & * & * & * \\ C_iG+D_iN & F_i & \alpha D_iK_1K_1^TB_i^T & -I+\alpha D_iK_1K_1^TD_i^T & * & * & * \\ K_2G & 0 & 0 & 0 & -\alpha I & * & * \\ 0 & 0 & (B_iK)^T & (D_iK)^T & K_2^T & -\eta_x I & * \\ 2\Delta_x G & 0 & 0 & 0 & 0 & 0 & -\eta_x I \end{bmatrix} < 0. \tag{2.53}$$

According to the online adjustment strategy in (2.49), the dynamic quantizer's parameter $\mu_x(k)$ can be adjusted by the scalar variable η_x obtained in Step 2. Then, through the online adjustment strategy (2.49), the given quantized feedback controller (2.37) can guarantee that the quantized fuzzy closed-loop system (2.38) is asymptotically stable with the prescribed $\mathcal{H}_\infty$ performance γ.

Remark 2.4 In Algorithm 2.2, according to the standard $\mathcal{H}_\infty$ controller design conditions of discrete-time systems, $\Lambda_i < 0$ in Step 1 can ensure the desired system performance, that is, the algorithm can provide a quantized state feedback control design with the same one without quantization.

Remark 2.5 In Step 1, the state feedback controller gain K is obtained by solving the standard LMI $\Lambda_i < 0$. In Step 2, matrix inequalities (2.47) and (2.53) determine the quantizer parameter η_x based on standard LMIs by using the matrix variables Q, G, and N obtained in Step 1. Compared with [5–8], the results show that the proposed quantized control design strategy can be implemented in the framework of LMI more simply and intuitively.

Remark 2.6 The quantized signal can be obtained by determining the dynamic quantizer's parameter $\mu_x(k)$ as a unique value and transformed it into a communication channel. In this section, the adjustment rule for updating the parameter $\mu_x(k)$ of the quantizer on both sides of the encoder/decoder is the same as [1]:

The adjusting rule:

$$\mu_x(k) = \begin{cases} \text{floor}\left(2\vartheta_x \|x(k)\| \times 10^{\upsilon}\right) \times 10^{-\upsilon}, & 0 \le \vartheta_x \|x(k)\| < \frac{1}{2}, \\ 1, & \frac{1}{2} \le \vartheta_x \|x(k)\| < 1, \\ \text{floor}\left(2\vartheta_x \|x(k)\|\right), & 1 \le \vartheta_x \|x(k)\|, \end{cases}$$

where $\upsilon = \min\left\{\upsilon \in \mathcal{N}^+ | \left(2\vartheta_x \|x(k)\| \times 10^{\upsilon}\right) > 1\right\}$ and function floor(θ) represents the maximum integer that does not exceed θ. It is worth noting that, in Algorithm 2.2, the method of updating $\mu_x(k)$ is to apply the above adjustment strategy together with $\vartheta_x = \frac{1}{\eta_x}$.

2.2.3 Simulation Example

In this section, a numerical example shows the effectiveness of the proposed design method. Consider a discrete-time system in the form of (2.34), then related parameters are given as follows:

$$A_1 = \begin{bmatrix} 1.1159 & -7.1340 & 0.0000 \\ 0.0808 & 1.0211 & -0.0368 \\ 0.0222 & 0.6245 & 1.0880 \end{bmatrix}, \quad A_2 = \begin{bmatrix} 1.3254 & -8.8267 & 0.0000 \\ 0.0808 & 1.1163 & -0.0510 \\ 0.0002 & 0.7534 & 0.9967 \end{bmatrix},$$

$$B_1 = \begin{bmatrix} 0.2415 & 0.0000 & 0.0000 \\ 0.0000 & 0.3444 & 0.0000 \\ 0.0000 & 0.0000 & 0.2169 \end{bmatrix}, \quad B_2 = \begin{bmatrix} 0.2415 & 0.0000 & 0.0000 \\ 0.0000 & 0.3444 & 0.0000 \\ 0.0000 & 0.0000 & 0.2169 \end{bmatrix},$$

$$C_1 = C_2 = \begin{bmatrix} 1 & 0 & 0 \end{bmatrix}, \quad D_1 = D_2 = \begin{bmatrix} 0.2 & 0 & 0 \end{bmatrix}, \quad E_1^T = E_2^T = \begin{bmatrix} 0 & 0 & 3 \end{bmatrix},$$

$$F_1 = F_2 = 0.1,$$

and the membership functions are considered as follows:

$$\rho_1(k) = \frac{|\sin(k)|}{8}, \quad \rho_2(k) = 1 - \rho_1(k).$$

In (2.37), the known parameters are assumed to be

$$K_1 = \begin{bmatrix} 2.0000 \\ 1.0000 \\ 0.1000 \end{bmatrix}, \quad K_2^T = \begin{bmatrix} 0.1000 \\ 2.0000 \\ 0.1000 \end{bmatrix}.$$

For this numerical example, choose $\gamma = 3$, and set the quantizer's range $M_x = 100$ and error bound $\Delta_x = 0.01$. Then, by applying the feedback controller and quantizer parameters designed in Algorithm 2.2, it can be acquired that the scalar variable $\eta_x = 92.8149$ and the quantized feedback controller gain

$$K = \begin{bmatrix} -4.9991 & 0.1191 & 0.0024 \\ -0.2334 & -1.9222 & 0.1293 \\ 0.0146 & -4.3863 & -4.6732 \end{bmatrix}.$$

In order to further illustrate the effectiveness of the designed quantized feedback controller, the simulation results are carried out by using MATLAB with the initial condition $x(0) = \begin{bmatrix} 0 & 0 & 0 \end{bmatrix}^T$, and the disturbance noise signal $w(k) = 0.5e^{-0.2k}\cos(0.2k)$. Then, Fig. 2.8 shows the dynamic response of the system state $x(k)$; Fig. 2.9 presents the response of the control input signal $u(k)$; Fig. 2.10 shows the real-time values of the ratio of $\sqrt{\sum_{j=0}^{k} z^T(j)z(j) / \sum_{j=0}^{k} w^T(j)w(j)}$, it can be observed from Fig. 2.10 that the convergence value of the ratio is 0.1512 which is below the given performance index $\gamma = 3$; Fig. 2.11 denotes the dynamic quantizer's

Fig. 2.8 Response of the system state $x(k)$

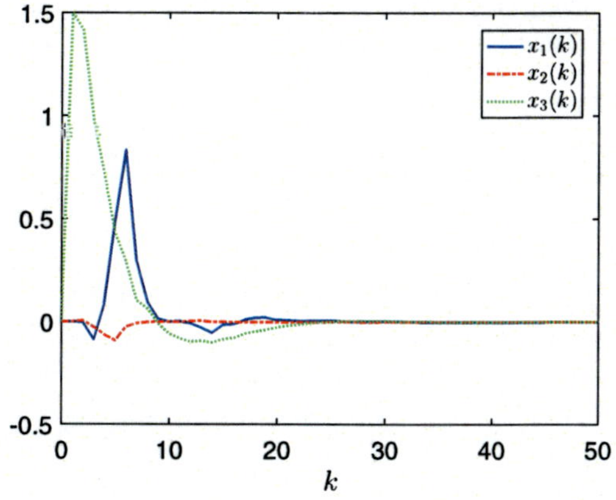

Fig. 2.9 Response of the control input $u(k)$

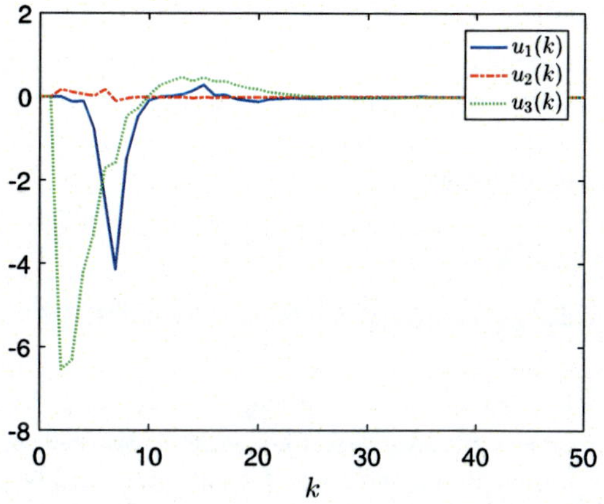

Fig. 2.10 History of $\sqrt{\sum_{j=0}^{k} z^T(j)z(j)/\sum_{j=0}^{k} w^T(j)w(j)}$

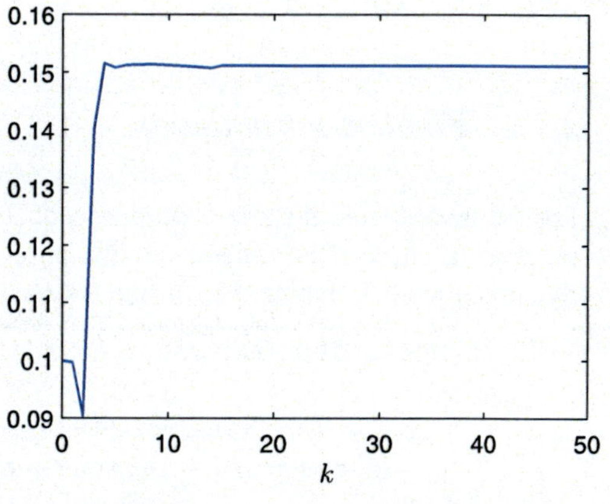

Fig. 2.11 Dynamic quantizer's parameter $\mu_x(k)$

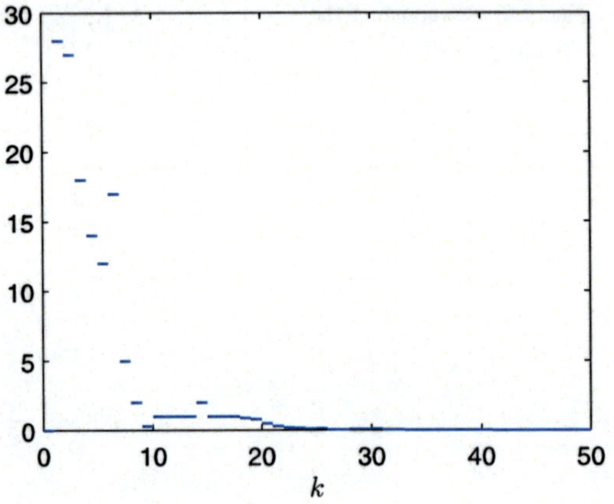

parameter $\mu_x(k)$, which is obtained by the adjusting rule in Remark 2.6. The simulation results of numerical example show that the two-step design method proposed in this section can ensure that the quantized fuzzy control system meets the specified performance.

2.3 Resilient Control with Input and Output Quantization

In this section, the resilient $\mathcal{H}_\infty$ controller for the quantized closed-loop system will be designed by descriptor representation approach in two cases. The measurement output signal and control input signal will be quantized by dynamic quantizers, and the conditions for the resilient controller and dynamic quantizers' parameters will be given by LMIs.

2.3.1 Problem Formulation

For the investigated quantized output feedback control problem in this section, we consider a class of nonlinear systems appropriated by the following T–S fuzzy dynamic model, in which the ith rule is described as follows:

Plant Rule i_{th}: **IF** $\varepsilon_1(t)$ is M_{1i}, $\varepsilon_2(t)$ is M_{2i}, and, ..., and $\varepsilon_d(t)$ is M_{di}, **THEN**

$$
\begin{aligned}
\dot{x}(t) &= (A_i + \Delta_{Ai})x(t) + B_i\mathbb{Q}(u(t)) + E_iw(t),\\
z(t) &= (C_i + \Delta_{Ci})x(t) + D_i\mathbb{Q}(u(t)) + F_iw(t),\\
y(t) &= (L_i + \Delta_{Li})x(t) + H_iw(t),
\end{aligned} \tag{2.54}
$$

where $x(t) \in \mathbb{R}^{n_x}$ is the state variable; $u(t) \in \mathbb{R}^{n_u}$ is the control input; $z(t) \in \mathbb{R}^{n_z}$ is the controlled output; $y(t) \in \mathbb{R}^{n_y}$ is the measurement output; $w(t) \in \mathbb{R}^{n_w}$ is the disturbance signal that is assumed to be the arbitrary signal in $\mathscr{L}_2[0, \infty)$. $\varepsilon(t) = [\varepsilon_1(t), \varepsilon_2(t), \ldots, \varepsilon_d(t)]$, where $\varepsilon_p(t)$, $p = 1, 2, \ldots, d$ are premise variables and measurable; M_{pi}, $i = 1, 2, \ldots, r$, $p = 1, 2, \ldots, d$ are the fuzzy sets; r is the number of fuzzy rules. $A_i \in \mathbb{R}^{n_x \times n_x}$, $B_i \in \mathbb{R}^{n_x \times n_u}$, $E_i \in \mathbb{R}^{n_x \times n_w}$, $C_i \in \mathbb{R}^{n_z \times n_x}$, $D_i \in \mathbb{R}^{n_z \times n_u}$, $F_i \in \mathbb{R}^{n_z \times n_w}$, $L_i \in \mathbb{R}^{n_y \times n_x}$, and $H_i \in \mathbb{R}^{n_y \times n_w}$ for $i = 1, 2, \ldots, r$ are known system matrices. $\Delta_{Ai} \in \mathbb{R}^{n_x \times n_x}$, $\Delta_{Ci} \in \mathbb{R}^{n_z \times n_x}$, and $\Delta_{Li} \in \mathbb{R}^{n_y \times n_x}$ for $i = 1, 2, \ldots, r$ are unknown real norm-bounded matrices, which represent time-varying parameter uncertainties. And the uncertainties can be given as the following form:

$$\begin{pmatrix} \Delta_{Ai} \\ \Delta_{Ci} \\ \Delta_{Li} \end{pmatrix} = \begin{pmatrix} X_{Ai} \\ X_{Ci} \\ X_{Li} \end{pmatrix} \Delta_x(t) Y, \tag{2.55}$$

where X_{Ai}, X_{Ci}, X_{Li}, and Y are known real constant matrices with appropriate dimensions, $\Delta_x(t)$ is an unknown real matrix satisfying $\Delta_x^T(t)\Delta_x(t) \leq I$.

Then according to Sect. 1.1, the T–S fuzzy model (2.54) is inferred as follows:

$$\begin{aligned} \dot{x}(t) &= \tilde{A}(\rho)x(t) + B(\rho)\mathbb{Q}(u(t)) + E(\rho)w(t), \\ z(t) &= \tilde{C}(\rho)x(t) + D(\rho)\mathbb{Q}(u(t)) + F(\rho)w(t), \\ y(t) &= \tilde{L}(\rho)x(t) + H(\rho)w(t), \end{aligned} \tag{2.56}$$

where

$$\tilde{A}(\rho) = A(\rho) + \Delta_A(\rho), \quad \tilde{C}(\rho) = C(\rho) + \Delta_C(\rho), \quad \tilde{L}(\rho) = L(\rho) + \Delta_L(\rho),$$

$$A(\rho) = \sum_{i=1}^{r} \rho_i(\varepsilon(t))A_i, \quad B(\rho) = \sum_{i=1}^{r} \rho_i(\varepsilon(t))B_i, \quad E(\rho) = \sum_{i=1}^{r} \rho_i(\varepsilon(t))E_i,$$

$$C(\rho) = \sum_{i=1}^{r} \rho_i(\varepsilon(t))C_i, \quad D(\rho) = \sum_{i=1}^{r} \rho_i(\varepsilon(t))D_i, \quad F(\rho) = \sum_{i=1}^{r} \rho_i(\varepsilon(t))F_i,$$

$$L(\rho) = \sum_{i=1}^{r} \rho_i(\varepsilon(t))L_i, \quad H(\rho) = \sum_{i=1}^{r} \rho_i(\varepsilon(t))H_i,$$

$$\Delta_A(\rho) = \sum_{i=1}^{r} \rho_i(\varepsilon(t))X_{Ai}\Delta_x(t)Y = X_A(\rho)\Delta_x(t)Y,$$

$$\Delta_C(\rho) = \sum_{i=1}^{r} \rho_i(\varepsilon(t))X_{Ci}\Delta_x(t)Y = X_C(\rho)\Delta_x(t)Y,$$

$$\Delta_L(\rho) = \sum_{i=1}^{r} \rho_i(\varepsilon(t))X_{Li}\Delta_x(t)Y = X_L(\rho)\Delta_x(t)Y.$$

The static output feedback controller considered in this chapter is given as follows

$$u(t) = (K + \Delta_K)\mathbb{Q}(y(t)), \tag{2.57}$$

where $\mathbb{Q}(y(t))$ is the input of the controller, K is the controller gain to be determined. Δ_K is parameter variation matrix and assumed to be the form

$$\Delta_K = R\Delta_k(t)W, \tag{2.58}$$

where R and W are known real constant matrices with appropriate dimensions, and $\Delta_k(t)$ describes the system uncertainty which is assumed to satisfy $\Delta_k^T(t)\Delta_k(t) \leq I$.

In (2.56) and (2.57), $\mathbb{Q}(y(t))$ and $\mathbb{Q}(u(t))$ represent quantized outputs of $y(t)$ and $u(t)$ by two dynamic quantizers, respectively, and are expressed as

$$\mathbb{Q}(y(t)) = \mu_y(t)q\left(\frac{y(t)}{\mu_y(t)}\right), \tag{2.59}$$

$$\mathbb{Q}(u(t)) = \mu_u(t)q\left(\frac{u(t)}{\mu_u(t)}\right). \tag{2.60}$$

In this chapter, for the quantized resilient control of the fuzzy system (2.56), the considered dynamic quantizer with general form is defined as the one given in Sect. 2.1, and assumed to be the one-parameter family of quantizer as $\mathbb{Q}(\omega(t)) = \mu_\omega(t)q\left(\frac{\omega(t)}{\mu_\omega(t)}\right)$, $\omega = y, u$, which comprises a dynamic parameter $\mu_\omega(t)$ and a static quantizer $q(\cdot)$ and satisfies the following conditions:

$$\|\mathbb{Q}(\omega(t)) - \omega(t)\| \leq \mu_\omega(t)\Delta_\omega, \quad if \ \|\omega(t)\| \leq \mu_\omega(t)M_\omega, \tag{2.61}$$

$$\|\mathbb{Q}(\omega(t))\| > \mu_\omega(t)M_\omega - \mu_\omega(t)\Delta_\omega, \quad if \ \|\omega(t)\| > \mu_\omega(t)M_\omega. \tag{2.62}$$

When the quantizer is unsaturated, the condition (2.61) provides a bound for the quantization error. The condition (2.62) gives a way for detecting the possibility of saturation. Δ_ω represents the quantization error bound and M_ω denotes the range of the quantizer $q(\cdot)$, respectively. In fact, it is not hard to notice that the range and the quantization error of this class of dynamic quantizers are $\mu_\omega(t)M_\omega$ and $\mu_\omega(t)\Delta_\omega$, respectively. That is to say, we can adjust the dynamic parameter to realize a suitable range and quantization error, and guarantee the desired system performance.

2.3.2 *Resilient $\mathcal{H}_\infty$ Controller Design with Quantization*

In this section, the output feedback resilient $\mathcal{H}_\infty$ control design problem for the fuzzy system (2.56) is addressed with the effect of quantized measurement output $\mathbb{Q}(y(t))$ and quantized control input $\mathbb{Q}(u(t))$. By structuring descriptor representation, this section will propose two different design conditions for the quantized output feedback $\mathcal{H}_\infty$ control.

Case A: Combining (2.56) and (2.57) can give

$$\begin{aligned}
\dot{x}(t) &= \tilde{A}(\rho)x(t) + B(\rho)\left(\mathbb{Q}(u(t)) - u(t)\right) + B(\rho)u(t) + E(\rho)w(t), \\
z(t) &= \tilde{C}(\rho)x(t) + D(\rho)\left(\mathbb{Q}(u(t)) - u(t)\right) + D(\rho)u(t) + F(\rho)w(t), \\
y(t) &= \tilde{L}(\rho)x(t) + H(\rho)w(t), \\
u(t) &= \tilde{K}(\mathbb{Q}(y(t)) - y(t)) + \tilde{K}y(t),
\end{aligned} \tag{2.63}$$

where $\tilde{K} = K + \Delta_K$. Then it can be rewritten into the following form based on the definitions of $\mathbb{Q}(y(t))$ and $\mathbb{Q}(u(t))$ in (2.59) and (2.60), respectively,

$$\begin{aligned}
\dot{x}(t) &= \tilde{A}(\rho)x(t) + B(\rho)\mu_u(t)\left(q\left(\frac{u(t)}{\mu_u(t)}\right) - \frac{u(t)}{\mu_u(t)}\right) + B(\rho)u(t) + E(\rho)w(t), \\
z(t) &= \tilde{C}(\rho)x(t) + D(\rho)\mu_u(t)\left(q\left(\frac{u(t)}{\mu_u(t)}\right) - \frac{u(t)}{\mu_u(t)}\right) + D(\rho)u(t) + F(\rho)w(t), \\
y(t) &= \tilde{L}(\rho)x(t) + H(\rho)w(t), \\
u(t) &= \tilde{K}\mu_y(t)\left(q\left(\frac{y(t)}{\mu_y(t)}\right) - \frac{y(t)}{\mu_y(t)}\right) + \tilde{K}y(t).
\end{aligned} \tag{2.64}$$

Further, it can be organized as

$$\begin{aligned}
\dot{x}(t) &= \tilde{A}(\rho)x(t) + B(\rho)e_u(t) + B(\rho)u(t) + E(\rho)w(t), \\
z(t) &= \tilde{C}(\rho)x(t) + D(\rho)e_u(t) + D(\rho)u(t) + F(\rho)w(t), \\
y(t) &= \tilde{L}(\rho)x(t) + H(\rho)w(t), \\
u(t) &= \tilde{K}e_y(t) + \tilde{K}y(t),
\end{aligned} \tag{2.65}$$

where $e_\omega(t) = \mu_\omega(t)\left(q\left(\frac{\omega(t)}{\mu_\omega(t)}\right) - \frac{\omega(t)}{\mu_\omega(t)}\right), \omega = y, u$. To cast the system into a descriptor form, the controller $u(t)$ in (2.57) is converted into following

$$0 \times \dot{u}(t) = \tilde{K}e_y(t) + \tilde{K}\left(\tilde{L}(\rho)x(t) + H(\rho)w(t)\right) - u(t). \tag{2.66}$$

As a result, from (2.65) and (2.66), we have the descriptor representation as follows:

$$\begin{aligned}
\mathbb{S}_A\dot{\eta}(t) &= \mathscr{A}(\rho)\eta(t) + \mathscr{M}e_y(t) + \mathscr{B}(\rho)e_u(t) + \mathscr{E}(\rho)w(t), \\
z(t) &= \mathscr{C}(\rho)\eta(t) + \mathscr{D}(\rho)e_u(t) + \mathscr{F}(\rho)w(t), \\
y(t) &= \mathscr{L}(\rho)\eta(t) + \mathscr{H}(\rho)w(t),
\end{aligned} \tag{2.67}$$

where $\eta(t) = \left[x^T(t) \;\; u^T(t)\right]^T$ and

$$\mathbb{S}_A = \begin{bmatrix} I & 0 \\ 0 & 0 \end{bmatrix}, \quad \mathcal{A}(\rho) = \begin{bmatrix} \tilde{A}(\rho) & B(\rho) \\ \tilde{K}\tilde{L}(\rho) & -I \end{bmatrix}, \quad \mathcal{M} = \begin{bmatrix} 0 \\ \tilde{K} \end{bmatrix},$$
$$\mathcal{B}(\rho) = \begin{bmatrix} B(\rho) \\ 0 \end{bmatrix}, \quad \mathcal{E}(\rho) = \begin{bmatrix} E(\rho) \\ \tilde{K}H(\rho) \end{bmatrix}, \quad \mathcal{C}(\rho) = \begin{bmatrix} \tilde{C}(\rho) & D(\rho) \end{bmatrix},$$
$$\mathcal{D}(\rho) = D(\rho), \quad \mathcal{F}(\rho) = F(\rho), \quad \mathcal{L}(\rho) = \begin{bmatrix} \tilde{L}(\rho) & 0 \end{bmatrix}, \quad \mathcal{H}(\rho) = H(\rho).$$

The quantized static output feedback $\mathcal{H}_\infty$ control problem considered in this section is to design the output feedback controller gain in (2.57) and the dynamic parameters in (2.59) and (2.60) such that

(1) The closed-loop system (2.67) is asymptotically stable when $w(t) = 0$.
(2) The closed-loop system (2.67) has a prescribed level γ of $\mathcal{H}_\infty$ noise attenuation, i.e., under the initial condition $x(0) = 0$, $\int_0^\infty z^T(t)z(t)dt < \gamma^2 \int_0^\infty w^T(t)w(t)dt$ is satisfied for any nonzero $w(t) \in \mathcal{L}_2[0, \infty)$.

To implement the quantized static output feedback $\mathcal{H}_\infty$ control design for the closed-loop system (2.67), the following condition is firstly presented to ensure the prescribed $\mathcal{H}_\infty$ performance.

Theorem 2.5 *Consider the quantized output feedback closed-loop system (2.67). For given scalar $\gamma > 0$, quantizers' ranges M_ω and error bounds Δ_ω, $\omega = y, u$, if there exist matrix P, positive scalars α_ω and δ_ω, $\omega = y, u$, such that the following matrix inequalities hold:*

$$P\mathbb{S}_A = \mathbb{S}_A^T P^T \geq 0, \tag{2.68}$$

$$M_\omega > \frac{1}{\delta_\omega}, \omega = y, u, \tag{2.69}$$

$$\begin{bmatrix} He\{P\mathcal{A}(\rho)\} & * & * \\ \Xi_1 & \Xi_2 & * \\ \Xi_3 & \Xi_4 & \Xi_5 \end{bmatrix} < 0, \tag{2.70}$$

where

$$\Xi_1 = \begin{bmatrix} P\mathcal{M} & P\mathcal{B}(\rho) & P\mathcal{E}(\rho) \end{bmatrix}^T, \quad \Xi_2 = diag\left\{ -\alpha_y I, \ -\alpha_u I, \ -\gamma^2 I \right\},$$
$$\Xi_3 = \begin{bmatrix} \mathcal{C}(\rho) \\ 2\alpha_y\delta_y\Delta_y\mathcal{L}(\rho) \\ 2\alpha_u\delta_u\Delta_u\mathcal{N} \end{bmatrix}, \quad \Xi_4 = \begin{bmatrix} 0 & \mathcal{D}(\rho) & \mathcal{F}(\rho) \\ 0 & 0 & 2\alpha_y\delta_y\Delta_y\mathcal{H}(\rho) \\ 0 & 0 & 0 \end{bmatrix},$$
$$\Xi_5 = diag\left\{ -I, -\alpha_y I, -\alpha_u I \right\}, \quad \mathcal{N} = \begin{bmatrix} 0 & I \end{bmatrix},$$

with on-line adjusting strategy for parameters $\mu_\omega(t)$ as:

$$\delta_\omega \|\omega(t)\| \leq \mu_\omega(t) \leq 2\delta_\omega \|\omega(t)\|, \ \omega = y, u. \tag{2.71}$$

Then, the controller (2.57) and the quantizers (2.59) and (2.60) with the adjusting strategy (2.71) can ensure that the closed-loop system (2.67) achieves the prescribed $\mathcal{H}_\infty$ performance γ.

Proof If the range M_y in (2.61) and dynamic parameter $\mu_y(t)$ of quantizer (2.59) satisfy (2.69) and (2.71), one has

$$\frac{1}{M_y}\|y(t)\| \leq \mu_y(t) \leq 2\delta_y\|y(t)\|, \tag{2.72}$$

i.e.,

$$\frac{1}{2\delta_y}\mu_y(t) \leq \|y(t)\| \leq M_y\mu_y(t). \tag{2.73}$$

For the measurement output $y(t)$, based on the property of the quantizer given in (2.73), it is effortless to obtain that $\left\|\frac{y(t)}{\mu_y(t)}\right\| \leq M_y$ from (2.61), then

$$\left\|q\left(\frac{y(t)}{\mu_y(t)}\right) - \frac{y(t)}{\mu_y(t)}\right\| \leq \Delta_y. \tag{2.74}$$

Considering the definition of $e_y(t)$ in (2.65) and the homogeneity property of Euclidean norm, and the inequalities (2.72) and (2.74), the following condition can be obtained

$$\begin{aligned}\|e_y(t)\| &= \left\|\mu_y(t)\left\{q\left(\frac{y(t)}{\mu_y(t)}\right) - \frac{y(t)}{\mu_y(t)}\right\}\right\| \\ &= \mu_y(t)\left\|q\left(\frac{y(t)}{\mu_y(t)}\right) - \frac{y(t)}{\mu_y(t)}\right\| \\ &\leq \mu_y(t)\Delta_y \\ &\leq 2\delta_y\Delta_y\|y(t)\|,\end{aligned} \tag{2.75}$$

i.e.,

$$e_y^T(t)e_y(t) \leq 4\delta_y^2\Delta_y^2 y^T(t)y(t). \tag{2.76}$$

Defining $\xi(t) = \begin{bmatrix}\eta^T(t) & e_y^T(t) & e_u^T(t) & w^T(t)\end{bmatrix}^T$, the above inequality can be rewritten as

$$\xi^T(t)\Big(Ha\{\mathcal{W}_2\} - diag\{0,\ I,\ 0,\ 0\}\Big)\xi(t) \geq 0, \tag{2.77}$$

where $\mathcal{W}_2 = \begin{bmatrix}2\delta_y\Delta_y\mathcal{L}(\rho) & 0 & 0 & 2\delta_y\Delta_y\mathcal{H}(\rho)\end{bmatrix}$. Similarly, combining (2.69) and (2.71), considering the definition of $e_u(t)$ in (2.65) and the homogeneity property of Euclidean norm, one gets

$$\|e_u(t)\| \leq 2\delta_u\Delta_u\|u(t)\|, \tag{2.78}$$

i.e.,

$$e_u^T(t)e_u(t) \leq 4\delta_u^2\Delta_u^2 u^T(t)u(t). \tag{2.79}$$

Further, the above inequality can be rewritten as

$$\xi^T(t)\Big(Ha\{\mathscr{W}_3\} - diag\{0,\ 0,\ I,\ 0\}\Big)\xi(t) \geq 0, \tag{2.80}$$

where $\mathscr{W}_3 = \begin{bmatrix} 2\delta_u \Delta_u \mathscr{N} & 0 & 0 & 0 \end{bmatrix}$. Now, choose a Lyapunov function as

$$V(\eta(t)) = \eta^T(t) P \mathbb{S}_A \eta(t),\ \ P\mathbb{S}_A = \mathbb{S}_A^T P^T \geq 0, \tag{2.81}$$

one gives

$$\begin{aligned}
&\dot{V}(\eta(t)) + z^T(t)z(t) - \gamma^2 w^T(t)w(t) \\
&\quad = \eta^T(t)P\mathbb{S}_A\dot{\eta}(t) + \dot{\eta}^T(t)\mathbb{S}_A^T P^T \eta(t) + z^T(t)z(t) - \gamma^2 w^T(t)w(t) \\
&\quad = \eta^T(t)P\Big(\mathscr{A}(\rho)\eta(t) + \mathscr{M}e_y(t) + \mathscr{B}(\rho)e_u(t) + \mathscr{E}(\rho)w(t)\Big) \\
&\qquad + \Big(\mathscr{A}(\rho)\eta(t) + \mathscr{M}e_y(t) + \mathscr{B}(\rho)e_u(t) + \mathscr{E}(\rho)w(t)\Big)^T P^T\eta(t) \\
&\qquad + Ha\{\mathscr{C}(\rho)\eta(t) + \mathscr{D}(\rho)e_u(t) + \mathscr{F}(\rho)w(t)\} - \gamma^2 w^T(t)w(t) \\
&\quad = \xi^T(t)\begin{bmatrix} He\{P\mathscr{A}(\rho)\} & * \\ \Xi_1 & diag\left\{0,\ 0,\ -\gamma^2 I\right\}\end{bmatrix}\xi(t) + \xi^T(t)Ha\{\mathscr{W}_1\}\xi(t),
\end{aligned}$$

where $\mathscr{W}_1 = \begin{bmatrix} \mathscr{C}(\rho) & 0 & \mathscr{D}(\rho) & \mathscr{F}(\rho) \end{bmatrix}$. Based on (2.70) and Schur complement (Lemma 1.1), it is easy to obtain

$$\begin{aligned}
&\begin{bmatrix} He\{P\mathscr{A}(\rho)\} & * \\ \Xi_1 & diag\left\{0,\ 0,\ -\gamma^2 I\right\}\end{bmatrix} + \alpha_y\Big(Ha\{\mathscr{W}_2\} - diag\{0,\ I,\ 0,\ 0\}\Big) \\
&\quad + \alpha_u\Big(Ha\{\mathscr{W}_3\} - diag\{0,\ 0,\ I,\ 0\}\Big) + Ha\{\mathscr{W}_1\} < 0.
\end{aligned} \tag{2.82}$$

Then based on inequalities (2.77) and (2.80), scalars $\alpha_y > 0$ and $\alpha_u > 0$, the following inequality

$$\xi^T(t)\left\{\begin{bmatrix} He\{P\mathscr{A}(\rho)\} & * \\ \Xi_1 & diag\left\{0,\ 0,\ -\gamma^2 I\right\}\end{bmatrix} + Ha\{\mathscr{W}_1\}\right\}\xi(t) < 0, \tag{2.83}$$

is obtained, which means that

$$\dot{V}(\eta(t)) + z^T(t)z(t) - \gamma^2 w^T(t)w(t) < 0, \tag{2.84}$$

can be guaranteed by the proposed conditions in Theorem 2.5. Further, considering zero initial condition and $V(\eta(\infty)) \geq 0$ with the definition of the Lyapunov function (2.81), one has

$$\int_0^\infty z^T(t)z(t)dt < \gamma^2 \int_0^\infty w^T(t)w(t)dt. \tag{2.85}$$

The proof is completed.

In the following, the design conditions for static output feedback controller gain and dynamic parameters will be given in the form of LMI based on the performance criterion proposed in Theorem 2.5. Let us define the matrix $P = \begin{bmatrix} P_1 & VP_2 \\ 0 & P_2 \end{bmatrix}$, where $P_1 > 0$ is necessary to make sure $P\mathbb{S}_A = \mathbb{S}_A^T P^T \geq 0$ and $V \in \mathbb{R}^{n_x \times n_u}$ is a dimension adjustment matrix. The inequality in (2.70) can be expressed as

$$\begin{bmatrix} \Phi_1 & * & * \\ \Phi_2 & \Phi_3 & * \\ \Phi_4 & \Phi_5 & \Phi_6 \end{bmatrix} < 0, \tag{2.86}$$

where

$$\Phi_1 = \begin{bmatrix} He\left\{P_1\tilde{A}(\rho) + VP_2\tilde{K}\tilde{L}(\rho)\right\} & * \\ P_2\tilde{K}\tilde{L}(\rho) + (P_1B(\rho) - VP_2)^T & -P_2 - P_2^T \end{bmatrix},$$

$$\Phi_2 = \begin{bmatrix} \left(VP_2\tilde{K}\right)^T & \left(P_2\tilde{K}\right)^T \\ (P_1B(\rho))^T & 0 \\ \left(P_1E(\rho) + VP_2\tilde{K}H(\rho)\right)^T & \left(P_2\tilde{K}H(\rho)\right)^T \end{bmatrix},$$

$$\Phi_4 = \begin{bmatrix} \tilde{C}(\rho) & D(\rho) \\ 2\alpha_y\delta_y\Delta_y\tilde{L}(\rho) & 0 \\ 0 & 2\alpha_u\delta_u\Delta_u I \end{bmatrix}, \quad \Phi_5 = \begin{bmatrix} 0 & D(\rho) & F(\rho) \\ 0 & 0 & 2\alpha_y\delta_y\Delta_y H(\rho) \\ 0 & 0 & 0 \end{bmatrix},$$

$$\Phi_3 = -diag\left\{\alpha_y I,\ \alpha_u I,\ \gamma^2 I\right\}, \quad \Phi_6 = -diag\left\{I,\ \alpha_y I,\ \alpha_u I\right\}.$$

It should be noted that there are still uncertain terms $\Delta_A(\rho)$, $\Delta_C(\rho)$, $\Delta_L(\rho)$, and Δ_K, therefore, in the following, we will eliminate these uncertainties. Firstly, according to (2.56), the above inequality can be rewritten as

$$\begin{bmatrix} \tilde{\Phi}_1 & * & * \\ \Phi_2 & \Phi_3 & * \\ \tilde{\Phi}_4 & \Phi_5 & \Phi_6 \end{bmatrix} + He\{\Pi_1\Delta_x(t)\Pi_2\} < 0, \tag{2.87}$$

where

$$\tilde{\Phi}_1 = \begin{bmatrix} He\left\{P_1A(\rho) + VP_2\tilde{K}L(\rho)\right\} & * \\ P_2\tilde{K}L(\rho) + (P_1B(\rho) - VP_2)^T & -P_2 - P_2^T \end{bmatrix},$$

$$\tilde{\Phi}_4 = \begin{bmatrix} C(\rho) & D(\rho) \\ 2\alpha_y\delta_y\Delta_y L(\rho) & 0 \\ 0 & 2\alpha_u\delta_u\Delta_u I \end{bmatrix},$$

$$\Pi_1 = \left[\left(P_1X_A(\rho) + VP_2\tilde{K}X_L(\rho)\right)^T \ \left(P_2\tilde{K}X_L(\rho)\right)^T \ 0\ 0\ 0\ X_C^T(\rho)\ \left(2\alpha_y\delta_y\Delta_y X_L(\rho)\right)^T\ 0\right]^T,$$

$$\Pi_2 = \begin{bmatrix} Y & 0 & 0 & 0 & 0 & 0 & 0 & 0 \end{bmatrix}.$$

Then, according to Lemma 1.2, the uncertain terms $\Delta_A(\rho)$, $\Delta_C(\rho)$, and $\Delta_L(\rho)$ can be eliminated, and we can obtain the following inequality

$$\begin{bmatrix} \tilde{\Phi}_1 & * & * & * \\ \Phi_2 & \Phi_3 & * & * \\ \tilde{\Phi}_4 & \Phi_5 & \Phi_6 & * \\ \Phi_7 & 0 & \Phi_8 & \Phi_9 \end{bmatrix} < 0, \tag{2.88}$$

where

$$\Phi_7 = \begin{bmatrix} \left(P_1 X_A(\rho) + V P_2 \tilde{K} X_L(\rho)\right)^T & \left(P_2 \tilde{K} X_L(\rho)\right)^T \\ \sigma_1 Y & 0 \end{bmatrix}^T,$$

$$\Phi_8 = \begin{bmatrix} X_C^T(\rho) & \left(2\alpha_y \delta_y \Delta_y X_L(\rho)\right)^T & 0 \\ 0 & 0 & 0 \end{bmatrix}, \quad \Phi_9 = -diag\{\sigma_1 I, \ \sigma_1 I\}.$$

Next, the uncertain term Δ_K will be eliminated. The above inequality can be rewritten as

$$\begin{bmatrix} \hat{\Phi}_1 & * & * & * \\ \hat{\Phi}_2 & \Phi_3 & * & * \\ \tilde{\Phi}_4 & \Phi_5 & \Phi_6 & * \\ \hat{\Phi}_7 & 0 & \Phi_8 & \Phi_9 \end{bmatrix} + He\{\Pi_3 \Delta_k(t) \Pi_4\} < 0, \tag{2.89}$$

where

$$\hat{\Phi}_1 = \begin{bmatrix} He\{P_1 A(\rho) + V P_2 K L(\rho)\} & * \\ P_2 K L(\rho) + (P_1 B(\rho) - V P_2)^T & -P_2 - P_2^T \end{bmatrix},$$

$$\hat{\Phi}_2 = \begin{bmatrix} (V P_2 K)^T & (P_2 K)^T \\ (P_1 B(\rho))^T & 0 \\ (P_1 E(\rho) + V P_2 K H(\rho))^T & (P_2 K H(\rho))^T \end{bmatrix},$$

$$\hat{\Phi}_7 = \begin{bmatrix} (P_1 X_A(\rho) + V P_2 K X_L(\rho))^T & (P_2 K X_L(\rho))^T \\ \sigma_1 Y & 0 \end{bmatrix}^T,$$

$$\Pi_3 = \begin{bmatrix} (V P_2 R)^T & (P_2 R)^T & 0 & 0 & 0 & 0 & 0 & 0 & 0 & 0 \end{bmatrix}^T,$$

$$\Pi_4 = \begin{bmatrix} W L(\rho) & 0 & W & 0 & W H(\rho) & 0 & 0 & 0 & W X_L(\rho) & 0 \end{bmatrix}.$$

Then, according to Lemma 1.2, the uncertain term Δ_K is eliminated, and the following inequality can be obtained

$$\begin{bmatrix} \hat{\Phi}_1 & * & * & * & * \\ \hat{\Phi}_2 & \Phi_3 & * & * & * \\ \tilde{\Phi}_4 & \Phi_5 & \Phi_6 & * & * \\ \hat{\Phi}_7 & 0 & \Phi_8 & \Phi_9 & * \\ \Phi_{10} & \Phi_{11} & 0 & \Phi_{12} & \Phi_{13} \end{bmatrix} < 0, \tag{2.90}$$

where

$$\Phi_{10} = \begin{bmatrix} (VP_2R)^T & (P_2R)^T \\ \sigma_2 WL(\rho) & 0 \end{bmatrix}, \quad \Phi_{11} = \begin{bmatrix} 0 & 0 & 0 \\ \sigma_2 W & 0 & \sigma_2 WH(\rho) \end{bmatrix},$$
$$\Phi_{12} = \begin{bmatrix} 0 & 0 \\ \sigma_2 WX_L(\rho) & 0 \end{bmatrix}, \quad \Phi_{13} = -diag\{\sigma_2 I, \ \sigma_2 I\}.$$

Defining $P_2K = T$, $\upsilon_\omega = \alpha_\omega \delta_\omega$, $\omega = y, \ u$ and based on the property of the membership function, the conditions (2.69) and (2.90) can be guaranteed by the following LMIs in Theorem 2.6, which gives the design conditions of the output feedback controller (2.57) and quantizers (2.59) and (2.60) to ensure the $\mathcal{H}_\infty$ performance γ for the fuzzy system (2.67).

Theorem 2.6 *For given scalar $\gamma > 0$, quantizers' ranges M_ω and error bounds Δ_ω, $\omega = y, u$, if there exist matrices $P_1 > 0$, P_2, and T, positive scalars α_ω and υ_ω, $\omega = y, u, \sigma_1, \sigma_2$ satisfying*

$$\upsilon_\omega M_\omega > \alpha_\omega, \ \omega = y, u, \tag{2.91}$$

$$\begin{bmatrix} \hat{\Phi}_{1i} & * & * & * & * \\ \hat{\Phi}_{2i} & \Phi_3 & * & * & * \\ \tilde{\Phi}_{4i} & \Phi_{5i} & \Phi_6 & * & * \\ \hat{\Phi}_{7i} & 0 & \Phi_{8i} & \Phi_9 & * \\ \Phi_{10i} & \Phi_{11i} & 0 & \Phi_{12i} & \Phi_{13} \end{bmatrix} < 0, \quad i = 1, 2, \ldots, r, \tag{2.92}$$

where

$$\hat{\Phi}_{1i} = \begin{bmatrix} He\{P_1A_i + VTL_i\} & * \\ TL_i + (P_1B_i - VP_2)^T & -P_2 - P_2^T \end{bmatrix}, \quad \hat{\Phi}_{2i} = \begin{bmatrix} (VT)^T & T^T \\ (P_1B_i)^T & 0 \\ (P_1E_i + VTH_i)^T & (TH_i)^T \end{bmatrix},$$
$$\tilde{\Phi}_{4i} = \begin{bmatrix} C_i & D_i \\ 2\upsilon_y\Delta_y L_i & 0 \\ 0 & 2\upsilon_u \Delta_u I \end{bmatrix}, \quad \Phi_{5i} = \begin{bmatrix} 0 & D_i & F_i \\ 0 & 0 & 2\upsilon_y\Delta_y H_i \\ 0 & 0 & 0 \end{bmatrix},$$
$$\hat{\Phi}_{7i} = \begin{bmatrix} (P_1X_{Ai} + VTX_{Li})^T & (TX_{Li})^T \\ \sigma_1 Y & 0 \end{bmatrix}, \quad \Phi_{8i} = \begin{bmatrix} X_{Ci}^T & (2\upsilon_y\Delta_y X_{Li})^T & 0 \\ 0 & 0 & 0 \end{bmatrix},$$
$$\Phi_{10i} = \begin{bmatrix} (VP_2R)^T & (P_2R)^T \\ \sigma_2 WL_i & 0 \end{bmatrix}, \quad \Phi_{11i} = \begin{bmatrix} 0 & 0 & 0 \\ \sigma_2 W & 0 & \sigma_2 WH_i \end{bmatrix}, \quad \Phi_{12i} = \begin{bmatrix} 0 & 0 \\ \sigma_2 WX_{Li} & 0 \end{bmatrix}.$$

Then, the $\mathcal{H}_\infty$ performance γ can be guaranteed for the fuzzy system (2.67). Moreover, the quantizers (2.59) and (2.60) are designed with on-line adjusting strategy for parameters $\mu_\omega(t)$, $\omega = y, u$ as (2.71) with $\delta_\omega = \upsilon_\omega/\alpha_\omega$, $\omega = y, u$, and the controller gain K in (2.57) can be obtained by

$$K = P_2^{-1}T. \tag{2.93}$$

Case B: Combining (2.56) and (2.57) yields

$$\begin{aligned}
\dot{x}(t) &= \tilde{A}(\rho)x(t) + B(\rho)\left(\mathbb{Q}(u(t)) - u(t)\right) + B(\rho)\tilde{K}\mathbb{Q}(y(t)) + E(\rho)w(t),\\
z(t) &= \tilde{C}(\rho)x(t) + D(\rho)\left(\mathbb{Q}(u(t)) - u(t)\right) + D(\rho)\tilde{K}\mathbb{Q}(y(t)) + F(\rho)w(t),\\
y(t) &= \tilde{L}(\rho)x(t) + H(\rho)w(t).
\end{aligned} \tag{2.94}$$

To cast the system into a descriptor form, the quantized output in (2.59) is converted into the following form

$$0 \times \dot{\mathbb{Q}}(y(t)) = \mu_y(t)\left(q\left(\frac{y(t)}{\mu_y(t)}\right) - \frac{y(t)}{\mu_y(t)}\right) + y(t) - \mathbb{Q}(y(t)). \tag{2.95}$$

Then with the same definitions of $e_y(t)$ and $e_u(t)$ in (2.65), one gets

$$\begin{aligned}
\dot{x}(t) &= \tilde{A}(\rho)x(t) + B(\rho)\tilde{K}\mathbb{Q}(y(t)) + B(\rho)e_u(t) + E(\rho)w(t),\\
z(t) &= \tilde{C}(\rho)x(t) + D(\rho)\tilde{K}\mathbb{Q}(y(t)) + D(\rho)e_u(t) + F(\rho)w(t),\\
y(t) &= \tilde{L}(\rho)x(t) + H(\rho)w(t),\\
0 \times \dot{\mathbb{Q}}(y(t)) &= e_y(t) + \tilde{L}(\rho)x(t) + H(\rho)w(t) - \mathbb{Q}(y(t)).
\end{aligned} \tag{2.96}$$

Further, from (2.96), the following descriptor representation can be obtained

$$\begin{aligned}
\mathbb{S}_B\dot{\varphi}(t) &= \mathbb{A}(\rho)\varphi(t) + \mathbb{M}e_y(t) + \mathbb{B}(\rho)e_u(t) + \mathbb{E}(\rho)w(t),\\
z(t) &= \mathbb{C}(\rho)\varphi(t) + \mathbb{D}(\rho)e_u(t) + \mathbb{F}(\rho)w(t),\\
y(t) &= \mathbb{L}(\rho)\varphi(t) + \mathbb{H}(\rho)w(t),
\end{aligned} \tag{2.97}$$

where $\varphi(t) = \left[x^T(t)\ \ \mathbb{Q}^T(y(t))\right]^T$ and

$$\begin{aligned}
&\mathbb{S}_B = \begin{bmatrix} I & 0 \\ 0 & 0 \end{bmatrix},\quad \mathbb{A}(\rho) = \begin{bmatrix} \tilde{A}(\rho) & B(\rho)\tilde{K} \\ \tilde{L}(\rho) & -I \end{bmatrix},\quad \mathbb{M} = \begin{bmatrix} 0 \\ I \end{bmatrix},\quad \mathbb{C}(\rho) = \left[\tilde{C}(\rho)\ \ D(\rho)\tilde{K}\right],\\
&\mathbb{D}(\rho) = D(\rho),\quad \mathbb{F}(\rho) = F(\rho),\quad \mathbb{B}(\rho) = \begin{bmatrix} B(\rho) \\ 0 \end{bmatrix},\quad \mathbb{E}(\rho) = \begin{bmatrix} E(\rho) \\ H(\rho) \end{bmatrix},\\
&\mathbb{L}(\rho) = \left[\tilde{L}(\rho)\ \ 0\right],\quad \mathbb{H}(\rho) = H(\rho).
\end{aligned}$$

Next, we will analyze the $\mathscr{H}_\infty$ performance of the closed-loop system (2.97) and derive the design conditions for the feedback controller (2.57) and quantizers (2.59) and (2.60).

Theorem 2.7 *Consider the quantized output feedback closed-loop system (2.97). For given scalar $\gamma > 0$, quantizers' ranges M_ω and error bounds Δ_ω, $\omega = y, u$, if there exist matrix P, positive scalars ζ_ω and λ_ω, $\omega = y, u$, such that the following matrix inequalities hold:*

$$P\,\mathbb{S}_B = \mathbb{S}_B^T P^T \geq 0, \tag{2.98}$$

$$M_{\omega} > \sqrt{\zeta_{\omega}},\ \omega = y, u, \tag{2.99}$$

$$\begin{bmatrix} He\{P\mathbb{A}(\rho)\} & * & * \\ \varsigma_1 & \varsigma_2 & * \\ \varsigma_3 & \varsigma_4 & \varsigma_5 \end{bmatrix} < 0, \tag{2.100}$$

where

$$\varsigma_1 = \begin{bmatrix} P\mathbb{M} & P\mathbb{B}(\rho) & P\mathbb{E}(\rho) \end{bmatrix}^T,\quad \varsigma_2 = diag\left\{ -\frac{\zeta_y}{4\lambda_y}I,\ -\frac{\zeta_u}{4\lambda_u}I,\ -\gamma^2 I \right\},$$

$$\varsigma_3 = \begin{bmatrix} \mathbb{C}(\rho) \\ \Delta_y\mathbb{L}(\rho) \\ \Delta_u\tilde{K}\mathcal{N} \end{bmatrix},\quad \varsigma_4 = \begin{bmatrix} 0 & \mathbb{D}(\rho) & \mathbb{F}(\rho) \\ 0 & 0 & \Delta_y\mathbb{H}(\rho) \\ 0 & 0 & 0 \end{bmatrix},$$

$$\varsigma_5 = diag\left\{ -I,\ -\lambda_y I,\ -\lambda_u I \right\},\quad \mathcal{N} = \begin{bmatrix} 0 & I \end{bmatrix},$$

with on-line adjusting strategy for parameters $\mu_{\omega}(t)$, $\omega = y, u$ as:

$$\frac{1}{\sqrt{\zeta_{\omega}}}\|\omega(t)\| \leq \mu_{\omega}(t) \leq \frac{2}{\sqrt{\zeta_{\omega}}}\|\omega(t)\|. \tag{2.101}$$

Then, the controller (2.57) and the quantizers (2.59) and (2.60) with the adjusting strategy (2.101) can ensure that the closed-loop system (2.97) achieves the prescribed $\mathcal{H}_{\infty}$ performance γ.

Proof With (2.99) and (2.101), it is easy to deduce that $\frac{1}{2}\sqrt{\zeta_y}\mu_y(t) \leq \|y(t)\| \leq M_y\mu_y(t)$. Then based on the definition of $e_y(t)$ in (2.65) and the homogeneity property of Euclidean norm, from (2.61), one has

$$\begin{aligned} \|e_y(t)\| &= \left\| \mu_y(t)\left\{ q\left(\frac{y(t)}{\mu_y(t)}\right) - \frac{y(t)}{\mu_y(t)} \right\} \right\| \\ &= \mu_y(t)\left\| q\left(\frac{y(t)}{\mu_y(t)}\right) - \frac{y(t)}{\mu_y(t)} \right\| \\ &\leq \mu_y(t)\Delta_y \\ &\leq \frac{2}{\sqrt{\zeta_y}}\Delta_y\|y(t)\|, \end{aligned} \tag{2.102}$$

i.e.,

$$\frac{\zeta_y}{4}e_y^T(t)e_y(t) \leq \Delta_y^2 y^T(t)y(t). \tag{2.103}$$

Defining $\phi^T(t) = \begin{bmatrix} \varphi^T(t) & e_y^T(t) & e_u^T(t) & w^T(t) \end{bmatrix}$ and considering (2.97), the above inequality can be rewritten as

$$\phi^T(t)\left(Ha\{\mathcal{W}_5\} - diag\left\{0,\ \frac{\zeta_y}{4}I,\ 0,\ 0\right\}\right)\phi(t) \geq 0, \tag{2.104}$$

where $\mathscr{W}_5 = \begin{bmatrix} \Delta_y \mathbb{L}(\rho) & 0 & 0 & \Delta_y \mathbb{H}(\rho) \end{bmatrix}$. Similarly, combining (2.99) and (2.101), considering the definition of $e_u(t)$ in (2.65) and the homogeneity property of Euclidean norm, one has

$$\|e_u(t)\| \leq \frac{2}{\sqrt{\zeta_u}} \Delta_u \|u(t)\| = \frac{2}{\sqrt{\zeta_u}} \Delta_u \left\| \tilde{K} \mathbb{Q}(y(t)) \right\|, \tag{2.105}$$

i.e.,

$$\frac{\zeta_u}{4} e_u^T(t) e_u(t) \leq \Delta_u^2 \mathbb{Q}^T(y(t)) \tilde{K}^T \tilde{K} \mathbb{Q}(y(t)). \tag{2.106}$$

Based on (2.97), the above inequality can be rewritten as

$$\phi^T(t) \left(Ha\{\mathscr{W}_6\} - diag\left\{ 0,\ 0,\ \frac{\zeta_u}{4} I,\ 0 \right\} \right) \phi(t) \geq 0, \tag{2.107}$$

where $\mathscr{W}_6 = \begin{bmatrix} \Delta_u \tilde{K} \mathscr{N} & 0 & 0 & 0 \end{bmatrix}$. Now, consider a Lyapunov function as

$$V(\varphi(t)) = \varphi^T(t) P \mathbb{S}_B \varphi(t), \quad P\mathbb{S}_B = \mathbb{S}_B^T P^T \geq 0, \tag{2.108}$$

Then one gets

$$\begin{aligned}
&\dot{V}(\varphi(t)) + z^T(t) z(t) - \gamma^2 w^T(t) w(t) \\
&= \varphi^T(t) P \mathbb{S}_B \dot{\varphi}(t) + \dot{\varphi}^T(t) \mathbb{S}_B^T P^T \varphi(t) + z^T(t) z(t) - \gamma^2 w^T(t) w(t) \\
&= \varphi^T(t) P \Big(\mathbb{A}(\rho)\varphi(t) + \mathbb{M} e_y(t) + \mathbb{B}(\rho) e_u(t) + \mathbb{E}(\rho) w(t) \Big) \\
&\quad + \Big(\mathbb{A}(\rho)\varphi(t) + \mathbb{M} e_y(t) + \mathbb{B}(\rho) e_u(t) + \mathbb{E}(\rho) w(t) \Big)^T P^T \varphi(t) \\
&\quad + Ha\{\mathbb{C}(\rho)\varphi(t) + \mathbb{D}(\rho) e_u(t) + \mathbb{F}(\rho) w(t)\} - \gamma^2 w^T(t) w(t) \\
&= \phi^T(t) \begin{bmatrix} He\{P\mathbb{A}(\rho)\} & * \\ \varsigma_1 & diag\{0,\ 0,\ -\gamma^2 I\} \end{bmatrix} \phi(t) + \phi^T(t) Ha\{\mathscr{W}_4\} \phi(t),
\end{aligned}$$

where $\mathscr{W}_4 = \begin{bmatrix} \mathbb{C}(\rho) & 0 & \mathbb{D}(\rho) & \mathbb{F}(\rho) \end{bmatrix}$. Based on (2.100) and Schur complement (Lemma 1.1), it is easy to obtain

$$\begin{aligned}
&\begin{bmatrix} He\{P\mathbb{A}(\rho)\} & * \\ \varsigma_1 & diag\{0,\ 0,\ -\gamma^2 I\} \end{bmatrix} + \frac{1}{\lambda_y} \left(Ha\{\mathscr{W}_5\} - diag\left\{ 0,\ \frac{\zeta_y}{4} I,\ 0,\ 0 \right\} \right) \\
&+ \frac{1}{\lambda_u} \left(Ha\{\mathscr{W}_6\} - diag\left\{ 0,\ 0,\ \frac{\zeta_u}{4} I,\ 0 \right\} \right) + Ha\{\mathscr{W}_4\} < 0.
\end{aligned} \tag{2.109}$$

Then based on inequalities (2.104) and (2.107), scalars $\lambda_y > 0$ and $\lambda_u > 0$, the following inequality is obtained,

$$\phi^T(t)\begin{bmatrix} He\{P\mathbb{A}(\rho)\} & * \\ \varsigma_1 & diag\left\{0,\ 0,\ -\gamma^2 I\right\} \end{bmatrix}\phi(t) + \phi^T(t)Ha\{\mathscr{W}_4\}\phi(t) < 0, \tag{2.110}$$

which means that $\dot{V}(\varphi(t)) + z^T(t)z(t) - \gamma^2 w^T(t)w(t) < 0$.

In the following, the design conditions are presented for static output feedback controller gain and dynamic parameters in the form of LMI based on the performance criterion proposed in Theorem 2.7. Firstly, defining $Z = P^{-T}$, pre- and post-multiplying the inequality (2.100) by $diag\{Z^T,\ 0,\ 0,\ 0,\ 0,\ 0,\ 0\}$ and its transpose, it can be rewritten as

$$\begin{bmatrix} He\{\mathbb{A}(\rho)Z\} & * & * \\ \tilde{\varsigma}_1 & \varsigma_2 & * \\ \tilde{\varsigma}_3 & \varsigma_4 & \varsigma_5 \end{bmatrix} < 0, \tag{2.111}$$

where

$$\tilde{\varsigma}_1 = \begin{bmatrix} \mathbb{M}^T \\ \mathbb{B}^T(\rho) \\ \mathbb{E}^T(\rho) \end{bmatrix}, \quad \tilde{\varsigma}_3 = \begin{bmatrix} \mathbb{C}(\rho)Z \\ \Delta_y\mathbb{L}(\rho)Z \\ \Delta_u\tilde{K}\mathcal{N}Z \end{bmatrix}.$$

As for the condition $P\mathbb{S}_B = \mathbb{S}_B^T P^T \geq 0$, it can be guaranteed by $\mathbb{S}_B Z = Z^T\mathbb{S}_B^T \geq 0$. Next, set the matrix $Z = \begin{bmatrix} Z_1 & 0 \\ Z_2 G & Z_2 \end{bmatrix}$ with $G \in \mathbb{R}^{n_y \times n_x}$ being a dimension adjustment matrix and $\frac{1}{\beta_\omega} = \frac{\zeta_\omega}{4\lambda_\omega}$, $\omega = y,\ u$. In order to ensure the establishment of the condition $\mathbb{S}_B Z = Z^T\mathbb{S}_B^T \geq 0$, $Z_1 > 0$ is necessary. Then the inequality (2.111) can be expressed as follows:

$$\begin{bmatrix} \Omega_1 & * & * \\ \Omega_2 & \Omega_3 & * \\ \Omega_4 & \Omega_5 & \Omega_6 \end{bmatrix} < 0, \tag{2.112}$$

where

$$\Omega_1 = \begin{bmatrix} He\left\{\tilde{A}(\rho)Z_1 + B(\rho)\tilde{K}Z_2G\right\} & * \\ \tilde{L}(\rho)Z_1 - Z_2G + \left(B(\rho)\tilde{K}Z_2\right)^T & -Z_2 - Z_2^T \end{bmatrix},$$

$$\Omega_2 = \begin{bmatrix} 0 & \beta_y I \\ \beta_u B^T(\rho) & 0 \\ E^T(\rho) & H^T(\rho) \end{bmatrix}, \quad \Omega_4 = \begin{bmatrix} \tilde{C}(\rho)Z_1 + D(\rho)\tilde{K}Z_2G & D(\rho)\tilde{K}Z_2 \\ \Delta_y\tilde{L}(\rho)Z_1 & 0 \\ \Delta_u\tilde{K}Z_2G & \Delta_u\tilde{K}Z_2 \end{bmatrix},$$

$$\Omega_5 = \begin{bmatrix} 0 & \beta_u D(\rho) & F(\rho) \\ 0 & 0 & \Delta_y H(\rho) \\ 0 & 0 & 0 \end{bmatrix}, \quad \Omega_3 = -diag\left\{\beta_y I,\ \beta_u I,\ \gamma^2 I\right\},$$

$$\Omega_6 = -diag\left\{I,\ \lambda_y I,\ \lambda_u I\right\}.$$

Similar to Case A, there are uncertain terms $\Delta_A(\rho)$, $\Delta_C(\rho)$, $\Delta_L(\rho)$, and Δ_K in above inequality, therefore, in the following, these uncertainties will be eliminated. Firstly, according to (2.56), the above inequality can be rewritten as

$$\begin{bmatrix} \tilde{\Omega}_1 & * & * \\ \Omega_2 & \Omega_3 & * \\ \tilde{\Omega}_4 & \Omega_5 & \Omega_6 \end{bmatrix} + He\{\Pi_5 \Delta_x(t) \Pi_6\} < 0, \tag{2.113}$$

where

$$\tilde{\Omega}_1 = \begin{bmatrix} He\left\{A(\rho)Z_1 + B(\rho)\tilde{K}Z_2G\right\} & * \\ L(\rho)Z_1 - Z_2G + \left(B(\rho)\tilde{K}Z_2\right)^T & -Z_2 - Z_2^T \end{bmatrix},$$

$$\tilde{\Omega}_4 = \begin{bmatrix} C(\rho)Z_1 + D(\rho)\tilde{K}Z_2G & D(\rho)\tilde{K}Z_2 \\ \Delta_y L(\rho)Z_1 & 0 \\ \Delta_u \tilde{K}Z_2G & \Delta_u \tilde{K}Z_2 \end{bmatrix},$$

$$\Pi_5 = \left[X_A^T(\rho)\ \ X_L^T(\rho)\ \ 0\ \ 0\ \ 0\ \ X_C^T(\rho)\ \ (\Delta_y X_L(\rho))^T\ \ 0\right]^T,$$

$$\Pi_6 = \left[YZ_1\ \ 0\ \ 0\ \ 0\ \ 0\ \ 0\ \ 0\ \ 0\right].$$

Then, according to Lemma 1.2, the uncertain terms $\Delta_A(\rho)$, $\Delta_C(\rho)$, and $\Delta_L(\rho)$ can be eliminated, and one can get the following inequality,

$$\begin{bmatrix} \tilde{\Omega}_1 & * & * & * \\ \Omega_2 & \Omega_3 & * & * \\ \tilde{\Omega}_4 & \Omega_5 & \Omega_6 & * \\ \Omega_7 & 0 & \Omega_8 & \Omega_9 \end{bmatrix} < 0, \tag{2.114}$$

where

$$\Omega_7 = \begin{bmatrix} X_A^T(\rho) & X_L^T(\rho) \\ \sigma_3 YZ_1 & 0 \end{bmatrix}, \quad \Omega_8 = \begin{bmatrix} X_C^T(\rho) & (\Delta_y X_L(\rho))^T & 0 \\ 0 & 0 & 0 \end{bmatrix},$$

$$\Omega_9 = -diag\{\sigma_3 I,\ \sigma_3 I\}.$$

Next, the uncertain term Δ_K will be eliminated. Above inequality can be rewritten as

$$\begin{bmatrix} \hat{\Omega}_1 & * & * & * \\ \Omega_2 & \Omega_3 & * & * \\ \hat{\Omega}_4 & \Omega_5 & \Omega_6 & * \\ \Omega_7 & 0 & \Omega_8 & \Omega_9 \end{bmatrix} + He\{\Pi_7 \Delta_k(t) \Pi_8\} < 0, \tag{2.115}$$

where

$$\hat{\Omega}_1 = \begin{bmatrix} He\{A(\rho)Z_1 + B(\rho)KZ_2G\} & * \\ L(\rho)Z_1 - Z_2G + (B(\rho)KZ_2)^T & -Z_2 - Z_2^T \end{bmatrix},$$

$$\hat{\Omega}_4 = \begin{bmatrix} C(\rho)Z_1 + D(\rho)KZ_2G & D(\rho)KZ_2 \\ \Delta_y L(\rho)Z_1 & 0 \\ \Delta_u KZ_2G & \Delta_u KZ_2 \end{bmatrix},$$

$$\Pi_7 = \begin{bmatrix}(B(\rho)R)^T & 0 & 0 & 0 & 0 & (D(\rho)R)^T & 0 & (\Delta_u R)^T & 0 & 0\end{bmatrix}^T,$$

$$\Pi_8 = \begin{bmatrix}WZ_2G & WZ_2 & 0 & 0 & 0 & 0 & 0 & 0 & 0 & 0\end{bmatrix}.$$

Then, according to Lemma 1.2, eliminating the uncertain term Δ_K can obtain the following inequality,

$$\begin{bmatrix} \hat{\Omega}_1 & * & * & * & * \\ \Omega_2 & \Omega_3 & * & * & * \\ \hat{\Omega}_4 & \Omega_5 & \Omega_6 & * & * \\ \Omega_7 & 0 & \Omega_8 & \Omega_9 & * \\ \Omega_{10} & 0 & \Omega_{11} & 0 & \Omega_{12} \end{bmatrix} < 0, \tag{2.116}$$

where

$$\Omega_{10} = \begin{bmatrix} \sigma_4 (B(\rho)R)^T & 0 \\ WZ_2G & WZ_2 \end{bmatrix}, \quad \Omega_{11} = \begin{bmatrix} \sigma_4 (D(\rho)R)^T & 0 & \sigma_4 (\Delta_u R)^T \\ 0 & 0 & 0 \end{bmatrix},$$

$$\Omega_{12} = -diag\{\sigma_4 I, \ \sigma_4 I\}.$$

Defining $KZ_2 = U$, based on the property of membership function, the conditions (2.99) and (2.100) can be guaranteed by the following LMIs in Theorem 2.8, which gives the design conditions of the output feedback controller (2.57) and quantizers (2.59) and (2.60) to ensure the $\mathscr{H}_\infty$ performance γ for the fuzzy system (2.97).

Theorem 2.8 *For given scalar $\gamma > 0$, quantizers' ranges M_ω and error bounds Δ_ω, $\omega = y, u$, if there exist matrices $Z_1 > 0$, Z_2, and U, positive scalars λ_ω and β_ω, $\omega = y, u, \sigma_3, \sigma_4$ satisfying*

$$\beta_\omega M_\omega^2 > 4\lambda_\omega, \ \omega = y, u, \tag{2.117}$$

$$\begin{bmatrix} \hat{\Omega}_{1i} & * & * & * & * \\ \Omega_{2i} & \Omega_3 & * & * & * \\ \hat{\Omega}_{4i} & \Omega_{5i} & \Omega_6 & * & * \\ \Omega_{7i} & 0 & \Omega_{8i} & \Omega_9 & * \\ \Omega_{10i} & 0 & \Omega_{11i} & 0 & \Omega_{12} \end{bmatrix} < 0, \ i = 1, 2, \ldots, r, \tag{2.118}$$

where

$$\hat{\Omega}_{1i} = \begin{bmatrix} He\{A_i Z_1 + B_i U G\} & * \\ L_i Z_1 - Z_2 G + (B_i U)^T & -Z_2 - Z_2^T \end{bmatrix}, \quad \Omega_{2i} = \begin{bmatrix} 0 & \beta_y I \\ \beta_u B_i^T & 0 \\ E_i^T & H_i^T \end{bmatrix},$$

$$\hat{\Omega}_{4i} = \begin{bmatrix} C_i Z_1 + D_i U G & D_i U \\ \Delta_y L_i Z_1 & 0 \\ \Delta_u U G & \Delta_u U \end{bmatrix}, \quad \Omega_{5i} = \begin{bmatrix} 0 & \beta_u D_i & F_i \\ 0 & 0 & \Delta_y H_i \\ 0 & 0 & 0 \end{bmatrix},$$

$$\Omega_{7i} = \begin{bmatrix} X_{Ai}^T & X_{Li}^T \\ \sigma_3 Y Z_1 & 0 \end{bmatrix}, \quad \Omega_{8i} = \begin{bmatrix} X_{Ci}^T & \Delta_y X_{Li}^T & 0 \\ 0 & 0 & 0 \end{bmatrix},$$

$$\Omega_{10i} = \begin{bmatrix} \sigma_4 (B_i R)^T & 0 \\ W Z_2 G & W Z_2 \end{bmatrix}, \quad \Omega_{11i} = \begin{bmatrix} \sigma_4 (D_i R)^T & 0 & \sigma_4 (\Delta_u R)^T \\ 0 & 0 & 0 \end{bmatrix}.$$

Then, the $\mathcal{H}_\infty$ performance γ can be guaranteed for the fuzzy system (2.67). Moreover, the quantizers (2.59) and (2.60) are designed with online adjusting strategy for parameters $\mu_\omega(t)$, $\omega = y, u$ as (2.71) with $\zeta_\omega = 4\lambda_\omega/\beta_\omega$, $\omega = y, u$, and the controller gain K in (2.57) can be obtained by

$$K = U Z_2^{-1}. \tag{2.119}$$

According to the view in [8], the dynamic parameters $\mu_\omega(t)$, $\omega = y, u$ also need to be adjusted. Referring to [8], for the on-line adjusting strategy of parameters $\mu_\omega(t)$, $\omega = y, u$ in Theorems 2.6 and 2.8, the adjusting rule is given as follows:

The adjusting rule:
For $\omega = y, u$

$$\mu_\omega(t) = \begin{cases} \text{floor}\left(2\vartheta_\omega \|\omega(t)\| \times 10^s\right) \times 10^{-s}, & 0 \le \vartheta_\omega \|\omega(t)\| < \frac{1}{2}, \\ 1, & \frac{1}{2} \le \vartheta_\omega \|\omega(t)\| < 1, \\ \text{floor}\left(2\vartheta_\omega \|\omega(t)\|\right), & 1 \le \vartheta_\omega \|\omega(t)\|, \end{cases} \tag{2.120}$$

where $s = \min\left\{s \in \mathcal{N}^+ | \left(2\vartheta_\omega \|\omega(t)\| \times 10^s\right) > 1\right\}$ and the function floor(θ) represents the maximum integer which does not exceed θ. More specifically, the $\vartheta_\omega(t)$ can be updated by applying the above adjusting strategy with $\vartheta_\omega = \delta_\omega$, $\omega = y, u$ for Theorem 2.6 and $\vartheta_\omega = \frac{1}{\sqrt{\zeta_\omega}}$, $\omega = y, u$ for Theorem 2.8.

Remark 2.7 Different from [8], the adjusting rule of this chapter gives the deterministic expression for quantizers' dynamic parameters $\vartheta_\omega(t)$, $\omega = y, u$, which can be determined as a unique value corresponding to each signal $\omega(t)$, $\omega = y, u$.

Remark 2.8 It is worth noting that by applying the descriptor representation approach to formulate the two augmented closed-loop systems (2.67) and (2.97), some nonlinear coupling terms between the system matrices and the controller gain matrix which are difficult to deal with, have been successfully avoided, such that we can easily handle the static output feedback control problem and derive strict LMI design conditions. Moreover, it avoids the appearance of coupling between quantization errors $e_y(t)$ and $e_u(t)$ and facilitates the design of $\mathcal{H}_\infty$ controller. The obtained conditions (2.91) and (2.92) in Theorem 2.6 or (2.117) and (2.118) in Theorem 2.8 can be used to deal with the output feedback $\mathcal{H}_\infty$ controller design problem for continuous-time T–S fuzzy systems with quantized measurement output and quantized control input if the dimension adjustment matrices $V \in \mathbb{R}^{n_x \times n_u}$ and $G \in \mathbb{R}^{n_y \times n_x}$ are known in advance. In this section, V and G can be chosen in the form as $V = \begin{bmatrix} I_{n_u \times n_u} & 0_{n_u \times (n_x - n_u)} \end{bmatrix}^T$ and $G = [\, I_{n_y \times n_y} \;\; 0_{n_y \times (n_x - n_y)} \,]$. Then Theorems 2.6 and 2.8 can be easily solved with the help of LMI control toolbox in MATLAB.

Remark 2.9 Based on the LMI conditions in Theorem 2.6 or Theorem 2.8, the static output feedback $\mathcal{H}_\infty$ control problem for continuous-time T–S fuzzy systems with quantized input and output signals can be solved. The $\mathcal{H}_\infty$ performance index γ can be optimized by

$$\begin{gathered} \min \ \iota \\ \text{subject to LMIs (2.91) and (2.92) or (2.117) and (2.118) with } \iota = \gamma^2. \end{gathered}$$

In this way, the corresponding optimal value of the $\mathcal{H}_\infty$ performance can be given as $\gamma_{\min} = \sqrt{\iota_{\min}}$.

Remark 2.10 Note that from the on-line adjusting strategies (2.71) and (2.101), quantizers' parameters $\mu_\omega(t)$, $\omega = y, u$ only depend on the quantized signal $\omega(t)$, $\omega = y, u$ and the scalars $\delta_\omega, \omega = y, u$ in Theorem 2.6 or $\zeta_\omega, \omega = y, u$ in Theorem 2.8, which can be obtained directly by the solution of LMIs (2.91) and (2.92) or (2.117) and (2.118), respectively, it implies that the computation of proposed algorithm for quantization design is relatively smaller than the design result in [8].

2.3.3 *Simulation Example*

An example for the inverted pendulum system of motor dynamics is used to illustrate the effectiveness of the proposed methods in this section, and it can be given as the following dynamic equation:

$$\begin{aligned} \dot{\rho}_p(t) &= \omega_p(t), \\ \dot{\omega}_p(t) &= \frac{g}{l}\sin(\rho_p(t)) + \frac{N(K_m + \Delta K_m)}{ml^2} I_a(t) + w(t), \\ \dot{I}_a(t) &= -\frac{K_b N}{L_a}\omega_p(t) - \frac{R_a}{L_a} I_a(t) + \frac{1}{L_a} U(t), \end{aligned} \tag{2.121}$$

where m, l, g, N, K_m, K_b, R_a, and L_a stand for the mass of pendulum, length of pendulum, gravity acceleration, gear ratio, motor torque coefficient, back EMF coefficient, resistance of resistor, and inductance of inductor, respectively. ΔK_m means the variation of motor torque coefficient caused by the friction. $U(t)$ represents the voltage input. $w(t)$ represents the disturbance signal. The outputs of this system are selected as

$$\begin{aligned} z(t) &= -\rho_p(t) + 0.1U(t), \\ y(t) &= \nu_1\rho_p(t) + \nu_2\omega_p(t) + \nu_3 I_a(t) + \nu_4 w(t), \end{aligned} \tag{2.122}$$

where ν_1, ν_2, ν_3, and ν_4 represent the sensor coefficient. Select the state of this system as $x(t) = \begin{bmatrix}\rho_p(t) & \omega_p(t) & I_a(t)\end{bmatrix}^T$. As for the nonlinear term $\sin(\rho_p(t))$, it can be approached by the T–S fuzzy model. Giving the physical parameters of the system as $m = 0.5\,\text{kg}$, $l = 1\,\text{m}$, $g = 9.8\,\text{m/s}^2$, $N = 5$, $K_m = 0.1\,\text{N}\cdot\text{m/A}$, $K_b = 0.1\,\text{V}\cdot\text{s/rad}$, $R_a = 0.8\,\Omega$, $L_a = 100\,\text{mH}$, $\nu_1 = 3$, $\nu_2 = 2$, $\nu_3 = 0.1$, and $\nu_4 = -0.2$, the T–S fuzzy model of this system can be given as

$$\begin{aligned} &R^i : if\ \rho_p(t)\ is\ M_{1i},\ then \\ &\dot{x}(t) = (A_i + \Delta_{Ai})x(t) + B_i\mathbb{Q}(u(t)) + E_i w(t), \\ &z(t) = (C_i + \Delta_{Ci})x(t) + D_i\mathbb{Q}(u(t)) + F_i w(t), \\ &y(t) = (L_i + \Delta_{Li})x(t) + H_i w(t), \end{aligned} \tag{2.123}$$

with $i = 1, 2$, where

$$A_1 = \begin{bmatrix} 0 & 1 & 0 \\ 9.8 & 0 & 1 \\ 0 & -5 & -8 \end{bmatrix},\quad A_2 = \begin{bmatrix} 0 & 1 & 0 \\ 0 & 0 & 1 \\ 0 & -5 & -8 \end{bmatrix},\quad B_1 = B_2 = \begin{bmatrix} 0 \\ 0 \\ 10 \end{bmatrix},$$

$$E_1 = E_2 = \begin{bmatrix} 0 \\ 1 \\ 0 \end{bmatrix},\quad C_1 = C_2 = [-1\ \ 0\ \ 0],\quad D_1 = D_2 = 0.1,\quad F_1 = F_2 = 0,$$

$$L_1 = L_2 = \begin{bmatrix} 3 & 2 & 0.1 \end{bmatrix},\quad H_1 = H_2 = -0.2,\quad X_{A1} = X_{A2} = \begin{bmatrix} 0 & 0 & 0 \\ 0 & 0 & 10 \\ 0 & 0 & 0 \end{bmatrix},$$

$$Y = \begin{bmatrix} 0.01 & 0 & 0 \\ 0 & 0.01 & 0 \\ 0 & 0 & 0.01 \end{bmatrix},\quad X_{C1} = X_{C2} = \begin{bmatrix} 0.1 & 0.1 & 0.1 \end{bmatrix},$$
$$X_{L1} = X_{L2} = \begin{bmatrix} 0.1 & 0.1 & 0.1 \end{bmatrix},\quad R = W = 0.01.$$

Moreover, the uncertain terms are selected as $\Delta_x(t) = \Delta_k(t) = 0.5\cos(0.1t)e^{-2t}$. For this fuzzy system, the membership function is assumed in the form as $\rho_1(\varepsilon(t)) = \sin^2(x_1(t))$, $\rho_2(\varepsilon(t)) = 1 - \rho_1(\varepsilon(t))$. And the quantization ranges and quantization error bounds are selected as $M_y = M_u = 50$, $\Delta_y = \Delta_u = 0.01$. By solving the LMIs (2.91) and (2.92) developed in Theorem 2.6 with $V = \begin{bmatrix}1 & 0 & 0\end{bmatrix}^T$, there is no a feasible solution due to the design conservatism, whereas solving the LMIs (2.117) and

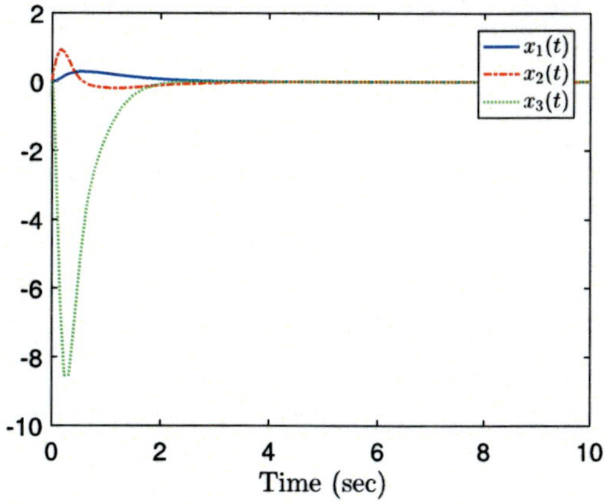

Fig. 2.12 Response of the system state $x(t)$

(2.118) developed in Theorem 2.8 of Case B with $G = \begin{bmatrix} 1 & 0 & 0 \end{bmatrix}$, a feasible solution is given as

$$Z_1 = \begin{bmatrix} 0.0704 & -0.1597 & 0.0116 \\ -0.1597 & 1.2717 & -5.6053 \\ 0.0116 & -5.6053 & 55.2213 \end{bmatrix}, \quad Z_2 = 0.1211.$$

Combining (2.119), it is calculated the output feedback controller gain as $K = -5.9242$. Moreover, it gives the optimized $\mathcal{H}_\infty$ performance index as $\gamma_{min} = 3.2766$.

Further, in order to demonstrate the validity of the algorithm more intuitively, the designed controller and quantizers are adopted to test the $\mathcal{H}_\infty$ performance $\gamma = 3.2766$ for the fuzzy system (2.123). Let us assume the initial condition to be $x(0) = \begin{bmatrix} 0 & 0 & 0 \end{bmatrix}^T$ and the disturbance signal to be $w(t) = 10\cos(0.1t)e^{-2t}$. Moreover, the dynamic parameters $\mu_y(t)$ and $\mu_u(t)$ are updated by the adjusting rule described in (2.120) with $\vartheta_y = \frac{1}{\sqrt{\zeta_y}} = 0.02$ and $\vartheta_u = \frac{1}{\sqrt{\zeta_u}} = 0.02$. Then the response of system state, measurement output, controlled output, and control input are shown in Figs. 2.12, 2.13, 2.14, and 2.15, respectively. The ratio of $\sqrt{\int_0^t z^T(i)z(i)di / \int_0^t w^T(i)w(i)di}$ showing the disturbance attenuation capability of the system is presented in Fig. 2.16. From this figure, it is easy to observe that the ratio tends to a constant value 1.0550 below the prescribed value $\gamma = 3.2766$, which implies that the asymptotical stability and the prescribed $\mathcal{H}_\infty$ performance of the fuzzy system (2.123) can be guaranteed by Theorem 2.8, and illustrates the feasibility and effectiveness of the proposed design strategy in this chapter. Besides, as shown in Figs. 2.17 and 2.18, the quantizers' dynamic parameters $\mu_y(t)$ and $\mu_u(t)$ are adaptively adjusted depending on the measurement output $y(t)$ and the control input $u(t)$, respectively, which is consistent with the proposed adjusting rule (2.120).

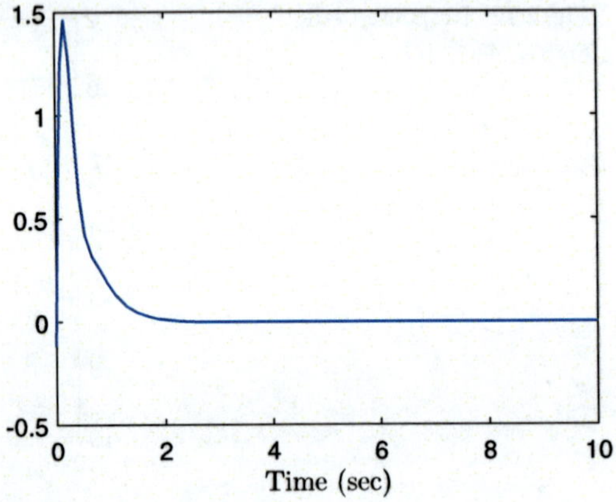

Fig. 2.13 Response of the measurement output $y(t)$

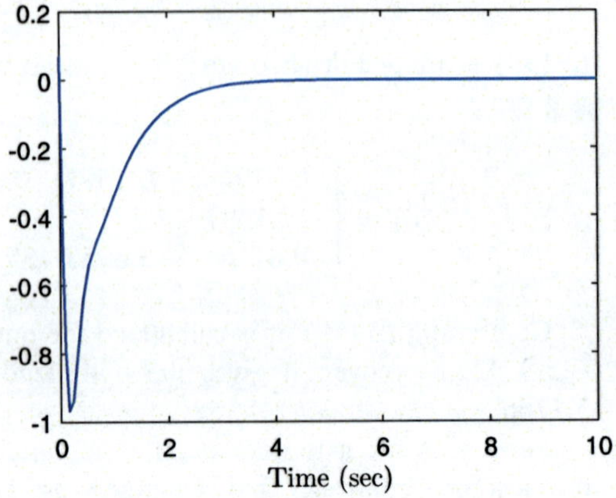

Fig. 2.14 Response of the controlled output $z(t)$

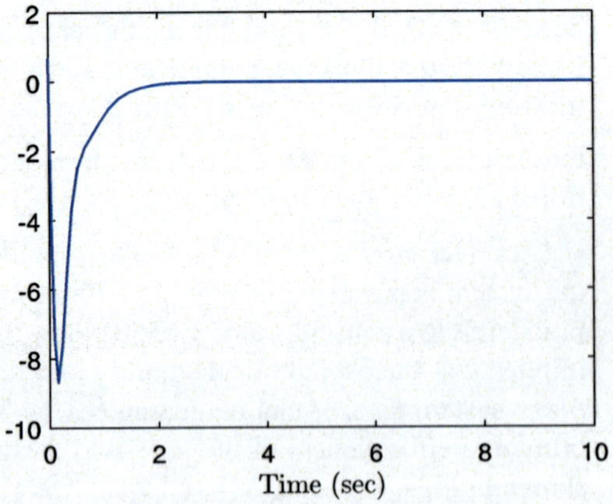

Fig. 2.15 Response of the control input $u(t)$

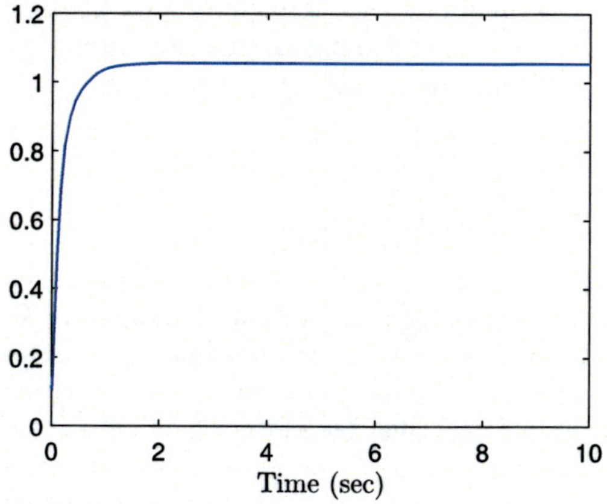

Fig. 2.16 History of $\sqrt{\int_0^t z^T(i)z(i)di / \int_0^t w^T(i)w(i)di}$

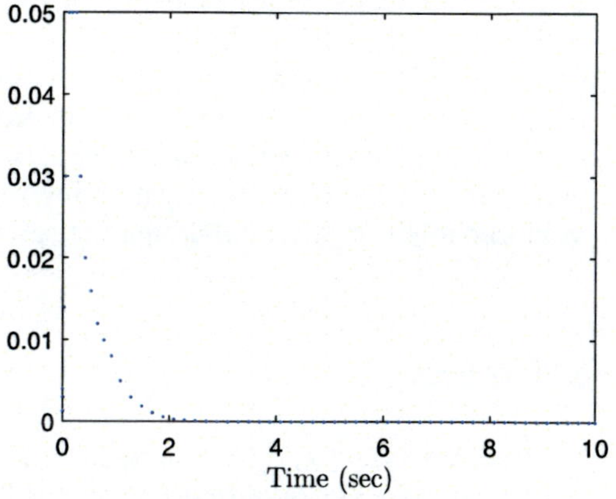

Fig. 2.17 Dynamic quantizer's parameter $\mu_y(t)$

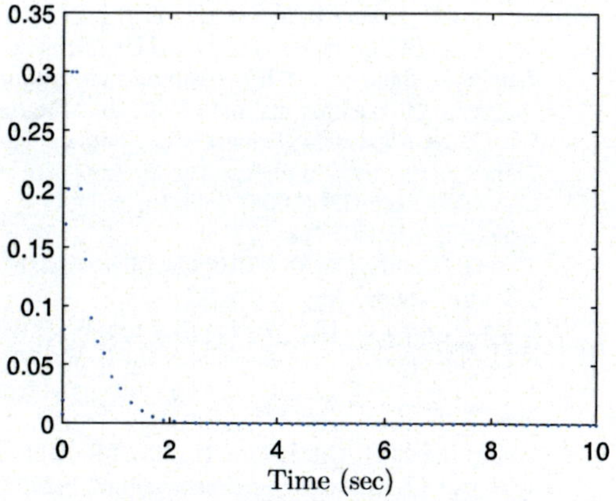

Fig. 2.18 Dynamic quantizer's parameter $\mu_u(t)$

From Figs. 2.12, 2.13, 2.14, 2.15, 2.16, 2.17, and 2.18, it can be concluded that both the quantized stabilization control and quantized $\mathscr{H}_\infty$ control design approaches developed in this chapter are effective.

2.4 Conclusion

In this chapter, the $\mathscr{H}_\infty$ control problem for T–S fuzzy systems with dynamic quantization has been studied sufficiently. For discrete-time fuzzy systems, via the two-step approach and an improved two-step approach, the robust controller and the non-fragile controller based on state feedback have been separately designed to maintain the same $\mathscr{H}_\infty$ performance with the one without quantization when the control input and the system state are quantized by dynamic quantizers, and the online adjusting rule for dynamic quantizers' parameters has also been formulated. For the continuous-time case, the descriptor representation approach has been adopted to design the resilient controller based on static output feedback to attenuate the effects of parameter uncertainties, controller gain perturbation, the control input, and measurement output quantization, so as to preserve the desired $\mathscr{H}_\infty$ performance. Moreover, some simulation results have been given to verify the effectiveness of these quantized robust $\mathscr{H}_\infty$ control design strategies.

References

1. Chang XH, Yang C, Xiong J (2019) Quantized fuzzy output feedback $\mathscr{H}_\infty$ control for nonlinear systems with adjustment of dynamic parameters. IEEE Trans Syst, Man, Cybern: Syst 49:2005–2015
2. Chang XH, Huang R, Wang H, Liu L (2020) Robust design strategy of quantized feedback control. IEEE Trans Circuits Syst II Exp Briefs 67:730–734
3. Tanaka K, Sano M (1994) A robust stabilization problem of fuzzy control systems and its application to backing up control of a truck–trailer. IEEE Trans Fuzzy Syst 2:119–134
4. Lo JC, Lin ML (2004) Robust $\mathscr{H}_\infty$ nonlinear control via fuzzy static output feedback. IEEE Trans Circuits Syst I Fundam Theory Appl 50:1494–1502
5. Che WW, Yang GH (2008) State feedback $\mathscr{H}_\infty$ control for quantized discrete-time systems. Asian J Control 10:718–723
6. Zhai G, Chen N, Gui W (2010) Quantizer design for interconnected feedback control systems. J Control Theory Appl 8:93–98
7. Chen N, Zhai G, Gui W, Yang C, Liu W (2010) Decentralised $\mathscr{H}_\infty$ quantisers design for uncertain interconnected networked systems. IET Control Theory Appl 4:177–185
8. Niu Y, Ho D W C (2014) Control strategy with adaptive quantizer's parameters under digital communication channels. Automatica 50:2665–2671
9. Wei Y, Park J H, Qiu J, Jung H (2018) Reliable output feedback control for piecewise affine systems with markov–type sensor failure. IEEE Trans Circuits Syst II Exp Briefs 65:913–917
10. Chang XH, Liu RR, Park J H (2019) A further study on output feedback $\mathscr{H}_\infty$ control for discrete-time systems. IEEE Trans Circuits Syst II Exp Briefs 67:305–309
11. Chang XH, Xiong J, Li ZM, Park J H (2018) Quantized static output feedback control for discrete-time systems. IEEE Trans Industr Inf 14:3426–3435

Chapter 3
Fuzzy Filtering with Multiple Signal Transmissions

Abstract This chapter investigates the filtering problem for T–S fuzzy systems, where multiple signals are considered to be transmitted through the communication channel. Firstly, the ℓ_2–ℓ_∞ filtering problem for T–S fuzzy systems with quantization is proposed, in which the measurement output and the performance output signals of the system are quantized by two static quantizers before being transmitted over the communication channel, respectively. Secondly, the induced ℓ_∞ filtering problem for T–S fuzzy systems is proposed, in which the measurement output and the performance output are taken into account the data packet dropout phenomenon modeled by two stochastic variables. Sufficient conditions are given to ensure that the filtering error system is not only stochastically stable but also has a prescribed induced ℓ_∞ performance. Thirdly, the $\mathscr{H}_\infty$ filtering problem for the T–S fuzzy systems is proposed, in which the measurement output and the performance output signals of the system are quantized by two dynamic quantizers. And the conditions are given to ensure the filtering error system is asymptotically stable with the prescribed $\mathscr{H}_\infty$ performance index. Finally, some examples are given to demonstrate the effectiveness of the proposed filtering methods for T–S fuzzy systems, respectively.

Keywords T–S fuzzy system · Filtering · Quantization · Data packet dropout

3.1 ℓ_2–ℓ_∞ Filtering with Static Quantization

In this section, the ℓ_2–ℓ_∞ filtering problem for nonlinear system with quantization is studied. The system with two quantized signals is investigated, namely, the measurement output signal and the performance output signal. And based on the fuzzy Lyapunov function approach, the ℓ_2–ℓ_∞ filter is designed for the uncertain T–S fuzzy system in terms of a set of strict LMIs.

X. Chang et al., *Control and Filtering of Fuzzy Systems Under Communication Channels*,
https://doi.org/10.1007/978-981-99-4346-3_3

3.1.1 Problem Formulation

Consider the following uncertain discrete-time T–S fuzzy model:

Plant Rule i_{th}: **IF** $h_1(k)$ is N_{1i}, $h_2(k)$ is N_{2i}, and, …, and $h_\theta(k)$ is $N_{\theta i}$, **THEN**

$$\begin{aligned} x(k+1) &= (A_i + \Delta_{Ai})x(k) + B_i w(k), \\ y(k) &= (C_i + \Delta_{Ci})x(k) + D_i w(k), \\ z(k) &= (L_i + \Delta_{Li})x(k) + E_i w(k), \end{aligned} \tag{3.1}$$

where $N_{\lambda i}$ are fuzzy sets with $\lambda = 1, \ldots, \theta$ and $i = 1, \ldots, r$, in which r is the number of fuzzy rules; $h(k) = [\, h_1(k), \ldots, h_\theta(k)\,]$ is premise variable vector; $x(k) \in \mathbb{R}^{n_x}$ is the state variable, $y(k) \in \mathbb{R}^{n_y}$ is the measurement output, $z(k) \in \mathbb{R}^{n_z}$ is the performance output, and $w(k) \in \mathbb{R}^{n_w}$ is the noise signal belonging to $\ell_2[0, \infty)$; $A_i \in \mathbb{R}^{n_x \times n_x}$, $B_i \in \mathbb{R}^{n_x \times n_w}$, $C_i \in \mathbb{R}^{n_y \times n_x}$, $D_i \in \mathbb{R}^{n_y \times n_w}$, $L_i \in \mathbb{R}^{n_z \times n_x}$, and $E_i \in \mathbb{R}^{n_z \times n_w}$ denote the system matrices which are known constant matrices; Δ_{Ai}, Δ_{Ci}, and Δ_{Li} are parameter uncertain matrices, which are considered as the following form in this section:

$$\begin{pmatrix} \Delta_{Ai} \\ \Delta_{Ci} \\ \Delta_{Li} \end{pmatrix} = \begin{pmatrix} F_{Ai} \\ F_{Ci} \\ F_{Li} \end{pmatrix} \Delta_x(k) N_x, \tag{3.2}$$

where $\Delta_x(k)$ is an uncertainty satisfying $\Delta_x^T(k)\Delta_x(k) \le I$; F_{Ai}, F_{Ci}, F_{Li}, and N_x are known constant matrices.

Denoting

$$s_i(h(k)) = \prod_{\lambda=1}^{\theta} N_{\lambda i}(h_\lambda(k)), \quad i = 1, \ldots, r,$$

where $N_{\lambda i}\big(h_\lambda(k)\big)$ is the grade of membership function of $h_\lambda(k)$ in $N_{\lambda i}$.

Assuming that

$$s_i(h(k)) > 0, \quad \sum_{i=1}^{r} s_i(h(k)) > 0, \quad i = 1, \ldots, r.$$

Let

$$\sigma_i(k) = \frac{s_i(h(k))}{\sum_{i=1}^{r} s_i(h(k))}, \quad i = 1, \ldots, r. \tag{3.3}$$

Then, we have

$$\sigma_i(k) \ge 0, \quad \sum_{i=1}^{r} \sigma_i(k) = 1, \quad i = 1, \ldots, r. \tag{3.4}$$

The uncertain T–S fuzzy model (3.1) can be described as follows:

$$\begin{aligned} x(k+1) &= (A_{\sigma_k} + \Delta_{A\sigma_k})x(k) + B_{\sigma_k}w(k), \\ y(k) &= (C_{\sigma_k} + \Delta_{C\sigma_k})x(k) + D_{\sigma_k}w(k), \\ z(k) &= (L_{\sigma_k} + \Delta_{L\sigma_k})x(k) + E_{\sigma_k}w(k), \end{aligned} \tag{3.5}$$

where

$$\begin{aligned} A_{\sigma_k} &= \sum_{i=1}^{r}\sigma_i(k)A_i, \quad B_{\sigma_k} = \sum_{i=1}^{r}\sigma_i(k)B_i, \quad C_{\sigma_k} = \sum_{i=1}^{r}\sigma_i(k)C_i, \\ D_{\sigma_k} &= \sum_{i=1}^{r}\sigma_i(k)D_i, \quad L_{\sigma_k} = \sum_{i=1}^{r}\sigma_i(k)L_i, \quad E_{\sigma_k} = \sum_{i=1}^{r}\sigma_i(k)E_i, \\ \Delta_{A\sigma_k} &= \sum_{i=1}^{r}\sigma_i(k)F_{Ai}\Delta_x(k)N_x, \quad \Delta_{C\sigma_k} = \sum_{i=1}^{r}\sigma_i(k)F_{Ci}\Delta_x(k)N_x, \\ \Delta_{L\sigma_k} &= \sum_{i=1}^{r}\sigma_i(k)F_{Li}\Delta_x(k)N_x. \end{aligned} \tag{3.6}$$

In order to transmit the measurement output $y(k)$ and the performance output $z(k)$ over the communication channel, we consider the static logarithmic quantizer [1]. The quantization levels have the following form:

$$U = \{\pm u_i,\ u_i = \rho^i u_0,\ i = 0,\ \pm 1,\ \pm 2, \ldots\} \cup \{\pm u_0\} \cup \{0\}, \tag{3.7}$$

where $0 < \rho < 1$ is quantization density and $u_0 > 0$. The quantizer function $Q(\cdot)$ is defined as

$$Q(\vartheta) = \begin{cases} u_m, & \frac{1}{1+\xi}u_m < \vartheta \le \frac{1}{1-\xi}u_m, \\ 0, & \vartheta = 0, \\ -Q(-\vartheta), & \vartheta < 0, \end{cases} \tag{3.8}$$

with

$$\xi = \frac{1-\rho}{1+\rho}. \tag{3.9}$$

To this end, the following two quantizers are employed in this section:

$$\begin{aligned} \bar{y}(k) &= Q(y(k)) = \begin{bmatrix} Q(y_1(k)) & \cdots & Q(y_{n_y}(k)) \end{bmatrix}^T, \\ \bar{z}(k) &= Q(z(k)) = \begin{bmatrix} Q(z_1(k)) & \cdots & Q(z_{n_z}(k)) \end{bmatrix}^T. \end{aligned} \tag{3.10}$$

For the quantization effects, we resort to the sector bound method proposed in [1] to deal with the quantization errors, the two quantizers described in (3.10) can be remodeled as

$$\begin{aligned}\bar{y}(k) - y(k) &= \Delta_y(k)y(k),\\ \bar{z}(k) - z(k) &= \Delta_z(k)z(k),\end{aligned} \tag{3.11}$$

which implies

$$\begin{aligned}\bar{y}(k) &= (I + \Delta_y(k))y(k),\\ \bar{z}(k) &= (I + \Delta_z(k))z(k),\end{aligned} \tag{3.12}$$

where

$$\begin{aligned}\|\Delta_y(k)\| &\le \Xi_y, \quad \Xi_y = diag\{\xi_y^1, \ldots, \xi_y^{n_y}\},\\ \|\Delta_z(k)\| &\le \Xi_z, \quad \Xi_z = diag\{\xi_z^1, \ldots, \xi_z^{n_z}\}.\end{aligned} \tag{3.13}$$

Then, the following filter form with quantized input is considered:

$$\begin{aligned}x_f(k+1) &= A_f x_f(k) + B_f \bar{y}(k),\\ z_f(k) &= C_f x_f(k) + D_f \bar{y}(k),\end{aligned} \tag{3.14}$$

where $x_f(k)$ and $z_f(k)$ are the state and the output of the filter, respectively; A_f, B_f, C_f and D_f are filter gain matrices to be determined.

By combining (3.5), (3.12), (3.14) and defining the augmented variable $\upsilon(k) = [\, x^T(k) \;\; x_f^T(k)\,]^T$ and the filtering error $e(k) = \bar{z}(k) - z_f(k)$, the filtering error system is given by

$$\begin{aligned}\upsilon(k+1) &= \mathcal{A}_{\sigma_k}\upsilon(k) + \mathcal{B}_{\sigma_k}w(k),\\ e(k) &= \mathcal{C}_{\sigma_k}\upsilon(k) + \mathcal{D}_{\sigma_k}w(k),\end{aligned} \tag{3.15}$$

where

$$\begin{aligned}\mathcal{A}_{\sigma_k} &= \begin{bmatrix} A_{\sigma_k} + \Delta_{A\sigma_k} & 0 \\ B_f(I + \Delta_y(k))(C_{\sigma_k} + \Delta_{C\sigma_k}) & A_f \end{bmatrix}, \quad \mathcal{B}_{\sigma_k} = \begin{bmatrix} B_{\sigma_k} \\ B_f(I + \Delta_y(k))D_{\sigma_k} \end{bmatrix},\\ \mathcal{C}_{\sigma_k} &= [\,(I + \Delta_z(k))(L_{\sigma_k} + \Delta_{L\sigma_k}) - D_f(I + \Delta_y(k))(C_{\sigma_k} + \Delta_{C\sigma_k}) \quad -C_f\,],\\ \mathcal{D}_{\sigma_k} &= (I + \Delta_z(k))E_{\sigma_k} - D_f(I + \Delta_y(k))D_{\sigma_k}.\end{aligned}$$

The purpose of this section is to determine the filter (3.14), such that the filtering error system (3.15) satisfies the following two conditions:

(1) When $w(k) = 0$, the filtering error system (3.15) is asymptotically stable;
(2) Under zero initial condition, for any nonzero $w(k) \in \ell_2[0, \infty)$ and given $\gamma > 0$, the error output satisfies

$$\|e(k)\|_\infty \le \gamma \|w(k)\|_2, \tag{3.16}$$

where $\|e(k)\|_\infty = \sup_{k\ge 0}\left\{\sqrt{e^T(k)e(k)}\right\}$ and $\|w(k)\|_2 = \sqrt{\sum_{k=0}^{\infty} w^T(k)w(k)}$.

3.1.2 Main Results

In this section, sufficient conditions are provided to design the ℓ_2–ℓ_∞ filter which also can ensure that filtering error system (3.15) is asymptotically stable with a given ℓ_2–ℓ_∞ performance γ.

Theorem 3.1 *For a given performance index $\gamma > 0$, the system (3.15) is asymptotically stable with the ℓ_2–ℓ_∞ performance index γ, if there exist matrices $P_{\sigma_k} > 0$, $P_{\sigma_{k+1}} > 0$ such that*

$$\begin{bmatrix} -P_{\sigma_k} + \mathcal{A}_{\sigma_k}^T P_{\sigma_{k+1}} \mathcal{A}_{\sigma_k} & * \\ \mathcal{B}_{\sigma_k}^T P_{\sigma_{k+1}} \mathcal{A}_{\sigma_k} & -I + \mathcal{B}_{\sigma_k}^T P_{\sigma_{k+1}} \mathcal{B}_{\sigma_k} \end{bmatrix} < 0, \tag{3.17}$$

$$\begin{bmatrix} -P_{\sigma_k} + \gamma^{-2} \mathcal{C}_{\sigma_k}^T \mathcal{C}_{\sigma_k} & * \\ \gamma^{-2} \mathcal{D}_{\sigma_k}^T \mathcal{C}_{\sigma_k} & -I + \gamma^{-2} \mathcal{D}_{\sigma_k}^T \mathcal{D}_{\sigma_k} \end{bmatrix} < 0, \tag{3.18}$$

where $P_{\sigma_k} = \sum_{i=1}^{r} \sigma_i(k) P_i$.

Proof By performing congruence transformation to (3.17) with $[\,\upsilon^T(k)\ w^T(k)]^T \neq 0$, one obtains

$$\begin{aligned} &(\mathcal{A}_{\sigma_k}\upsilon(k) + \mathcal{B}_{\sigma_k} w(k))^T P_{\sigma_{k+1}} (\mathcal{A}_{\sigma_k}\upsilon(k) + \mathcal{B}_{\sigma_k} w(k)) \\ &- \upsilon^T(k) P_{\sigma_k} \upsilon(k) < w^T(k) w(k). \end{aligned} \tag{3.19}$$

Construct Lyapunov function as

$$V(k) = \upsilon^T(k) P_{\sigma_k} \upsilon(k). \tag{3.20}$$

Then from (3.15), one has

$$V(k+1) = (\mathcal{A}_{\sigma_k}\upsilon(k) + \mathcal{B}_{\sigma_k} w(k))^T P_{\sigma_{k+1}} (\mathcal{A}_{\sigma_k}\upsilon(k) + \mathcal{B}_{\sigma_k} w(k)). \tag{3.21}$$

Then, (3.19) can be rewritten as

$$V(k+1) - V(k) < w^T(k) w(k). \tag{3.22}$$

For $w(k) = 0$, we can find that

$$\Delta V(k) = V(k+1) - V(k) < 0, \tag{3.23}$$

which implies that the filtering error system (3.15) is asymptotically stable.

Then by taking the sum on both sides of the inequality (3.22) from 0 to $k-1$, one has

$$\sum_{s=0}^{k-1} \Big(V(s+1) - V(s) \Big) < \sum_{s=0}^{k-1} w^T(s) w(s). \tag{3.24}$$

After simply calculating, one can be obtained

$$V(k) - V(0) < \sum_{s=0}^{k-1} w^T(s)w(s). \tag{3.25}$$

Under zero initial condition, the following inequality can be obtained:

$$V(k) < \sum_{s=0}^{k-1} w^T(s)w(s). \tag{3.26}$$

Similarly, performing congruence transformation to (3.18) with $[\,\upsilon^T(k)\ w^T(k)]^T$ and its transpose, we have

$$\gamma^{-2}e^T(k)e(k) < V(k) + w^T(k)w(k). \tag{3.27}$$

Combining (3.26) with (3.27) yields

$$e^T(k)e(k) < \gamma^2\sum_{s=0}^{k-1} w^T(s)w(s) + \gamma^2 w^T(k)w(k) = \gamma^2\sum_{s=0}^{k} w^T(s)w(s), \tag{3.28}$$

which means that $\|e(k)\|_\infty < \gamma\|w(k)\|_2$. The proof is completed.

Note that the inequalities (3.17) and (3.18) contain the uncertainty terms and the nonlinear terms which need to be handled to obtain the LMI-based filter design conditions. In the following, the filter design scheme is proposed in terms of a set of strict LMIs.

Theorem 3.2 *For a given performance index $\gamma > 0$, the system (3.15) is asymptotically stable and satisfies the ℓ_2–ℓ_∞ performance γ, if there exist scalars $\varepsilon_l > 0$, $\eta_l > 0$, $\beta > 0$, $\delta > 0$ and matrices $P_{1i} > 0$, P_{2i}, $P_{3i} > 0$, G_1, G_2, G_3, $\bar{A}_f$, $\bar{B}_f$, $\bar{C}_f$, $\bar{D}_f$ such that*

$$\Theta_{il} = \begin{bmatrix} -P_{1i} & * & * & * & * & * & * & * & * \\ -P_{2i} & -P_{3i} & * & * & * & * & * & * & * \\ 0 & 0 & -I & * & * & * & * & * & * \\ {\Theta_{14}^{i}}^T & \bar{A}_f & {\Theta_{34}^{i}}^T & \Theta_{44}^{l} & * & * & * & * & * \\ {\Theta_{15}^{i}}^T & \bar{A}_f & {\Theta_{35}^{i}}^T & {\Theta_{45}^{l}}^T & \Theta_{55}^{l} & * & * & * & * \\ 0 & 0 & 0 & \Xi_y^T\bar{B}_f^T & \Xi_y^T\bar{B}_f^T & -\varepsilon_l I & * & * & * \\ \varepsilon_l C_i & 0 & \varepsilon_l D_i & 0 & 0 & 0 & -\varepsilon_l I & * & * \\ 0 & 0 & 0 & {\Theta_{48}^{i}}^T & {\Theta_{58}^{i}}^T & 0 & \varepsilon_l F_{Ci}^T & -\eta_l I & * \\ \eta_l N_x & 0 & 0 & 0 & 0 & 0 & 0 & 0 & -\eta_l I \end{bmatrix} < 0, \tag{3.29}$$

$$\Sigma_i = \begin{bmatrix} -P_{1i} & * & * & * & * & * & * & * & * & * \\ -P_{2i} & -P_{3i} & * & * & * & * & * & * & * & * \\ 0 & 0 & -I & * & * & * & * & * & * & * \\ {\Sigma_{14}^{i}}^T & -\bar{C}_f & {\Sigma_{34}^{i}}^T & -\gamma^2 I & * & * & * & * & * & * \\ 0 & 0 & 0 & \Xi_z^T & -\beta I & * & * & * & * & * \\ 0 & 0 & 0 & -\Xi_y^T \bar{D}_f^T & 0 & -\beta I & * & * & * & * \\ \beta L_i & 0 & \beta E_i & 0 & 0 & 0 & -\beta I & * & * & * \\ \beta C_i & 0 & \beta D_i & 0 & 0 & 0 & 0 & -\beta I & * & * \\ 0 & 0 & 0 & {\Sigma_{49}^{i}}^T & 0 & 0 & \beta F_{Li}^T & \beta F_{Ci}^T & -\delta I & * \\ \delta N_x & 0 & 0 & 0 & 0 & 0 & 0 & 0 & 0 & -\delta I \end{bmatrix} < 0, \tag{3.30}$$

for $i, l = 1, \ldots, r$, *where*

$$\begin{aligned} &\Theta_{14}^i = (G_1 A_i + \bar{B}_f C_i)^T,\ \Theta_{15}^i = (G_3 A_i + \bar{B}_f C_i)^T, \\ &\Theta_{34}^i = (G_1 B_i + \bar{B}_f D_i)^T,\ \Theta_{35}^i = (G_3 B_i + \bar{B}_f D_i)^T, \\ &\Theta_{44}^l = -G_1 - G_1^T + P_{1l},\ \Theta_{45}^l = -G_2 - G_3^T + P_{2l}, \\ &\Theta_{48}^i = G_1 F_{Ai} + \bar{B}_f F_{Ci},\ \Theta_{55}^l = -G_2 - G_2^T + P_{3l}, \\ &\Theta_{58}^i = G_3 F_{Ai} + \bar{B}_f F_{Ci},\ \Sigma_{14}^i = (L_i - \bar{D}_f C_i)^T, \\ &\Sigma_{34}^i = (E_i - \bar{D}_f D_i)^T,\ \Sigma_{49}^i = F_{Li} - \bar{D}_f F_{Ci}. \end{aligned} \tag{3.31}$$

In addition, the ℓ_2*–*ℓ_∞ *filter gain matrices in (3.14) can be obtained by*

$$A_f = G_2^{-1} \bar{A}_f,\quad B_f = G_2^{-1} \bar{B}_f,\quad C_f = \bar{C}_f,\quad D_f = \bar{D}_f. \tag{3.32}$$

Proof By applying Schur complement (Lemma 1.1) to the matrix inequality (3.17), one has

$$\begin{bmatrix} -P_{\sigma_k} & * & * \\ 0 & -I & * \\ \mathcal{A}_{\sigma_k} & \mathcal{B}_{\sigma_k} & -P_{\sigma_{k+1}}^{-1} \end{bmatrix} < 0. \tag{3.33}$$

Let us consider a nonsingular matrix G with appropriate dimension and pre- and post-multiply $diag\{I,\ I,\ G\}$ and its transpose to (3.33), one yields

$$\begin{bmatrix} -P_{\sigma_k} & * & * \\ 0 & -I & * \\ G\mathcal{A}_{\sigma_k} & G\mathcal{B}_{\sigma_k} & -GP_{\sigma_{k+1}}^{-1} G^T \end{bmatrix} < 0. \tag{3.34}$$

From the fact that $-(G - P_{\sigma_{k+1}})^T P_{\sigma_{k+1}}^{-1} (G - P_{\sigma_{k+1}}) \leq 0$ with $P_{\sigma_{k+1}} > 0$ means $-GP_{\sigma_{k+1}}^{-1} G^T \leq -G - G^T + P_{\sigma_{k+1}}$, we know that (3.34) can be verified by

$$\begin{bmatrix} -P_{\sigma_k} & * & * \\ 0 & -I & * \\ G\mathcal{A}_{\sigma_k} & G\mathcal{B}_{\sigma_k} & \Theta_p \end{bmatrix} < 0, \tag{3.35}$$

with $\Theta_p = -G - G^T + P_{\sigma_{k+1}}$.

Then, we need to isolate the uncertainties in the matrix inequality (3.35) before they are eliminated. According to the definitions in (3.9) and (3.13), we affirm that the matrix Ξ_y is nonsingular, then (3.35) can be rewritten as follows:

$$\begin{aligned} &\begin{bmatrix} -P_{\sigma_k} & * & * \\ 0 & -I & * \\ GA_{11\sigma_k} & GB_{11\sigma_k} & \Theta_p \end{bmatrix} + \begin{bmatrix} 0 \\ 0 \\ G\begin{bmatrix} 0 \\ B_f \end{bmatrix}\Xi_y \end{bmatrix} \frac{\Delta_y(k)}{\Xi_y}[[C_{\sigma_k} + \Delta_{C\sigma_k} \ \ 0] \ \ D_{\sigma_k} \ \ 0] \\ &+ \begin{bmatrix} \begin{bmatrix} C_{\sigma_k}^T + \Delta_{C\sigma_k}^T \\ 0 \end{bmatrix} \\ D_{\sigma_k}^T \\ 0 \end{bmatrix} \frac{\Delta_y^T(k)}{\Xi_y^T}[0 \ \ 0 \ \ \Xi_y^T[0 \ B_f^T]G^T] < 0, \end{aligned} \tag{3.36}$$

where

$$A_{11\sigma_k} = \begin{bmatrix} A_{\sigma_k} + \Delta_{A\sigma_k} & 0 \\ B_f(C_{\sigma_k} + \Delta_{C\sigma_k}) & A_f \end{bmatrix}, \quad B_{11\sigma_k} = [\, B_{\sigma_k}^T \ \ D_{\sigma_k}^T B_f^T \,]^T.$$

From (3.13), one has $\left(\frac{\Delta_y(k)}{\Xi_y}\right)^T \frac{\Delta_y(k)}{\Xi_y} \le I$. By using Lemma 1.2 to (3.36), there will be a sufficient condition guaranteeing that the above inequality holds

$$\begin{aligned} &\begin{bmatrix} -P_{\sigma_k} & * & * \\ 0 & -I & * \\ GA_{11\sigma_k} & GB_{11\sigma_k} & \Theta_p \end{bmatrix} + \frac{1}{\varepsilon_{\sigma_{k+1}}} \begin{bmatrix} 0 \\ 0 \\ G\begin{bmatrix} 0 \\ B_f \end{bmatrix}\Xi_y \end{bmatrix} [0 \ \ 0 \ \ \Xi_y^T[0 \ B_f^T]G^T] \\ &+ \frac{1}{\varepsilon_{\sigma_{k+1}}} \begin{bmatrix} \varepsilon_{\sigma_{k+1}}\begin{bmatrix} C_{\sigma_k}^T + \Delta_{C\sigma_k}^T \\ 0 \end{bmatrix} \\ \varepsilon_{\sigma_{k+1}} D_{\sigma_k}^T \\ 0 \end{bmatrix} [\,\varepsilon_{\sigma_{k+1}}[C_{\sigma_k} + \Delta_{C\sigma_k} \ \ 0] \ \ \varepsilon_{\sigma_{k+1}} D_{\sigma_k} \ \ 0] < 0. \end{aligned} \tag{3.37}$$

Using Schur complement (Lemma 1.1) gives

$$\begin{bmatrix} -P_{\sigma_k} & * & * & * & * \\ 0 & -I & * & * & * \\ GA_{11\sigma_k} & GB_{11\sigma_k} & \Theta_p & * & * \\ 0 & 0 & \Xi_y^T[0 \ B_f^T]G^T & -\varepsilon_{\sigma_{k+1}} I & * \\ \varepsilon_{\sigma_{k+1}}[C_{\sigma_k} + \Delta_{C\sigma_k} \ 0] & \varepsilon_{\sigma_{k+1}} D_{\sigma_k} & 0 & 0 & -\varepsilon_{\sigma_{k+1}} I \end{bmatrix} < 0. \tag{3.38}$$

Notice that the uncertainty $\Delta_y(k)$ dose not exist in (3.38) by using Lemma 1.2, and only the uncertainty $\Delta_x(k)$ is left. Next, the same method as eliminating $\Delta_y(k)$ is used to eliminate $\Delta_x(k)$, we gain

$$\begin{bmatrix} -P_{\sigma_k} & * & * & * & * & * & * \\ 0 & -I & * & * & * & * & * \\ GA_{12\sigma_k} & GB_{11\sigma_k} & \Theta_p & * & * & * & * \\ 0 & 0 & \Xi_y^T\begin{bmatrix}0 & B_f^T\end{bmatrix}G^T & -\varepsilon_{\sigma_{k+1}}I & * & * & * \\ \begin{bmatrix}\varepsilon_{\sigma_{k+1}}C_{\sigma_k}^T \\ 0\end{bmatrix}^T & \varepsilon_{\sigma_{k+1}}D_{\sigma_k} & 0 & 0 & -\varepsilon_{\sigma_{k+1}}I & * & * \\ 0 & 0 & \begin{bmatrix}F_{A\sigma_k} \\ B_fF_{C\sigma_k}\end{bmatrix}^T G^T & 0 & \varepsilon_{\sigma_{k+1}}F_{C\sigma_k}^T & -\eta_{\sigma_{k+1}}I & * \\ \eta_{\sigma_{k+1}}\begin{bmatrix}N_x^T \\ 0\end{bmatrix}^T & 0 & 0 & 0 & 0 & 0 & -\eta_{\sigma_{k+1}}I \end{bmatrix} < 0, \tag{3.39}$$

where

$$A_{12\sigma_k} = \begin{bmatrix} A_{\sigma_k} & 0 \\ B_fC_{\sigma_k} & A_f \end{bmatrix}.$$

In order to eliminate the coupling items in (3.39), constructing two matrix variables as

$$G = \begin{bmatrix} G_1 & G_2 \\ G_3 & G_2 \end{bmatrix}, \ P_{\sigma_k} = \begin{bmatrix} P_{1\sigma_k} & * \\ P_{2\sigma_k} & P_{3\sigma_k} \end{bmatrix} = \sum_{i=1}^{r} \sigma_i(k) \begin{bmatrix} P_{1i} & * \\ P_{2i} & P_{3i} \end{bmatrix}, \tag{3.40}$$

and according to the definition $\bar{A}_f = G_2A_f$, $\bar{B}_f = G_2B_f$ in (3.32). We know that (3.39) can be rewritten as

$$\begin{bmatrix} -P_{1\sigma_k} & * & * & * & * & * & * & * & * \\ -P_{2\sigma_k} & -P_{3\sigma_k} & * & * & * & * & * & * & * \\ 0 & 0 & -I & * & * & * & * & * & * \\ \Theta_{14}{}^T & \bar{A}_f & \Theta_{34}{}^T & \Theta_{44} & * & * & * & * & * \\ \Theta_{15}{}^T & \bar{A}_f & \Theta_{35}{}^T & \Theta_{45}^T & \Theta_{55} & * & * & * & * \\ 0 & 0 & 0 & \Xi_y^T\bar{B}_f^T & \Xi_y^T\bar{B}_f^T & -\varepsilon_{\sigma_{k+1}}I & * & * & * \\ \varepsilon_{\sigma_{k+1}}C_{\sigma_k} & 0 & \varepsilon_{\sigma_{k+1}}D_{\sigma_k} & 0 & 0 & 0 & -\varepsilon_{\sigma_{k+1}}I & * & * \\ 0 & 0 & 0 & \Theta_{48}^T & \Theta_{58}^T & 0 & \varepsilon_{\sigma_{k+1}}F_{C\sigma_k}^T & -\eta_{\sigma_{k+1}}I & * \\ \eta_{\sigma_{k+1}}N_x & 0 & 0 & 0 & 0 & 0 & 0 & 0 & -\eta_{\sigma_{k+1}}I \end{bmatrix} < 0, \tag{3.41}$$

where

$$\begin{aligned} &\Theta_{14} = (G_1A_{\sigma_k} + \bar{B}_fC_{\sigma_k})^T, \quad \Theta_{15} = (G_3A_{\sigma_k} + \bar{B}_fC_{\sigma_k})^T, \\ &\Theta_{34} = (G_1B_{\sigma_k} + \bar{B}_fD_{\sigma_k})^T, \quad \Theta_{35} = (G_3B_{\sigma_k} + \bar{B}_fD_{\sigma_k})^T, \\ &\Theta_{44} = -G_1 - G_1^T + P_{1\sigma_{k+1}}, \quad \Theta_{45} = -G_2 - G_3^T + P_{2\sigma_{k+1}}, \\ &\Theta_{48} = G_1F_{A\sigma_k} + \bar{B}_fF_{C\sigma_k}, \quad \Theta_{55} = -G_2 - G_2^T + P_{3\sigma_{k+1}}, \\ &\Theta_{58} = G_3F_{A\sigma_k} + \bar{B}_fF_{C\sigma_k}. \end{aligned} \tag{3.42}$$

Analogous to the derivation process for the condition (3.17) and definition (3.32), then the matrix inequality (3.18) holds if the following condition is true:

$$\begin{bmatrix} -P_{1\sigma_k} & * & * & * & * & * & * & * & * & * \\ -P_{2\sigma_k} & -P_{3\sigma_k} & * & * & * & * & * & * & * & * \\ 0 & 0 & -I & * & * & * & * & * & * & * \\ \Sigma_{14}^T & -\bar{C}_f & \Sigma_{34}^T & -\gamma^2 I & * & * & * & * & * & * \\ 0 & 0 & 0 & \Xi_z^T & -\beta I & * & * & * & * & * \\ 0 & 0 & 0 & -\Xi_y^T \bar{D}_f^T & 0 & -\beta I & * & * & * & * \\ \beta L_{\sigma_k} & 0 & \beta E_{\sigma_k} & 0 & 0 & 0 & -\beta I & * & * & * \\ \beta C_{\sigma_k} & 0 & \beta D_{\sigma_k} & 0 & 0 & 0 & 0 & -\beta I & * & * \\ 0 & 0 & 0 & \Sigma_{49}^T & 0 & 0 & \beta F_{L\sigma_k}^T & \beta F_{C\sigma_k}^T & -\delta I & * \\ \delta N_x & 0 & 0 & 0 & 0 & 0 & 0 & 0 & 0 & -\delta I \end{bmatrix} < 0, \quad (3.43)$$

where

$$\Sigma_{14} = (L_{\sigma_k} - \bar{D}_f C_{\sigma_k})^T, \quad \Sigma_{34} = (E_{\sigma_k} - \bar{D}_f D_{\sigma_k})^T, \quad \Sigma_{49} = F_{L\sigma_k} - \bar{D}_f F_{C\sigma_k}. \quad (3.44)$$

From (3.41) and (3.43), we have

$$\sum_{i=1}^{r} \sum_{l=1}^{r} \sigma_i(k)\sigma_l(k+1)\Theta_{il} < 0, \quad (3.45)$$

and

$$\sum_{i=1}^{r} \sigma_i(k)\Sigma_i < 0. \quad (3.46)$$

By recalling the property of the membership function described in (3.4), we can know that if (3.29) and (3.30) are established, the inequalities (3.45) and (3.46) are true. The proof is completed.

3.1.3 Simulation Example

In this section, we consider a factual situation that there are the parameter perturbation and disturbance signal in the DC motor system [2]. The dynamic equations of this system are given by

$$\begin{aligned} L\frac{di(t)}{dt} &= -u_R(t) - (K_b + \Delta_b(t))\omega(t), \\ J\frac{d\omega(t)}{dt} &= K_a i(t) - B\omega(t) + w(t), \end{aligned} \quad (3.47)$$

where $i(t)$ is armature current, $\omega(t)$ is rotational speed, $w(t)$ is the torque disturbance, L is armature inductor, $u_R(t)$ is the voltage across the armature resistor, $\Delta_b(t)$ is an uncertainty satisfying $\|\Delta_b(t)\| < K_b$, J is moment of inertia of the mechanical rotating part, K_a is the torque constant, B is viscous friction coefficient of the rotating part, respectively.

Assume $x_1(t)=i(t)$, $x_2(t)=\omega(t)$, and select the parameters as $J=0.082\,\text{kg}\cdot\text{m}^2$, $B=0.3\,\text{N}\cdot\text{m}\cdot\text{s/rad}$, $L=1000\,\text{mH}$, $K_a=0.576\,\text{N}\cdot\text{m/A}$, $K_b=0.612\,\text{V}\cdot\text{s/rad}$ and $\Delta_b(t)=0.06b(t)$ $(-1\le b(t)\le 1)$, respectively. As in [2], the system can be described as

$$\begin{aligned}\dot{x}_1(t) &= -4x_1(t)-x_1^3(t)-\ (0.612+0.06b(t))x_2(t),\\ \dot{x}_2(t) &= (0.576/0.082)x_1(t)-0.3/0.082x_2(t)+1/0.082w(t).\end{aligned} \tag{3.48}$$

Let us rewrite (3.48) as

$$\dot{x}(t)=\begin{bmatrix}-4-x_1^2(t) & -(0.612+0.06b(t))\\ 0.576/0.082 & -0.3/0.082\end{bmatrix}x(t)+\begin{bmatrix}0\\ 1/0.082\end{bmatrix}w(t). \tag{3.49}$$

For the nonlinear system (3.49), one can assume that $-3\le x_1(t)\le 3$, that is, $-13\le -4-x_1^2(t)\le -4$. By fuzzy modeling for the nonlinear system, (3.49) can be represented by the following T–S model with two fuzzy rules:

$$\dot{x}(t)=\sum_{i=1}^{2}\sigma_i(t)\big(A_{ib}x(t)+B_{ib}w(t)\big), \tag{3.50}$$

where $x^T(t)=[\,x_1^T(t)\quad x_2^T(t)\,]$ and

$$\begin{aligned}A_{1b}&=\begin{bmatrix}-4 & -(0.612+0.06b(t))\\ 0.576/0.082 & -0.3/0.082\end{bmatrix}, & B_{1b}&=\begin{bmatrix}0\\ 1/0.082\end{bmatrix},\\ A_{2b}&=\begin{bmatrix}-13 & -(0.612+0.06b(t))\\ 0.576/0.082 & -0.3/0.082\end{bmatrix}, & B_{2b}&=\begin{bmatrix}0\\ 1/0.082\end{bmatrix}.\end{aligned}$$

The membership function can be given by

$$\sigma_1(t)=x_1^2(t)/9,\quad \sigma_2(t)=1-x_1^2(t)/9. \tag{3.51}$$

In the following, we will study the ℓ_2–ℓ_∞ filtering problem for the system (3.50). By taking a sampling time $T=0.05$ s, we get a discrete-time T–S fuzzy system as (3.1) and the corresponding system parameter matrices are

$$A_1 = \begin{bmatrix} 0.8 & -0.0306 \\ 0.3512 & 0.8171 \end{bmatrix}, \quad B_1 = \begin{bmatrix} 0 \\ 0.05/0.082 \end{bmatrix},$$
$$A_2 = \begin{bmatrix} 0.35 & -0.0306 \\ 0.3512 & 0.8171 \end{bmatrix}, \quad B_2 = \begin{bmatrix} 0 \\ 0.05/0.082 \end{bmatrix}, \tag{3.52}$$
$$C_1 = [-3 \quad 0.5], \quad D_1 = 2.5, \quad L_1 = [0.2 \quad 1], \quad E_1 = 1,$$
$$C_2 = [1.5 \quad 2], \quad D_2 = -2.3, \quad L_2 = [0.3 \quad 1], \quad E_2 = -0.5.$$

And the following parameters are also considered in (3.2):

$$F_{A1} = F_{A2} = [-0.6 \quad 0]^T, \quad F_{C1} = F_{C2} = 0.01,$$
$$F_{L1} = F_{L2} = 3.5, \quad N_x = [0 \quad 0.1]. \tag{3.53}$$

We adopt the other parameters $\rho_y = \rho_z = 0.7$, the maximum ℓ_2-ℓ_∞ attenuation level can be obtained as 1.6855 by solving LMIs (3.29) and (3.30). In this case, the filter matrices can be calculated as the follows:

$$A_f = \begin{bmatrix} 1.2160 & -0.2963 \\ 0.6312 & 0.3836 \end{bmatrix}, \quad B_f = \begin{bmatrix} -0.2008 \\ -0.4673 \end{bmatrix},$$
$$C_f = [0.3349 \quad -0.2740], \quad D_f = 0.4213.$$

Assume that the initial condition is $x(0) = x_f(0) = [0 \quad 0]^T$, the uncertainties are $\Delta_x(k) = 0.6\cos(0.8k)$, $\Delta_y(k) = 0.1\sin(k)$, $\Delta_z(k) = 0.1\sin(0.8k)$, and the external disturbance is $w(k) = e^{-2k}\cos(0.1k)$. The response results of $x(k)$ and $x_f(k)$ are displayed in Figs. 3.1 and 3.2, respectively. Figure 3.3 describes the response of filtering error $e(k)$. It can be perceived that the error curve tends to 0. The ratio of $\sqrt{e^T(k)e(k)/\sum_{k=0}^{\infty} w^T(k)w(k)}$ is shown in Fig. 3.4. It can be known that the maximum value of the ratio is less than 1.6855. From the simulation results of this example, we can get that the design method of the ℓ_2-ℓ_∞ filter proposed in this section is effective.

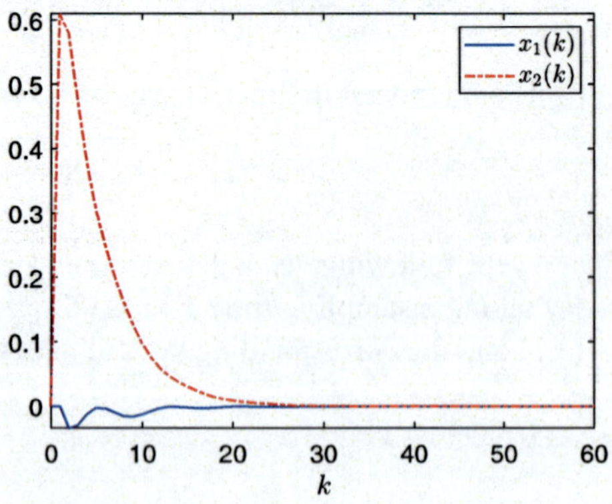

Fig. 3.1 Response of the system state $x(k)$

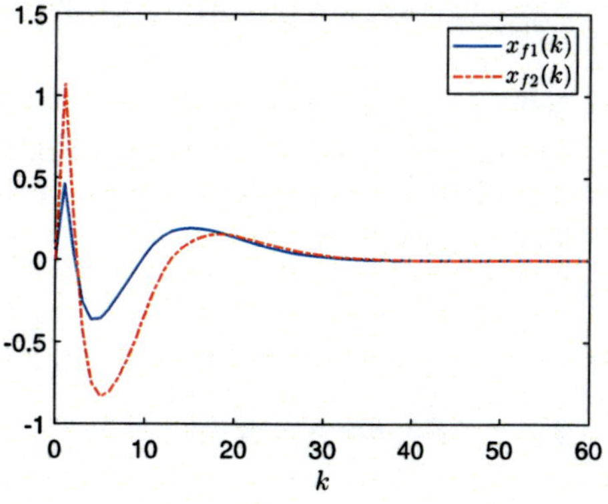

Fig. 3.2 Response of the filter state $x_f(k)$

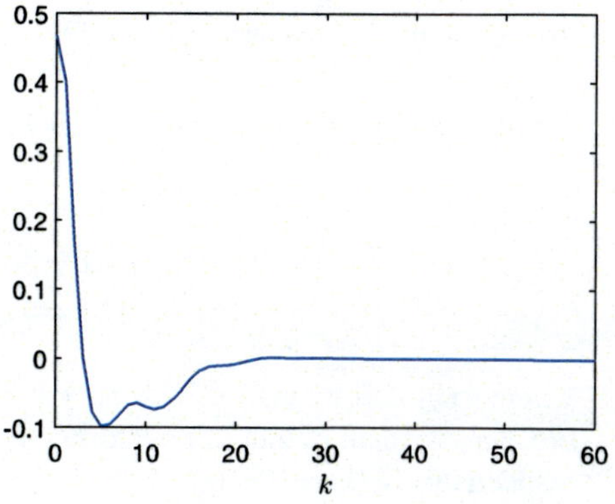

Fig. 3.3 Filtering error $e(k)$

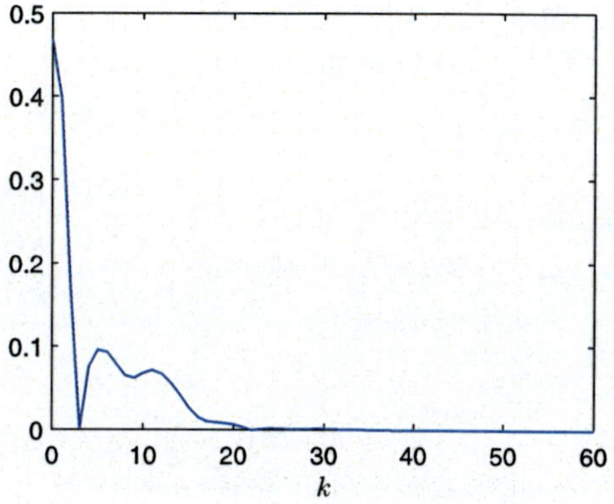

Fig. 3.4 History of $\sqrt{e^T(k)e(k)/\sum_{k=0}^{\infty} w^T(k)w(k)}$

3.2 Induced ℓ_∞ Filtering with Data Packet Dropouts

In this section, the induced ℓ_∞ filtering problem for a class of nonlinear systems based on T–S fuzzy model is considered. In addition, both the measurement output and the performance output are considered to be suffered from data packet dropout when they are transmitted through the communication channel. Finally, the conditions to design induced ℓ_∞ filter are given in the form of LMIs.

3.2.1 Problem Formulation

For the induced ℓ_∞ filtering problem in this section, we consider a nonlinear discrete-time system which is approximated by a T–S fuzzy model as follows:

Plant Rule i_{th}: **IF** $h_1(k)$ is N_{1i}, $h_2(k)$ is N_{2i}, and, …, and $h_\theta(k)$ is $N_{\theta i}$, **THEN**

$$\begin{aligned} x(k+1) &= A_i x(k) + B_i w(k), \\ y(k) &= C_i x(k) + D_i w(k), \\ z(k) &= L_i x(k) + E_i w(k), \end{aligned} \tag{3.54}$$

where $x(k) \in \mathbb{R}^{n_x}$, $y(k) \in \mathbb{R}^{n_y}$, $z(k) \in \mathbb{R}^{n_z}$, and $w(k) \in \mathbb{R}^{n_w}$ are the state variable, the measurement output, the performance output, and the noise signal in $\ell_\infty[0, \infty)$, respectively; $h(k) = [\, h_1(k), \ \ldots, \ h_\theta(k)\,]$ is premise variable vector; $N_{\lambda i}$ with $\lambda = 1, \ldots, \theta$ and $i = 1, \ldots, r$ are the fuzzy sets with r being the number of fuzzy rules; $A_i \in \mathbb{R}^{n_x \times n_x}$, $B_i \in \mathbb{R}^{n_x \times n_w}$, $C_i \in \mathbb{R}^{n_y \times n_x}$, $D_i \in \mathbb{R}^{n_y \times n_w}$, $L_i \in \mathbb{R}^{n_z \times n_x}$, and $E_i \in \mathbb{R}^{n_z \times n_w}$ are known constant matrices with appropriate dimensions.

As in Sect. 3.1, the T–S fuzzy model (3.54) can be deduced as follows:

$$\begin{aligned} x(k+1) &= A(\upsilon(k))x(k) + B(\upsilon(k))w(k), \\ y(k) &= C(\upsilon(k))x(k) + D(\upsilon(k))w(k), \\ z(k) &= L(\upsilon(k))x(k) + E(\upsilon(k))w(k), \end{aligned} \tag{3.55}$$

where

$$\begin{aligned} A(\upsilon(k)) &= \sum_{i=1}^{r} \sigma_i(k) A_i, \quad B(\upsilon(k)) = \sum_{i=1}^{r} \sigma_i(k) B_i, \\ C(\upsilon(k)) &= \sum_{i=1}^{r} \sigma_i(k) C_i, \quad D(\upsilon(k)) = \sum_{i=1}^{r} \sigma_i(k) D_i, \\ L(\upsilon(k)) &= \sum_{i=1}^{r} \sigma_i(k) L_i, \quad E(\upsilon(k)) = \sum_{i=1}^{r} \sigma_i(k) E_i. \end{aligned}$$

In this section, the following fuzzy filter is provided to estimate the performance output $z(k)$ in (3.55):

$$\begin{aligned} x_f(k+1) &= A_f(\upsilon(k))x_f(k) + B_f(\upsilon(k))\bar{y}(k), \\ z_f(k) &= C_f(\upsilon(k))x_f(k) + D_f(\upsilon(k))\bar{y}(k), \end{aligned} \tag{3.56}$$

in which $x_f(k)$ and $z_f(k)$ denote the state and the output of the fuzzy filter, respectively; $A_f(\upsilon(k))$, $B_f(\upsilon(k))$, $C_f(\upsilon(k))$, and $D_f(\upsilon(k))$ are fuzzy filter matrices which can be described as

$$\begin{aligned} A_f(\upsilon(k)) &= \sum_{d=1}^{r} \sigma_d(k)A_{fd}, \quad B_f(\upsilon(k)) = \sum_{d=1}^{r} \sigma_d(k)B_{fd}, \\ C_f(\upsilon(k)) &= \sum_{d=1}^{r} \sigma_d(k)C_{fd}, \quad D_f(\upsilon(k)) = \sum_{d=1}^{r} \sigma_d(k)D_{fd}. \end{aligned}$$

In this section, suppose that the data packet may be lost, which means that the measurement output $y(k)$ and the performance output $z(k)$ in system (3.55) are transmitted by unreliable communication channel. Then, by using a stochastic method, the data packet dropout phenomenon can be modeled:

$$\begin{aligned} \bar{y}(k) &= \delta_y(k)y(k), \\ \bar{z}(k) &= \delta_z(k)z(k), \end{aligned} \tag{3.57}$$

where $\{\delta_y(k)\}$ and $\{\delta_z(k)\}$ obey Bernoulli process. If $y(k)$ or $z(k)$ fails to be transmitted (i.e., data packet dropout), then $\delta_y(k) = 0$ or $\delta_z(k) = 0$. And $\delta_y(k) = 1$ or $\delta_z(k) = 1$ means the successful transmission of related signals. A nature assumption on $\{\delta_y(k)\}$ and $\{\delta_z(k)\}$ can be obtained:

$$\begin{aligned} Prob\{\delta_y(k) = 0\} &= 1 - \bar{\delta}_y, \quad Prob\{\delta_y(k) = 1\} = \Gamma\{\delta_y(k)\} = \bar{\delta}_y, \\ Prob\{\delta_z(k) = 0\} &= 1 - \bar{\delta}_z, \quad Prob\{\delta_z(k) = 1\} = \Gamma\{\delta_z(k)\} = \bar{\delta}_z. \end{aligned} \tag{3.58}$$

In the light of (3.54), (3.56), and (3.57), we have

$$\begin{aligned} x_f(k+1) &= A_f(\upsilon(k))x_f(k) + B_f(\upsilon(k))\delta_y(k)y(k), \\ z_f(k) &= C_f(\upsilon(k))x_f(k) + D_f(\upsilon(k))\delta_y(k)y(k), \\ \bar{z}(k) &= \delta_z(k)L(\upsilon(k))x(k) + \delta_z(k)E(\upsilon(k))w(k). \end{aligned} \tag{3.59}$$

Combining (3.55) and (3.59), the following filtering error system is acquired:

$$\begin{aligned} r(k+1) &= (\mathscr{A}_1 + \tilde{\delta}_y(k)\mathscr{A}_2)r(k) + (\mathscr{B}_1 + \tilde{\delta}_y(k)\mathscr{B}_2)w(k), \\ e(k) &= (\mathscr{C}_1 + \tilde{\delta}_z(k)\mathscr{C}_2 + \tilde{\delta}_y(k)\mathscr{C}_3)r(k) + (\mathscr{D}_1 + \tilde{\delta}_z(k)\mathscr{D}_2 + \tilde{\delta}_y(k)\mathscr{D}_3)w(k), \end{aligned} \tag{3.60}$$

where

$$
\begin{aligned}
r(k) &= [\,x^T(k) \quad x_f^T(k)\,]^T, \quad e(k) = \bar{z}(k) - z_f(k), \\
\mathscr{A}_1 &= \begin{bmatrix} A(\upsilon(k)) & 0 \\ \bar{\delta}_y B_f(\upsilon(k))C(\upsilon(k)) & A_f(\upsilon(k)) \end{bmatrix}, \quad \mathscr{A}_2 = \begin{bmatrix} 0 & 0 \\ B_f(\upsilon(k))C(\upsilon(k)) & 0 \end{bmatrix}, \\
\mathscr{B}_1 &= \begin{bmatrix} B(\upsilon(k)) \\ \bar{\delta}_y B_f(\upsilon(k))D(\upsilon(k)) \end{bmatrix}, \quad \mathscr{B}_2 = \begin{bmatrix} 0 \\ B_f(\upsilon(k))D(\upsilon(k)) \end{bmatrix}, \\
\mathscr{C}_1 &= [\,\bar{\delta}_z L(\upsilon(k)) - \bar{\delta}_y D_f(\upsilon(k))C(\upsilon(k)) \quad -C_f(\upsilon(k))\,], \\
\mathscr{C}_2 &= [\,L(\upsilon(k)) \quad 0\,], \quad \mathscr{C}_3 = [\,-D_f(\upsilon(k))C(\upsilon(k)) \quad 0\,], \\
\mathscr{D}_1 &= \bar{\delta}_z E(\upsilon(k)) - \bar{\delta}_y D_f(\upsilon(k))D(\upsilon(k)), \\
\mathscr{D}_2 &= E(\upsilon(k)), \quad \mathscr{D}_3 = -D_f(\upsilon(k))D(\upsilon(k)), \\
\tilde{\delta}_y(k) &= \delta_y(k) - \bar{\delta}_y, \quad \tilde{\delta}_z(k) = \delta_z(k) - \bar{\delta}_z.
\end{aligned}
$$

It is clear that $\Gamma\{\tilde{\delta}_y(k)\} = 0$, $\Gamma\{\tilde{\delta}_z(k)\} = 0$, $\Gamma\{\tilde{\delta}_y^2(k)\} = \bar{\delta}_y(1 - \bar{\delta}_y)$, and $\Gamma\{\tilde{\delta}_z^2(k)\} = \bar{\delta}_z(1 - \bar{\delta}_z)$.

Before starting the next section, let us explicitly state the issue of the induced ℓ_∞ filtering to be investigated:

(1) In the data packet dropout situation, determine a filter (3.56) guaranteeing filtering error system (3.60) to be stochastically stable in the mean square when $w(k) = 0$, that is, if there exists a finite $W > 0$ such that

$$
\Gamma\left\{\sum_{k=0}^{\infty} \|r(k)\|^2 \Big| r(0)\right\} < r^T(0)Wr(0). \tag{3.61}
$$

(2) For any nonzero $w(k) \in \ell_\infty[0, \infty)$, under zero initial condition, the filtering error system meets the following a predefined induced ℓ_∞ disturbance attenuation level $\gamma > 0$:

$$
\|e(k)\|_\infty \le \gamma \|w(k)\|_\infty. \tag{3.62}
$$

3.2.2 Main Results

In this section, sufficient conditions are given in form of linear matrix inequalities to ensure that filtering error system (3.60) is stochastically stable with a given induced ℓ_∞ performance γ [3, 4].

Theorem 3.3 *For a given scalar $\gamma > 0$, if there exist matrix $P > 0$ and scalars $\alpha \in (0, 1)$, $\eta \in (0, 1)$ such that the following two matrix inequalities hold:*

$$
\begin{bmatrix} \Omega_1 & * \\ \mathscr{B}_1^T P \mathscr{A}_1 + \mu_y^2 \mathscr{B}_2^T P \mathscr{A}_2 & \Omega_2 \end{bmatrix} < 0, \tag{3.63}
$$

$$\begin{bmatrix} \Omega_3 & * \\ \gamma^{-2}(\mathcal{D}_1^T\mathcal{C}_1 + \mu_z^2\mathcal{D}_2^T\mathcal{C}_2 + \mu_y^2\mathcal{D}_3^T\mathcal{C}_3) & \Omega_4 \end{bmatrix} < 0, \tag{3.64}$$

where

$$\Omega_1 = -P + \alpha P + \mathcal{A}_1^T P\mathcal{A}_1 + \mu_y^2\mathcal{A}_2^T P\mathcal{A}_2,\quad \Omega_2 = -\eta I + \mathcal{B}_1^T P\mathcal{B}_1 + \mu_y^2\mathcal{B}_2^T P\mathcal{B}_2,$$
$$\Omega_3 = -\alpha P + \gamma^{-2}(\mathcal{C}_1^T\mathcal{C}_1 + \mu_z^2\mathcal{C}_2^T\mathcal{C}_2 + \mu_y^2\mathcal{C}_3^T\mathcal{C}_3),\quad \mu_y = \sqrt{\bar{\delta}_y(1-\bar{\delta}_y)},$$
$$\Omega_4 = -(1-\eta)I + \gamma^{-2}(\mathcal{D}_1^T\mathcal{D}_1 + \mu_z^2\mathcal{D}_2^T\mathcal{D}_2 + \mu_y^2\mathcal{D}_3^T\mathcal{D}_3),\quad \mu_z = \sqrt{\bar{\delta}_z(1-\bar{\delta}_z)},$$

the filtering error system in (3.60) is stochastically stable with the induced ℓ_∞ performance index γ.

Proof Choose a Lyapunov function as

$$V(r(k)) = r^T(k)Pr(k).$$

Then, along the trajectories of filtering error system in (3.60) with $w(k) = 0$, one has

$$\begin{aligned} \Gamma\{\Delta V(r(k))\} &= \Gamma\{V(r(k+1))|r(k)\} - V(r(k)) \\ &= \Gamma\{r^T(k)(\mathcal{A}_1 + \tilde{\delta}_y(k)\mathcal{A}_2)^T P(\mathcal{A}_1 + \tilde{\delta}_y(k)\mathcal{A}_2)r(k)|r(k)\} - r^T(k)Pr(k) \\ &= r^T(k)(\mathcal{A}_1^T P\mathcal{A}_1 + \mu_y^2\mathcal{A}_2^T P\mathcal{A}_2 - P)r(k). \end{aligned} \tag{3.65}$$

By pre- and post-multiplying $[\,r^T(k) \quad 0\,]$ and its transpose to (3.63) with $r(k) \neq 0$, one obtains

$$r^T(k)(\mathcal{A}_1^T P\mathcal{A}_1 + \mu_y^2\mathcal{A}_2^T P\mathcal{A}_2 - P)r(k) + \alpha r^T(k)Pr(k) < 0.$$

Since $0 < \alpha < 1$, the above inequality can be expressed as

$$r^T(k)(\mathcal{A}_1^T P\mathcal{A}_1 + \mu_y^2\mathcal{A}_2^T P\mathcal{A}_2 - P)r(k) < 0,$$

which implies $\Gamma\{\Delta V(r(k))\} < 0$.

Define $\bar{P} = \mathcal{A}_1^T P\mathcal{A}_1 + \mu_y^2\mathcal{A}_2^T P\mathcal{A}_2 - P$. Then one can get

$$\Gamma\{V(r(k+1))|r(k)\} - V(r(k)) \leq -\lambda_{\min}(-\bar{P})\Gamma\{\|r(k)\|^2\}. \tag{3.66}$$

For any $m \geq 1$, by taking mathematical expectation of (3.66) on both sides and summing up the inequality on both sides from 0 to m, it is obtained that

$$\Gamma\{V(r(m+1))\} - V(r(0)) \leq -\lambda_{\min}(-\bar{P})\sum_{k=0}^{m}\left(\Gamma\{\|r(k)\|^2\}\right),$$

which means that

$$\sum_{k=0}^{m} \left(\Gamma\{\|r(k)\|^2\} \right) \leq (\lambda_{\min}(-\bar{P}))^{-1}(V(r(0)) - \Gamma\{V(r(m+1))\}). \tag{3.67}$$

For all $k \geq 0$, in the light of (3.67) and $\Gamma\{V(r(k))\} \geq 0$, one can derive that

$$\begin{aligned} \Gamma\left\{\sum_{k=0}^{\infty} \|r(k)\|^2 | r(0)\right\} &\leq (\lambda_{\min}(-\bar{P}))^{-1} r^T(0) \lambda_{\max}(P) r(0) \\ &= r^T(0)(\lambda_{\min}(-\bar{P}))^{-1} \lambda_{\max}(P) r(0) \\ &= r^T(0) W r(0), \end{aligned} \tag{3.68}$$

where $r(0)$ is the initial condition and $W = (\lambda_{\min}(-\bar{P}))^{-1}\lambda_{\max}(P)$. Therefore, the filtering error system is stochastically stable in the mean square.

In what follows, the induced ℓ_∞ performance of the filtering error system in (3.60) with $w(k) \neq 0$ will be established.

Define $\xi_k = [r^T(k) \quad w^T(k)]^T$ and $\Lambda(k) = \Gamma\{V(r(k))\}$. Then by pre- and post-multiplying ξ_k^T and its transpose to (3.63), and based on the fact that $\Lambda(k+1) - \Lambda(k) = \Gamma\{V(r(k+1))\} - \Gamma\{V(r(k))\}$, one can deduce that

$$\Lambda(k+1) - \Lambda(k) + \alpha\Lambda(k) < \eta w^T(k) w(k). \tag{3.69}$$

Set $\tau = 1 - \alpha$, which implies $0 < \tau < 1$. And multiply (3.69) by τ^{-k}, one gets

$$\tau^{-k}\Lambda(k+1) - \tau^{-k+1}\Lambda(k) < \tau^{-k}\eta w^T(k) w(k). \tag{3.70}$$

By taking the sum on both sides of the matrix inequality (3.70), from 0 to $k-1$, it can be derived that

$$\begin{aligned} \sum_{j=0}^{k-1}\left(\tau^{-j}\Lambda(j+1) - \tau^{-j+1}\Lambda(j)\right) &< \sum_{j=0}^{k-1}\left(\tau^{-j}\eta w^T(j) w(j)\right) \\ &< \eta \sup_{k\geq 0}\|w(k)\|^2 \sum_{j=0}^{k-1}\tau^{-j}. \end{aligned} \tag{3.71}$$

After summing up for (3.71), it can be obtained that

$$\tau^{-k+1}\Lambda(k) - \tau\Lambda(0) < \eta \sup_{k\geq 0}\|w(k)\|^2 \frac{1-\tau^{-k}}{1-\tau^{-1}}. \tag{3.72}$$

Under zero initial condition and $0 < \tau < 1$, one can conclude from (3.72) that

$$\Lambda(k) < \frac{\eta}{\alpha}\sup_{k\geq 0}\|w(k)\|^2. \tag{3.73}$$

On the other hand, (3.64) can be rewritten as

$$\begin{bmatrix} -\alpha P & 0 \\ 0 & -(1-\eta)I \end{bmatrix} + \gamma^{-2}\begin{bmatrix} \mathscr{C}_1^T \\ \mathscr{D}_1^T \end{bmatrix}\begin{bmatrix} \mathscr{C}_1 & \mathscr{D}_1 \end{bmatrix} + \gamma^{-2}\begin{bmatrix} \mu_z\mathscr{C}_2^T \\ \mu_z\mathscr{D}_2^T \end{bmatrix}\begin{bmatrix} \mu_z\mathscr{C}_2 & \mu_z\mathscr{D}_2 \end{bmatrix}$$
$$+ \gamma^{-2}\begin{bmatrix} \mu_y\mathscr{C}_3^T \\ \mu_y\mathscr{D}_3^T \end{bmatrix}\begin{bmatrix} \mu_y\mathscr{C}_3 & \mu_y\mathscr{D}_3 \end{bmatrix} < 0. \tag{3.74}$$

Pre- and post-multiplying ξ_k^T and its transpose to (3.74), one has

$$\gamma^{-2}\Gamma\{\|e(k)\|^2\} < \alpha\Lambda(k) + (1-\eta)\Gamma\{\|w(k)\|^2\}, \forall k \in \mathbf{Z}_+. \tag{3.75}$$

Substituting (3.73) into (3.75), it yields

$$\gamma^{-2}\Gamma\{\|e(k)\|^2\} < \eta\sup_{k\geq 0}\|w(k)\|^2 + (1-\eta)\Gamma\{\|w(k)\|^2\},$$

and by taking the supremum over $k \in \mathbf{Z}_+$, one can obtain that $\|e(k)\|_\infty < \gamma\|w(k)\|_\infty$, which means that the filtering error satisfies the induced ℓ_∞ performance. The proof is completed.

In order to remove the coupling items in (3.63) and (3.64), the following theorem is proposed.

Theorem 3.4 *For given constants $\gamma > 0$, $\bar{\delta}_y$, and $\bar{\delta}_z$, if there exist scalars $\eta \in (0, 1)$, $\beta > 0$, and $\sigma > 0$, matrices M_1, M_2, M_3, M_4, G_1, G_2, G_3, $\bar{A}_{fd}$, $\bar{B}_{fd}$, $\bar{C}_{fd}$, and $\bar{D}_{fd}$ satisfying the following LMIs:*

$$\sigma - \beta < 0, \tag{3.76}$$

$$\bar{\Upsilon}_{id} < 0, \quad i, d = 1, \ldots, r, \tag{3.77}$$

$$\bar{\Psi}_{id} < 0, \quad i, d = 1, \ldots, r, \tag{3.78}$$

where

$$\bar{\Upsilon}_{id} = \begin{bmatrix} \Upsilon_{11id} & * \\ \Upsilon_{12}^T & \Upsilon_3 \end{bmatrix}, \quad \Upsilon_{11id} = \begin{bmatrix} \mathscr{M} & * \\ \mathscr{E}_{id}^T & \mathscr{G} \end{bmatrix}, \quad \Upsilon_3 = diag\{-\beta I, -\sigma I, -\sigma I\},$$

$$\mathscr{M} = diag\{-M - M^T + \sigma I, -\eta I\}, \quad \mathscr{G} = diag\{-G - G^T, -G - G^T\},$$

$$G = \begin{bmatrix} G_1 & G_2 \\ G_3 & G_2 \end{bmatrix}, \quad \mathscr{E}_{id} = \begin{bmatrix} \mathscr{A}_{1id}^T & \mu_y\mathscr{A}_{2id}^T \\ \mathscr{B}_{1id}^T & \mu_y\mathscr{B}_{2id}^T \end{bmatrix}, \quad \mathscr{A}_{1id} = \begin{bmatrix} G_1A_i + \bar{\delta}_y\bar{B}_{fd}C_i & \bar{A}_{fd} \\ G_3A_i + \bar{\delta}_y\bar{B}_{fd}C_i & \bar{A}_{fd} \end{bmatrix},$$

$$\mathscr{A}_{2id} = \begin{bmatrix} \bar{B}_{fd}C_i & 0 \\ \bar{B}_{fd}C_i & 0 \end{bmatrix}, \quad \mathscr{B}_{1id} = \begin{bmatrix} G_1B_i + \bar{\delta}_y\bar{B}_{fd}D_i \\ G_3B_i + \bar{\delta}_y\bar{B}_{fd}D_i \end{bmatrix}, \quad \mathscr{B}_{2id} = \begin{bmatrix} \bar{B}_{fd}D_i \\ \bar{B}_{fd}D_i \end{bmatrix},$$

$$\Upsilon_{12} = \begin{bmatrix} \begin{bmatrix} M & 0 \end{bmatrix} & * \\ 0 & diag\{M, M\} \end{bmatrix}^T, \quad M = \begin{bmatrix} M_1 & M_2 \\ M_3 & M_4 \end{bmatrix}, \quad \bar{\Psi}_{id} = \begin{bmatrix} \Psi_{1id} & * \\ \Psi_{2id}^T & \Psi_3 \end{bmatrix},$$

$$\Psi_{1id} = \begin{bmatrix} \tilde{\Psi}_{11} & * \\ \tilde{\Psi}_{12id} & -\gamma^2 I \end{bmatrix}, \quad \tilde{\Psi}_{11} = diag\{-M - M^T + \beta I, (\eta - 1)I\},$$

$$\tilde{\Psi}_{12id} = [\,\tilde{\Psi}^1_{12id} \quad \bar{\delta}_z E_i - \bar{\delta}_y \bar{D}_{fd} D_i\,],\quad \tilde{\Psi}^1_{12id} = [\,\bar{\delta}_z L_i - \bar{\delta}_y \bar{D}_{fd} C_i \quad -\bar{C}_{fd}\,],$$

$$\Psi_{2id} = \begin{bmatrix} [\mu_z L_i \;\; 0]^T & [-\mu_y \bar{D}_{fd} C_i \;\; 0]^T \\ \mu_z E_i^T & -\mu_y D_i^T \bar{D}_{fd}^T \\ 0 & 0 \end{bmatrix},\quad \Psi_3 = diag\{-\gamma^2 I,\ -\gamma^2 I\},$$

the filtering error system in (3.60) is stochastically stable with the induced ℓ_∞ performance index γ, and the induced ℓ_∞ filter gain matrices in (3.56) can be obtained by the following equations:

$$A_{fd} = G_2^{-1}\bar{A}_{fd},\quad B_{fd} = G_2^{-1}\bar{B}_{fd},\quad C_{fd} = \bar{C}_{fd},\quad D_{fd} = \bar{D}_{fd}. \tag{3.79}$$

Proof By Schur complement (Lemma 1.1), the matrix inequality (3.63) can be rewritten as

$$\begin{bmatrix} \mathcal{F} & * \\ \mathcal{E}^T & \mathcal{G}_1 \end{bmatrix} < 0, \tag{3.80}$$

where

$$\mathcal{F} = diag\{-P + \alpha P,\ -\eta I\},\quad \mathcal{E} = \begin{bmatrix} \mathcal{A}_1^T & \mu_y \mathcal{A}_2^T \\ \mathcal{B}_1^T & \mu_y \mathcal{B}_2^T \end{bmatrix},\quad \mathcal{G}_1 = diag\{-P^{-1},\ -P^{-1}\}.$$

Define $G_I = diag\{I,\ I,\ G,\ G\}$ with G being a nonsingular matrix. Then, by performing congruence transformations with G_I to (3.80), one has

$$\begin{bmatrix} \mathcal{F} & * \\ \mathcal{E}_G^T & \mathcal{G}_2 \end{bmatrix} < 0, \tag{3.81}$$

where

$$\mathcal{E}_G = \begin{bmatrix} \mathcal{A}_1^T G^T & \mu_y \mathcal{A}_2^T G^T \\ \mathcal{B}_1^T G^T & \mu_y \mathcal{B}_2^T G^T \end{bmatrix},\quad \mathcal{G}_2 = -diag\{GP^{-1}G^T,\ GP^{-1}G^T\}.$$

Based on the fact that $-(G-P)P^{-1}(G-P)^T \le 0$ with $P > 0$, which implies $-GP^{-1}G^T \le -G - G^T + P$, (3.81) can be verified by

$$\begin{bmatrix} \mathcal{F} & * \\ \mathcal{E}_G^T & \mathcal{G}_3 \end{bmatrix} < 0, \tag{3.82}$$

where

$$\mathcal{G}_3 = diag\{-G - G^T + P,\ -G - G^T + P\}.$$

Next, to eliminate the coupling items in (3.82), such as, αP, $\mathcal{A}_1^T G^T$, $\mu_y \mathcal{A}_2^T G^T$, $\mathcal{B}_1^T G^T$, and $\mu_y \mathcal{B}_2^T G^T$, assuming that the Lyapunov matrix has the following form

$$P = \frac{1}{\sigma} M^T M, \tag{3.83}$$

where M is a nonsingular matrix and $\sigma > 0$.

Substituting (3.83) into (3.82) and according to $-\frac{1}{\sigma} M^T M \leq -M - M^T + \sigma I$, (3.82) can be rewritten as

$$\begin{bmatrix} \breve{\mathcal{M}} & * \\ \mathcal{E}_G^T & \mathcal{G} \end{bmatrix} + \begin{bmatrix} \mathcal{M}_1 & * \\ 0 & \mathcal{M}_4 \end{bmatrix} < 0, \tag{3.84}$$

where

$$\breve{\mathcal{M}} = diag\left\{-M - M^T + \sigma I, \ -\eta I\right\}, \quad \mathcal{M}_1 = diag\left\{\frac{\alpha}{\sigma} M^T M, \ 0\right\},$$
$$\mathcal{M}_4 = diag\left\{\frac{1}{\sigma} M^T M, \ \frac{1}{\sigma} M^T M\right\}.$$

Then based on Schur complement (Lemma 1.1), (3.84) can be rewritten as

$$\begin{bmatrix} \breve{\Upsilon}_1 & * \\ \breve{\Upsilon}_2^T & \breve{\Upsilon}_3 \end{bmatrix} < 0, \tag{3.85}$$

where

$$\breve{\Upsilon}_1 = \begin{bmatrix} \breve{\mathcal{M}} & * \\ \mathcal{E}_G^T & \mathcal{G} \end{bmatrix}, \quad \breve{\Upsilon}_2 = \begin{bmatrix} [M \ 0] & * \\ 0 & \breve{\Upsilon}_{22} \end{bmatrix}^T, \quad \breve{\Upsilon}_{22} = diag\{M, \ M\},$$
$$\breve{\Upsilon}_3 = diag\left\{-\frac{\sigma}{\alpha} I, \ -\sigma I, \ -\sigma I\right\}.$$

In order to further eliminate the coupling items in (3.85), constructing the following matrices:

$$G = \begin{bmatrix} G_1 & G_2 \\ G_3 & G_2 \end{bmatrix}, \quad \bar{A}_f(\upsilon(k)) = G_2 A_f(\upsilon(k)),$$
$$M = \begin{bmatrix} M_1 & M_2 \\ M_3 & M_4 \end{bmatrix}, \quad \bar{B}_f(\upsilon(k)) = G_2 B_f(\upsilon(k)),$$

where G_2 is nonsingular without loss of generality.

Then (3.85) can be ensured by

$$\begin{bmatrix} \tilde{\Upsilon}_1 & * \\ \breve{\Upsilon}_2^T & \breve{\Upsilon}_3 \end{bmatrix} < 0, \tag{3.86}$$

where

$$\tilde{\Upsilon}_1 = \begin{bmatrix} \breve{\mathcal{M}} & * \\ \tilde{\Upsilon}_G^T & \mathcal{G} \end{bmatrix}, \quad \tilde{\Upsilon}_G = \begin{bmatrix} \mathcal{A}_{1G}^T & \mu_y \mathcal{A}_{2G}^T \\ \mathcal{B}_{1G}^T & \mu_y \mathcal{B}_{2G}^T \end{bmatrix}, \quad \mathcal{A}_{2G} = \begin{bmatrix} \bar{B}_f(\upsilon(k))C(\upsilon(k)) & 0 \\ \bar{B}_f(\upsilon(k))C(\upsilon(k)) & 0 \end{bmatrix},$$

$$\mathcal{A}_{1G} = \begin{bmatrix} G_1 A(\upsilon(k)) + \bar{\delta}_y \bar{B}_f(\upsilon(k))C(\upsilon(k)) & \bar{A}_f(\upsilon(k)) \\ G_3 A(\upsilon(k)) + \bar{\delta}_y \bar{B}_f(\upsilon(k))C(\upsilon(k)) & \bar{A}_f(\upsilon(k)) \end{bmatrix},$$

$$\mathcal{B}_{1G} = \begin{bmatrix} G_1 B(\upsilon(k)) + \bar{\delta}_y \bar{B}_f(\upsilon(k))D(\upsilon(k)) \\ G_3 B(\upsilon(k)) + \bar{\delta}_y \bar{B}_f(\upsilon(k))D(\upsilon(k)) \end{bmatrix}, \quad \mathcal{B}_{2G} = \begin{bmatrix} \bar{B}_f(\upsilon(k))D(\upsilon(k)) \\ \bar{B}_f(\upsilon(k))D(\upsilon(k)) \end{bmatrix},$$

which means that (3.86) can guarantee (3.63).

Along the same line as the proof in the condition (3.63), by setting $\bar{C}_f(\upsilon(k)) = C_f(\upsilon(k))$ and $\bar{D}_f(\upsilon(k)) = D_f(\upsilon(k))$, it can be concluded that the matrix inequality (3.64) holds if the following condition is satisfied:

$$\begin{bmatrix} \breve{\Psi}_1 & * \\ \breve{\Psi}_2^T & \Psi_3 \end{bmatrix} < 0, \tag{3.87}$$

with

$$\breve{\Psi}_1 = \begin{bmatrix} \breve{\Psi}_{11} & * \\ \breve{\Psi}_{12} & -\gamma^2 I \end{bmatrix}, \quad \breve{\Psi}_{11} = diag\left\{-M - M^T + \frac{\sigma}{\alpha} I,\ (\eta - 1)I\right\},$$

$$\breve{\Psi}_{12} = [\breve{\Psi}_{12}^1 \quad \bar{\delta}_z E(\upsilon(k)) - \bar{\delta}_y \bar{D}_f(\upsilon(k))D(\upsilon(k))\ ],$$

$$\breve{\Psi}_{12}^1 = [\ \bar{\delta}_z L(\upsilon(k)) - \bar{\delta}_y \bar{D}_f(\upsilon(k))C(\upsilon(k)) \quad - \bar{C}_f(\upsilon(k))\ ],$$

$$\breve{\Psi}_2 = \begin{bmatrix} [\mu_z L(\upsilon(k)) \quad 0] & \mu_z E(\upsilon(k)) & 0 \\ [-\mu_y \bar{D}_f(\upsilon(k))C(\upsilon(k)) \quad 0] & -\mu_y \bar{D}_f(\upsilon(k))D(\upsilon(k)) & 0 \end{bmatrix}^T.$$

Define $\beta = \frac{\sigma}{\alpha}$. Then, in the light of condition (3.76) with $\sigma > 0$ and $\beta > 0$, it follows from (3.86) and (3.87) that

$$\sum_{i=1}^{r}\sum_{d=1}^{r} \sigma_i(k)\sigma_d(k)\tilde{\Upsilon}_{id} < 0, \tag{3.88}$$

and

$$\sum_{i=1}^{r}\sum_{d=1}^{r} \sigma_i(k)\sigma_d(k)\bar{\Psi}_{id} < 0. \tag{3.89}$$

Therefore, the inequalities (3.88) and (3.89) can be guaranteed by the conditions (3.77) and (3.78), respectively. The proof is completed.

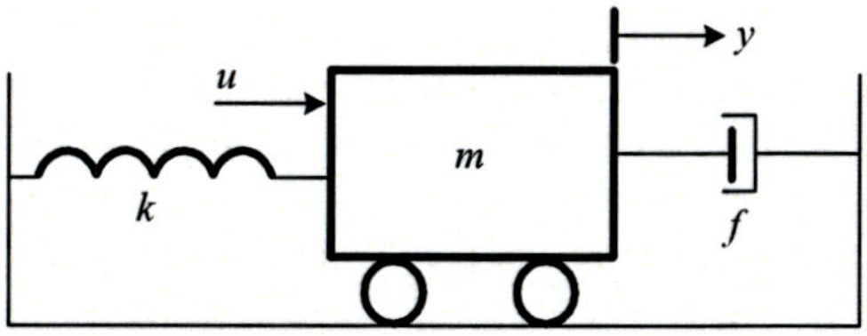

Fig. 3.5 Mass-spring-damper system

Table 3.1 The related parameters of the mass-spring-damper system

m	Mass of car	$m = 1.5$ kg
k	Stiffness coefficient of spring	$k = 341.3$ N/m
f	Viscous friction coefficient of damper	$f = 18.67$ N · s/m

3.2.3 Simulation Example

In this section, an example of the mass-spring-damper mechanical system [5] is given to demonstrate the effectiveness of the designed method. As shown in Fig. 3.5, we consider a car m vibrating on a horizontal surface within a bounded region and attached to a vertical surface through a spring and a damper. The mass is subject to an external force $u = F_u(t)$. Assume that friction force does not exist due to static and Coulomb friction. Then Newton's law of motion given as

$$m\ddot{y}(t) = F_u(t) - F_s(t) - F_d(t) + w(t), \tag{3.90}$$

where $y(t)$ is the displacement from a reference position, $F_u(t)$ is the restoring force of the spring, $F_d(t)$ is a resistive force of the damper, and $w(t)$ is the external disturbance. Supposing that $F_s(t) = ky(t) + y^3(t)$ with k being a real constant known as the stiffness coefficient, which signifies that the spring is nonlinear one. $F_d(t) = f\dot{y}(t)$ is the dynamic property of the viscous damper with f being the viscous friction coefficient. Table 3.1 gives some related parameters of the mechanical system. Notice that when the external force $F_u(t) = 0$, the mass-spring-damper mechanical system in (3.90) is stable. Thus, we can choose $F_u(t) = 0$ for studying the filtering problem. Set the displacement $y(t)$ and its derivative $\dot{y}(t)$ as the state variables of system (3.90), that is

$$x_1(t) = y(t), \quad x_2(t) = \dot{y}(t). \tag{3.91}$$

From (3.91), system (3.90) can be controlled

$$\begin{aligned} \dot{x}_1(t) &= x_2(t), \\ m\dot{x}_2(t) &= -kx_1(t) - x_1^3(t) - fx_2(t) + w(t). \end{aligned} \tag{3.92}$$

Based on parameters in Table 3.1, (3.92) can be rewritten

$$\dot{x}_1(t) = x_2(t),$$
$$\dot{x}_2(t) = -(341.3/1.5)x_1(t) - (1/1.5)x_1^3(t) - (18.67/1.5)x_2(t) + (1/1.5)w(t). \tag{3.93}$$

Suppose that $-1 \le x_1(t) \le 1$, i.e., $0 \le x_1^2(t) \le 1$. Then a continuous-time T–S system can be obtained to approximate the mass-spring-damper system as follows:

$$\dot{x}(t) = \sum_{i=1}^{2} \sigma_i(t)\big(A_{ib}x(t) + B_{ib}w(t)\big), \tag{3.94}$$

where

$$A_{1b} = \begin{bmatrix} 0 & 1 \\ -341.3/1.5 & -18.67/1.5 \end{bmatrix}, \quad B_{1b} = \begin{bmatrix} 0 \\ 1/1.5 \end{bmatrix},$$
$$A_{2b} = \begin{bmatrix} 0 & 1 \\ -342.3/1.5 & -18.67/1.5 \end{bmatrix}, \quad B_{2b} = \begin{bmatrix} 0 \\ 1/1.5 \end{bmatrix}.$$

Define the premise variable $h_1(t) = x_1^2(t)$, which implies that the maximum and minimum values of $h_1(t)$ are 1 and 0, respectively. Then, $h_1(t)$ can be rewritten as

$$h_1(t) = \sigma_1(t) \times 1 + \sigma_2(t) \times 0, \quad \sigma_1(t) + \sigma_2(t) = 1.$$

And, the membership functions can be calculated as

$$\sigma_1(t) = x_1^2(t), \quad \sigma_2(t) = 1 - \sigma_1(t). \tag{3.95}$$

Based on (3.95), we will address the induced ℓ_∞ filtering problem of system (3.93). Set sampling time $T = 0.5$ s. Then discrete-time system (3.55) is obtained and the corresponding system parameters are given as

$$A_1 = \begin{bmatrix} 0.0482 & 0.0018 \\ -0.4084 & 0.0259 \end{bmatrix}, \quad B_1 = \begin{bmatrix} 0.0028 \\ 0.0012 \end{bmatrix},$$
$$A_2 = \begin{bmatrix} -0.0231 & 0.0017 \\ -0.5752 & -0.0440 \end{bmatrix}, \quad B_2 = \begin{bmatrix} 0.0020 \\ 0.0011 \end{bmatrix},$$
$$C_1 = [1 \ \ 0], \quad D_1 = 0.5, \quad E_1 = 0.1,$$
$$C_2 = [1 \ \ 0], \quad D_2 = 0.5, \quad E_2 = 0.1,$$
$$L_1 = [1 \ \ 0.6], \quad L_2 = [1 \ \ 0.6].$$

Choose $\Gamma\{\delta_y(k)\} = \bar{\delta}_y = 0.9$ and $\Gamma\{\delta_z(k)\} = \bar{\delta}_z = 0.85$. Then, by solving LMIs (3.76), (3.77), and (3.78) in Theorem 3.4, we can obtain $\gamma_{\min} = 0.0466$ and the following related matrix variables:

$$\begin{aligned}
\bar{A}_{f1} &= \begin{bmatrix} 79.8213 & 410.8237 \\ 60.0165 & 38.6745 \end{bmatrix}, \quad \bar{B}_{f1} = \begin{bmatrix} -24.4117 \\ -8.3932 \end{bmatrix}, \\
\bar{A}_{f2} &= \begin{bmatrix} 79.8213 & 410.8237 \\ 60.0165 & 38.6745 \end{bmatrix}, \quad \bar{B}_{f2} = \begin{bmatrix} -24.4117 \\ -8.3932 \end{bmatrix}, \\
\bar{C}_{f1} &= [-0.8511 \quad -1.0968], \quad \bar{D}_{f1} = 0.1715, \\
\bar{C}_{f2} &= [-0.8511 \quad -1.0968], \quad \bar{D}_{f2} = 0.1715, \\
G_2 &= \begin{bmatrix} 3886.1 & 1202.1 \\ 644.2 & 1904.5 \end{bmatrix}.
\end{aligned}$$

Substituting the above results into (3.79), the corresponding gains can be obtained

$$\begin{aligned}
A_{f1} &= \begin{bmatrix} 0.0121 & 0.1111 \\ 0.0274 & -0.0173 \end{bmatrix}, \quad B_{f1} = \begin{bmatrix} -0.0055 \\ -0.0025 \end{bmatrix}, \\
A_{f2} &= \begin{bmatrix} 0.0121 & 0.1111 \\ 0.0274 & -0.0173 \end{bmatrix}, \quad B_{f2} = \begin{bmatrix} -0.0055 \\ -0.0025 \end{bmatrix}, \\
C_{f1} &= \bar{C}_{f1}, \quad D_{f1} = \bar{D}_{f1}, \quad C_{f2} = \bar{C}_{f2}, \quad D_{f2} = \bar{D}_{f2}.
\end{aligned}$$

To show the performance of the designed filter, set the initial conditions $x(0) = x_f(0) = [0 \quad 0]^T$ and the disturbance $w(k) = 12e^{-2k}\cos(0.1k)$. Then, the response of $x(k)$ and its estimate $x_f(k)$ are displayed in Figs. 3.6 and 3.7, respectively. The simulation results of signals $z(k)$ and $z_f(k)$ are shown in Fig. 3.8. Figure 3.9 displays the response of error signal $e(k)$, which can be perceived that the error curve tends to 0. The simulation ratio of $\sqrt{e^T(k)e(k)/\|w(k)\|_\infty^2}$ is shown in Fig. 3.10, from which we can see that the maximum value of the ratio is 0.0142. It is less than $\gamma_{\min} = 0.0466$.

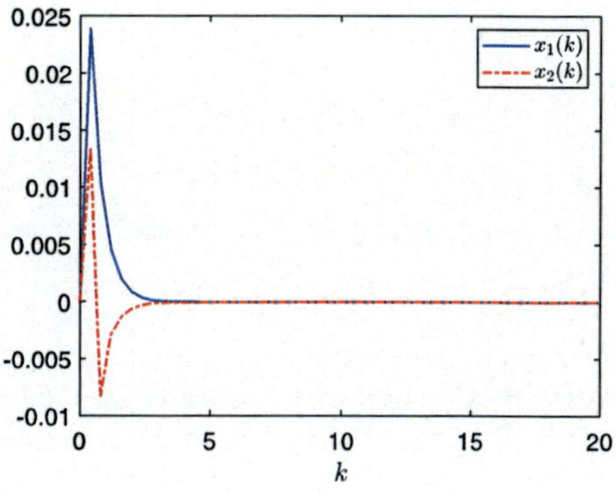

Fig. 3.6 Response of the system state $x(k)$

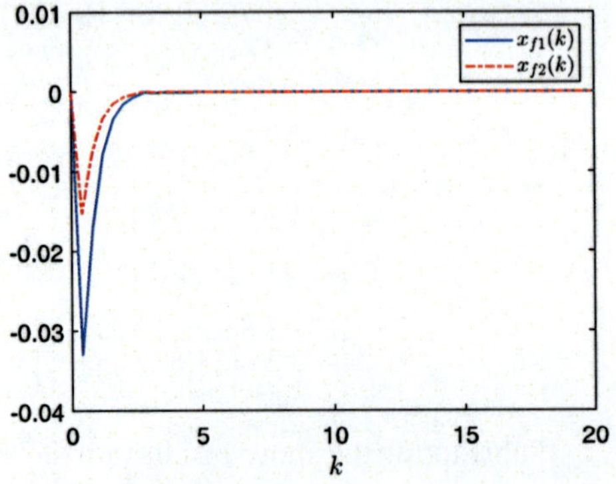

Fig. 3.7 Response of the filter state $x_f(k)$

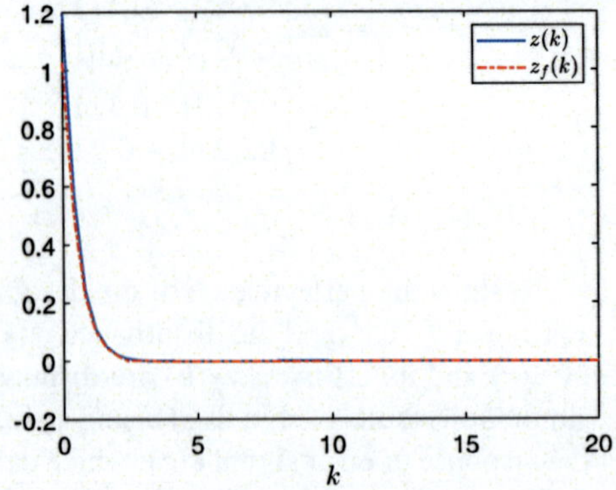

Fig. 3.8 Responses of $z(k)$ and $z_f(k)$

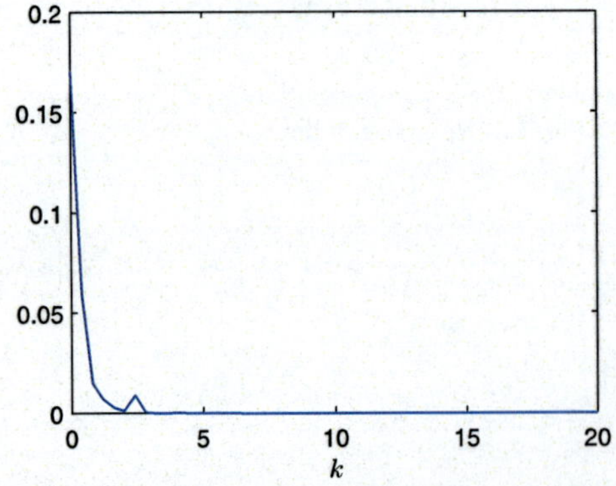

Fig. 3.9 Filtering error $e(k)$

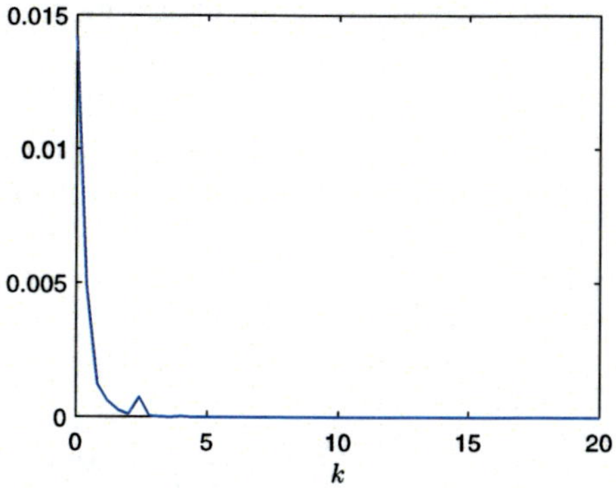

Fig. 3.10 History of $\sqrt{e^T(k)e(k)/\|w(k)\|_\infty^2}$

3.3 $\mathscr{H}_\infty$ Filtering with Dynamic Quantization

In this section, the $\mathscr{H}_\infty$ filtering problem for T–S fuzzy system with multi-signal dynamic quantization is considered, i.e., the measurement output and the performance output are quantized by dynamic quantizers in this section. And the conditions to design $\mathscr{H}_\infty$ filter with dynamic quantization are given in the form of a set of LMIs.

3.3.1 *Problem Formulation*

For the investigated $\mathscr{H}_\infty$ filtering problem in this section, we consider the following T–S fuzzy model:

Plant Rule i_{th}: **IF** $h_1(k)$ is N_{1i}, $h_2(k)$ is N_{2i}, and, ..., and $h_\theta(k)$ is $N_{\theta i}$, **THEN**

$$\begin{aligned} x(k+1) &= A_i x(k) + B_i w(k), \\ y(k) &= C_i x(k), \\ z(k) &= L_i x(k), \end{aligned} \tag{3.96}$$

where $h(k) = [h_1(k), h_2(k), \ldots, h_\theta(k)]$ is the vector of premise variables. N_{ji} is the fuzzy sets with $j = 1, 2, \ldots, \theta$, $i = 1, 2, \ldots, r$. $x(k) \in \mathbb{R}^{n_x}$ stands for the state variable; $y(k) \in \mathbb{R}^{n_y}$ stands for the measurement output of the system; $z(k) \in \mathbb{R}^{n_z}$ stands for the performance output of the system; $w(k) \in \mathbb{R}^{n_w}$ refers to the noise input which belongs to $\ell_2[0, \infty)$; A_i, B_i, C_i and L_i are given system matrices with appropriate dimensions.

As in Sect. 3.1, the T–S fuzzy model can be deduced as follows:

$$\begin{aligned} x(k+1) &= A(\rho)x(k) + B(\rho)w(k), \\ y(k) &= C(\rho)x(k), \\ z(k) &= L(\rho)x(k), \end{aligned} \tag{3.97}$$

where

$$A(\rho) = \sum_{i=1}^{r} \sigma_i(k)A_i, \quad B(\rho) = \sum_{i=1}^{r} \sigma_i(k)B_i,$$
$$C(\rho) = \sum_{i=1}^{r} \sigma_i(k)C_i, \quad L(\rho) = \sum_{i=1}^{r} \sigma_i(k)L_i.$$

Unlike static quantizer that require an infinite number of quantization levels, we construct the dynamic quantizer which will be defined as the one given in [6], for quantized signal $\tau \in \mathbb{R}^l$, the relationship between range and error bound of dynamic quantizer is given as

$$\begin{cases} \|f(\tau) - \tau\| \le \Delta, & if\ \|\tau\| \le M, \\ \|f(\tau)\| > M - \Delta, & if\ \|\tau\| > M, \end{cases} \tag{3.98}$$

where M and Δ are used to represent the quantizer's range and error bound, respectively. In the control strategy to be developed below, we consider the one-parameter family of quantizers as

$$f_\mu(\tau) = \mu f\left(\frac{\tau}{\mu}\right), \quad \mu > 0, \tag{3.99}$$

where μ is dynamic parameter of the quantizer. Note that if μ is zero, this will mean that the output of the quantizer is set to zero.

Due to the limitation of bandwidth and other factors of the communication channel, the output of the system may be subject to some unpredictable situations. In order to reduce the possibility of uncertainties, it is necessary to quantify the measurement and performance outputs. In this section, we have the following signal quantization expression:

$$\begin{aligned} f_\mu\left(y(k)\right) &= \mu_y f\left(\frac{y(k)}{\mu_y}\right), \\ f_\mu\left(z(k)\right) &= \mu_z f\left(\frac{z(k)}{\mu_z}\right). \end{aligned} \tag{3.100}$$

Therefore, a filter with measurement output quantization is designed to estimate the state of the system, which is given as follows:

$$\begin{aligned} x_f(k+1) &= A_f x_f(k) + B_f \tilde{y}(k), \\ z_f(k) &= C_f x_f(k) + D_f \tilde{y}(k), \end{aligned} \tag{3.101}$$

where $x_f(k) \in \mathbb{R}^{n_x}$ and $z_f(k) \in \mathbb{R}^{n_z}$ stand for the state variable and the output of filter, respectively; $\tilde{y}(k) \in \mathbb{R}^{n_y}$ is the quantized measurement output with the form of (3.100); A_f, B_f, C_f and D_f are the matrices of the filter system with the appropriate dimensions.

Further, the following equation can be obtained:

$$\begin{aligned} x_f(k+1) &= A_f x_f(k) + B_f \mu_y f\left(\frac{y(k)}{\mu_y}\right), \\ z_f(k) &= C_f x_f(k) + D_f \mu_y f\left(\frac{y(k)}{\mu_y}\right). \end{aligned} \tag{3.102}$$

Combining (3.99), (3.100) and (3.102), the following equation can be obtained:

$$\begin{aligned} x_f(k+1) &= A_f x_f(k) + B_f y(k) + B_f \mu_y \left(f\left(\frac{y(k)}{\mu_y}\right) - \frac{y(k)}{\mu_y}\right), \\ z_f(k) &= C_f x_f(k) + D_f y(k) + D_f \mu_y \left(f\left(\frac{y(k)}{\mu_y}\right) - \frac{y(k)}{\mu_y}\right). \end{aligned} \tag{3.103}$$

By defining $e(k) = \tilde{z}(k) - z_f(k)$ where $\tilde{z}(k)$ is the quantized performance output with the form of (3.100), the filtering error system can be obtained as follows:

$$\begin{aligned} \Theta(k+1) &= \mathcal{A}_1(\rho)\Theta(k) + \mathcal{B}(\rho)w(k) + \mathcal{E}_1\mathcal{V}_y(k), \\ e(k) &= C_{L1}(\rho)\Theta(k) + D_{L1}\mathcal{V}_y(k) + \mathcal{V}_z(k), \\ y(k) &= C_y(\rho)\Theta(k), \\ z(k) &= L_z(\rho)\Theta(k), \end{aligned} \tag{3.104}$$

where

$$\begin{aligned} \Theta(k) &= \left[x^T(k)\; x_f^T(k)\right]^T, \\ \mathcal{A}_1(\rho) &= \begin{bmatrix} A(\rho) & 0 \\ B_f C(\rho) & A_f \end{bmatrix}, \quad \mathcal{B}(\rho) = \begin{bmatrix} B(\rho) \\ 0 \end{bmatrix}, \quad \mathcal{E}_1 = \begin{bmatrix} 0 \\ B_f \end{bmatrix}, \\ C_{L1}(\rho) &= \left[L(\rho) - D_f C(\rho) \;\; -C_f\right], \quad D_{L1} = -D_f, \\ \mathcal{V}_y(k) &= \mu_y \left(f\left(\frac{y(k)}{\mu_y}\right) - \frac{y(k)}{\mu_y}\right), \quad C_y(\rho) = \left[C(\rho) \;\; 0\right], \\ \mathcal{V}_z(k) &= \mu_z \left(f\left(\frac{z(k)}{\mu_z}\right) - \frac{z(k)}{\mu_z}\right), \quad L_z(\rho) = \left[L(\rho) \;\; 0\right]. \end{aligned}$$

The purpose of this section is to design filter given in (3.102) with multi-signal quantization, such that the filtering error system (3.104) satisfies the following conditions:

(1) The filtering error system in (3.104) is said to be asymptotically stable when $w(k) = 0$;
(2) Under zero initial condition, for any $w(k) \neq 0$ and a given $\gamma > 0$, the filtering error $e(k)$ satisfies

$$\sum_{k=0}^{\infty} e^T(k)e(k) < \gamma^2 \sum_{k=0}^{\infty} w^T(k)w(k).$$

3.3.2 Main Results

Theorem 3.5 *For given quantizer's ranges M_δ and quantization errors Δ_δ, $\delta = y,\ z$, the filtering error system (3.104) is asymptotically stable with the prescribed $\mathcal{H}_\infty$ performance index $\gamma > 0$, if there exists matrix $P > 0$, positive scalars ρ_δ and $\hbar_\delta$, $\delta = y,\ z$ such that the following matrix inequalities hold:*

$$M_y > 1/\rho_y, \quad M_z > 1/\rho_z, \tag{3.105}$$

$$\begin{bmatrix} -P & * & * & * & * & * & * & * \\ 0 & -\gamma^2 I & * & * & * & * & * & * \\ 0 & 0 & -\hbar_y I & * & * & * & * & * \\ 0 & 0 & 0 & -\hbar_z I & * & * & * & * \\ \mathcal{A}_1(\rho) & \mathcal{B}(\rho) & \mathcal{E}_1 & 0 & -P^{-1} & * & * & * \\ C_{L1}(\rho) & 0 & D_{L1} & I & 0 & -I & * & * \\ 2\hbar_y\rho_y\Delta_y C_y(\rho) & 0 & 0 & 0 & 0 & 0 & -\hbar_y I & * \\ 2\hbar_z\rho_z\Delta_z L_z(\rho) & 0 & 0 & 0 & 0 & 0 & 0 & -\hbar_z I \end{bmatrix} < 0, \tag{3.106}$$

where the online adjusting strategy for quantizer's parameters μ_δ as

$$\rho_\delta \|\delta(k)\| \leq \mu_\delta \leq 2\rho_\delta \|\delta(k)\|, \quad \delta = y,\ z. \tag{3.107}$$

Proof Firstly, we will confirm the truth that the filtering error system in (3.104) is asymptotically stable with $w(k) = 0$ if the conditions in (3.105) and (3.106) can be guaranteed. Combining (3.105) and (3.107), the following inequality can be obtained:

$$\frac{1}{2\rho_\delta}\mu_\delta \leq \|\delta(k)\| \leq M_\delta \mu_\delta. \tag{3.108}$$

It is clear that the definition of (3.98) is satisfied, so

$$\left\| \frac{\delta(k)}{\mu_\delta} \right\| \leq M_\delta, \quad \delta = y,\ z, \tag{3.109}$$

then

$$\left\| f\left(\frac{\delta(k)}{\mu_\delta}\right) - \frac{\delta(k)}{\mu_\delta} \right\| \leq \Delta_\delta. \tag{3.110}$$

Based on the homogeneity property of Euclidean norm and the definition of $\mathscr{V}_\delta(k)$ in (3.104), one gives

$$\begin{aligned} \|\mathscr{V}_\delta(k)\| &= \left\| \mu_\delta \left(f\left(\frac{\delta(k)}{\mu_\delta}\right) - \frac{\delta(k)}{\mu_\delta} \right) \right\| \\ &= \mu_\delta \left\| f\left(\frac{\delta(k)}{\mu_\delta}\right) - \frac{\delta(k)}{\mu_\delta} \right\| \leq \mu_\delta \Delta_\delta. \end{aligned} \tag{3.111}$$

Obviously, combining (3.107), (3.111) can be rewritten as

$$\|\mathscr{V}_\delta(k)\| \leq 2\rho_\delta \Delta_\delta \|\delta(k)\|, \tag{3.112}$$

i.e.,

$$\begin{aligned} \mathscr{V}_y^T(k)\mathscr{V}_y(k) &\leq 4\rho_y^2\Delta_y^2(C_y(\rho)\Theta(k))^T C_y(\rho)\Theta(k), \\ \mathscr{V}_z^T(k)\mathscr{V}_z(k) &\leq 4\rho_z^2\Delta_z^2(L_z(\rho)\Theta(k))^T L_z(\rho)\Theta(k). \end{aligned} \tag{3.113}$$

Then let's define $\Omega^T(k) = \left[\, \Theta^T(k)\ w^T(k)\, \mathscr{V}_y^T(k)\, \mathscr{V}_z^T(k) \,\right]$, according to (3.104), it yields

$$\begin{aligned} \Omega^T(k)\Upsilon_1\Omega(k) &\geq 0, \\ \Omega^T(k)\Upsilon_2\Omega(k) &\geq 0, \end{aligned} \tag{3.114}$$

where

$$\begin{aligned} \Upsilon_1 &= Ha\left\{\left[2\rho_y\Delta_y C_y(\rho)\ 0\ 0\ 0\right]\right\} - \mathrm{diag}\{0,\ 0,\ I,\ 0\}, \\ \Upsilon_2 &= Ha\left\{\left[2\rho_z\Delta_z L_z(\rho)\ 0\ 0\ 0\right]\right\} - \mathrm{diag}\{0,\ 0,\ 0,\ I\}. \end{aligned}$$

Then, by using Schur complement (Lemma 1.1), the condition (3.106) can be rewritten as

$$\Upsilon_0 + \Upsilon_3 + \hbar_y\Upsilon_1 + \hbar_z\Upsilon_2 < 0, \tag{3.115}$$

where

$$\begin{aligned} \Upsilon_0 &= \left[\mathscr{A}_1(\rho)\ \mathscr{B}(\rho)\ \mathscr{E}_1\ 0\right]^T P \left[\mathscr{A}_1(\rho)\ \mathscr{B}(\rho)\ \mathscr{E}_1\ 0\right] - \mathrm{diag}\{P, 0, 0, 0\}, \\ \Upsilon_3 &= Ha\{\left[C_{L1}(\rho)\ 0\ D_{L1}\ I\right]\} - \mathrm{diag}\{0, \gamma^2 I, 0, 0\}. \end{aligned}$$

Thus, from Lemma 1.3 and (3.114), we have

$$\Omega^T(k)(\Upsilon_0 + \Upsilon_3)\Omega(k) < 0. \tag{3.116}$$

Then, the Lyapunov function is adopted as

$$V(\Theta(k)) = \Theta^T(k)P\Theta(k),\ P > 0, \tag{3.117}$$

and we can obtain

$$\begin{aligned} &V(\Theta(k+1)) - V(\Theta(k)) \\ &= \left(\mathscr{A}_1(\rho)\Theta(k) + \mathscr{B}(\rho)w(k) + \mathscr{E}_1\mathscr{V}_y(k)\right)^T P \\ &\quad \times \left(\mathscr{A}_1(\rho)\Theta(k) + \mathscr{B}(\rho)w(k) + \mathscr{E}_1\mathscr{V}_y(k)\right) - \Theta^T(k)P\Theta(k) \\ &= \Omega^T(k)\Upsilon_0\Omega(k). \end{aligned} \tag{3.118}$$

Combining (3.116) and (3.118) can obtain that

$$V(\Theta(k+1)) - V(\Theta(k)) + e^T(k)e(k) - \gamma^2 w^T(k)w(k) < 0. \tag{3.119}$$

Under the condition of external disturbance $w(k) = 0$, we have

$$\Delta V = V(\Theta(k+1)) - V(\Theta(k)) < V(\Theta(k+1)) - V(\Theta(k)) + e^T(k)e(k) < 0, \tag{3.120}$$

which means the filtering error system (3.104) is asymptotically stable.

Next, in order to analyze the $\mathscr{H}_\infty$ performance, taking the summation of the both sides of (3.119) from 0 to ∞, one can be obtained that

$$V(\Theta(\infty)) - V(\Theta(0)) + \sum_{k=0}^{\infty} e^T(k)e(k) - \sum_{k=0}^{\infty} \gamma^2 w^T(k)w(k) < 0. \tag{3.121}$$

With zero condition $V(\Theta(0)) = 0$ and $V(\Theta(\infty)) \geq 0$, we obtain

$$\sum_{k=0}^{\infty} e^T(k)e(k) < \sum_{k=0}^{\infty} \gamma^2 w^T(k)w(k), \tag{3.122}$$

which implies that the filtering error system has a prescribed $\mathscr{H}_\infty$ performance γ. The proof is completed.

In this section, the $\mathscr{H}_\infty$ filter design conditions, where the gain matrices of filter (3.102) can be determined, will be given. In fact, the existence of the inverse matrix which makes the conditions in Theorem 3.5 cannot be directly solved by the Matlab LMI toolbox. Therefore, we will eliminate the inverse matrix and other couple terms in this section, then based on the stability analysis criterion in Theorem 3.5, the corresponding design conditions are provided by Theorem 3.6 in form of LMIs, which guarantee that the filtering error system (3.104) is asymptotically stable with the prescribed $\mathscr{H}_\infty$ performance index γ.

Theorem 3.6 *For given quantizer's ranges M_δ and quantization errors Δ_δ, $\delta = y,\ z$, the filtering error system (3.104) is asymptotically stable with the prescribed*

$\mathscr{H}_\infty$ performance index $\gamma > 0$, if there exist matrices $P_1 > 0$, P_2, $P_3 > 0$, G_1, G_2, G_3 and positive scalars η, σ_δ, $\hbar_\delta$, $\delta = y,\ z$ such that the following linear matrix inequalities hold:

$$\sigma_y M_y > \hbar_y, \quad \sigma_z M_z > \hbar_z, \tag{3.123}$$

$$\begin{bmatrix} -P_1 & * & * & * & * & * & * & * & * & * \\ -P_2 & -P_3 & * & * & * & * & * & * & * & * \\ 0 & 0 & -\gamma^2 I & * & * & * & * & * & * & * \\ 0 & 0 & 0 & -\hbar_y I & * & * & * & * & * & * \\ 0 & 0 & 0 & 0 & -\hbar_z I & * & * & * & * & * \\ G_1 A_i + \bar{B}_f C_i & \bar{A}_f & G_1 B_i & \bar{B}_f & 0 & -G_1 - G_1^T + P_1 & * & * & * & * \\ G_2 A_i + \bar{B}_f C_i & \bar{A}_f & G_2 B_i & \bar{B}_f & 0 & -G_2 - G_3^T + P_2 & -G_3 - G_3^T + P_3 & * & * & * \\ L_i - \bar{D}_f C_i & -\bar{C}_f & 0 & -\bar{D}_f & I & 0 & 0 & -I & * & * \\ 2\sigma_y \Delta_y C_i & 0 & 0 & 0 & 0 & 0 & 0 & 0 & -\hbar_y I & * \\ 2\sigma_z \Delta_z L_i & 0 & 0 & 0 & 0 & 0 & 0 & 0 & 0 & -\hbar_z I \end{bmatrix} < 0,\ i = 1, \ldots, r. \tag{3.124}$$

Moreover, the online adjusting strategy for quantizer's parameters μ_δ are given in (3.107) and the filtering matrices are given as

$$A_f = G_3^{-1} \bar{A}_f, \quad B_f = G_3^{-1} \bar{B}_f, \quad C_f = \bar{C}_f, \quad D_f = \bar{D}_f. \tag{3.125}$$

Proof First, by defining $\rho_\delta = \sigma_\delta / \hbar_\delta$, $\delta = y,\ z$, (3.105) can be rewritten as

$$M_y > \hbar_y / \sigma_y, \quad M_z > \hbar_z / \sigma_z. \tag{3.126}$$

Therefore, (3.123) can be obtained from (3.126) by inequality transformation.

Then we perform congruence transformation to (3.106) by $\mathscr{G} = \mathrm{diag}\{I, I, I, I, G^T, I, I, I\}$ and its transpose, respectively, one has

$$\begin{bmatrix} -P & * & * & * & * & * & * & * \\ 0 & -\gamma^2 I & * & * & * & * & * & * \\ 0 & 0 & -\hbar_y I & * & * & * & * & * \\ 0 & 0 & 0 & -\hbar_z I & * & * & * & * \\ G^T \mathscr{A}_1(\rho) & G^T \mathscr{B}(\rho) & G^T \mathscr{E}_1 & 0 & -G^T P^{-1} G & * & * & * \\ C_{L1}(\rho) & 0 & D_{L1} & I & 0 & -I & * & * \\ 2\hbar_y \rho_y \Delta_y C_y(\rho) & 0 & 0 & 0 & 0 & 0 & -\hbar_y I & * \\ 2\hbar_z \rho_z \Delta_z L_z(\rho) & 0 & 0 & 0 & 0 & 0 & 0 & -\hbar_z I \end{bmatrix} < 0. \tag{3.127}$$

By using $-G^T P^{-1} G < -G - G^T + P$, the following inequality can be obtained:

$$\begin{bmatrix} -P & * & * & * & * & * & * & * \\ 0 & -\gamma^2 I & * & * & * & * & * & * \\ 0 & 0 & -\hbar_y I & * & * & * & * & * \\ 0 & 0 & 0 & -\hbar_z I & * & * & * & * \\ G^T \mathscr{A}_1(\rho) & G^T \mathscr{B}(\rho) & G^T \mathscr{E}_1 & 0 & -G - G^T + P & * & * & * \\ C_{L1}(\rho) & 0 & D_{L1} & I & 0 & -I & * & * \\ 2\hbar_y \rho_y \Delta_y C_y(\rho) & 0 & 0 & 0 & 0 & 0 & -\hbar_y I & * \\ 2\hbar_z \rho_z \Delta_z L_z(\rho) & 0 & 0 & 0 & 0 & 0 & 0 & -\hbar_z I \end{bmatrix} < 0. \tag{3.128}$$

Furthermore, let's assume that $P = \begin{bmatrix} P_1 & * \\ P_2 & P_3 \end{bmatrix}$, $G^T = \begin{bmatrix} G_1 & G_3 \\ G_2 & G_3 \end{bmatrix}$, one has

$$\begin{bmatrix} -P_1 & * & * & * & * & * & * & * & * & * \\ -P_2 & -P_3 & * & * & * & * & * & * & * & * \\ 0 & 0 & -\gamma^2 I & * & * & * & * & * & * & * \\ 0 & 0 & 0 & -\hbar_y I & * & * & * & * & * & * \\ 0 & 0 & 0 & 0 & -\hbar_z I & * & * & * & * & * \\ \Gamma_1(\rho) & G_3 A_f & G_1 B(\rho) & G_3 B_f & 0 & -G_1 - G_1^T + P_1 & * & * & * & * \\ \Gamma_2(\rho) & G_3 A_f & G_2 B(\rho) & G_3 B_f & 0 & -G_2 - G_3^T + P_2 & -G_3 - G_3^T + P_3 & * & * & * \\ \Gamma_3(\rho) & -C_f & 0 & -D_f & I & 0 & 0 & -I & * & * \\ \Gamma_4(\rho) & 0 & 0 & 0 & 0 & 0 & 0 & 0 & -\hbar_y I & * \\ \Gamma_5(\rho) & 0 & 0 & 0 & 0 & 0 & 0 & 0 & 0 & -\hbar_z I \end{bmatrix} < 0, \tag{3.129}$$

where

$$\Gamma_1(\rho) = G_1 A(\rho) + G_3 B_f C(\rho), \quad \Gamma_2(\rho) = G_2 A(\rho) + G_3 B_f C(\rho),$$
$$\Gamma_3(\rho) = L(\rho) - D_f C(\rho), \quad \Gamma_4(\rho) = 2\sigma_y \Delta_y C(\rho),$$
$$\Gamma_5(\rho) = 2\sigma_z \Delta_z L(\rho).$$

By defining $\bar{A}_f = G_3 A_f, \bar{B}_f = G_3 B_f, \bar{C}_f = C_f$ and $\bar{D}_f = D_f$, then the inequality (3.129) can be rewritten as

$$\begin{bmatrix} -P_1 & * & * & * & * & * & * & * & * & * \\ -P_2 & -P_3 & * & * & * & * & * & * & * & * \\ 0 & 0 & -\gamma^2 I & * & * & * & * & * & * & * \\ 0 & 0 & 0 & -\hbar_y I & * & * & * & * & * & * \\ 0 & 0 & 0 & 0 & -\hbar_z I & * & * & * & * & * \\ G_1 A(\rho) + \bar{B}_f C(\rho) & \bar{A}_f & G_1 B(\rho) & \bar{B}_f & 0 & -G_1 - G_1^T + P_1 & * & * & * & * \\ G_2 A(\rho) + \bar{B}_f C(\rho) & \bar{A}_f & G_2 B(\rho) & \bar{B}_f & 0 & -G_2 - G_3^T + P_2 & -G_3 - G_3^T + P_3 & * & * & * \\ L(\rho) - \bar{D}_f C(\rho) & -\bar{C}_f & 0 & -\bar{D}_f & I & 0 & 0 & -I & * & * \\ 2\sigma_y \Delta_y C(\rho) & 0 & 0 & 0 & 0 & 0 & 0 & 0 & -\hbar_y I & * \\ 2\sigma_z \Delta_z L(\rho) & 0 & 0 & 0 & 0 & 0 & 0 & 0 & 0 & -\hbar_z I \end{bmatrix} < 0. \tag{3.130}$$

Therefore, one can obtain that (3.124) can be guaranteed if the condition in (3.130) is satisfied. The proof is completed.

Remark 3.1 In fact, according to the work studied in [7], the dynamic quantizer parameters μ_δ should also be determined as a unique value by a specific tuning strategy. Such a dynamic adjustment strategy may yield smaller quantization errors compared to the general static adjustment strategy [8–10]. From the online adjusting strategy in (3.107), we know that the dynamic parameters μ_δ are related to quantized signals $\delta(k)$ and the scalars ρ_δ, $\delta = y,\ z$. Noted that $\rho_\delta = \sigma_\delta / \hbar_\delta$, $\delta = y,\ z$ which can be obtained by solving the conditions in Theorem 3.6.

Remark 3.2 In this section, the proposed conditions in Theorem 3.6 are strict LMIs. Therefore, the filter gain matrices can be obtained by solving (3.123) and (3.124) in Matlab LMI toolbox. Then, the following optimization problem is utilized:

$$\min \gamma \text{ subject to LMIs (3.123) and (3.124).} \tag{3.131}$$

3.3.3 Simulation Example

A numerical example will be shown to demonstrate that the designed $\mathscr{H}_\infty$ filter strategy is effective. And parameter matrices of the system (3.97) will be given as follows:

$$\begin{aligned} A_1 &= \begin{bmatrix} 0.7 & -0.05 \\ 0.1 & 0.8 \end{bmatrix}, \quad B_1 = \begin{bmatrix} 0.1 \\ 0.1 \end{bmatrix}, \\ A_2 &= \begin{bmatrix} 0.94 & 0.3 \\ -0.05 & 0.6 \end{bmatrix}, \quad B_2 = \begin{bmatrix} 0.3 \\ 0.1 \end{bmatrix}, \\ C_1 &= \begin{bmatrix} 0.2 & 0.1 \end{bmatrix}, \quad L_1 = \begin{bmatrix} 1 & 0.1 \end{bmatrix}, \\ C_2 &= \begin{bmatrix} 1.2 & 0 \end{bmatrix}, \quad L_2 = \begin{bmatrix} 2 & 0.5 \end{bmatrix}. \end{aligned}$$

In addition, member function is set to

$$\sigma_1(k) = \sin^2(0.05x_1(k)), \quad \sigma_2(k) = 1 - \sigma_1(k). \tag{3.132}$$

Furthermore, we adopt the quantization errors $\Delta_y = \Delta_z = 0.01$, quantization ranges $M_y = M_z = 50$, the other parameters $\hbar_y = 5$, $\hbar_z = 5$ and select the $\mathscr{H}_\infty$ performance index $\gamma = 3$. According to the form of the LMIs of Theorem 3.6, the unknown parameters in inequalities (3.123) and (3.124) can be obtained by using the LMI toolbox, then we have

$$\begin{aligned} A_f &= \begin{bmatrix} 0.5175 & 0.2063 \\ 0.0778 & 0.5582 \end{bmatrix}, \quad B_f = \begin{bmatrix} -0.3696 \\ 0.0512 \end{bmatrix}, \\ C_f &= \begin{bmatrix} -0.9188 & -0.0438 \end{bmatrix}, \quad D_f = 0.8616. \end{aligned} \tag{3.133}$$

Moreover, we assume that $x(0) = [0 \ \ 0]^T$ and $w(k) = \sin(k)e^{-k}$, the simulation results are given in Figs. 3.11, 3.12, 3.13, 3.14, 3.15, 3.16 and 3.17. Figure 3.11 plots the response of system state $x(k)$. Figure 3.12 plots the response of filter state $x_f(k)$. Figure 3.13 plots the responses of $z(k)$ and $z_f(k)$. Figure 3.14 plots the filtering error $e(k)$. Figure 3.15 plots the response of quantizer's parameter μ_y.

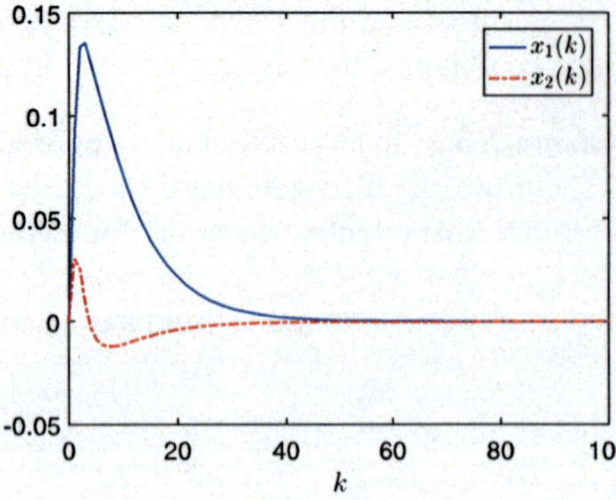

Fig. 3.11 Response of the system state $x(k)$

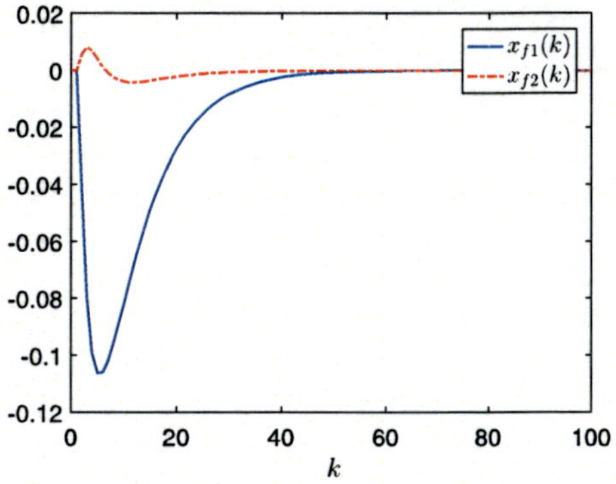

Fig. 3.12 Response of the filter state $x_f(k)$

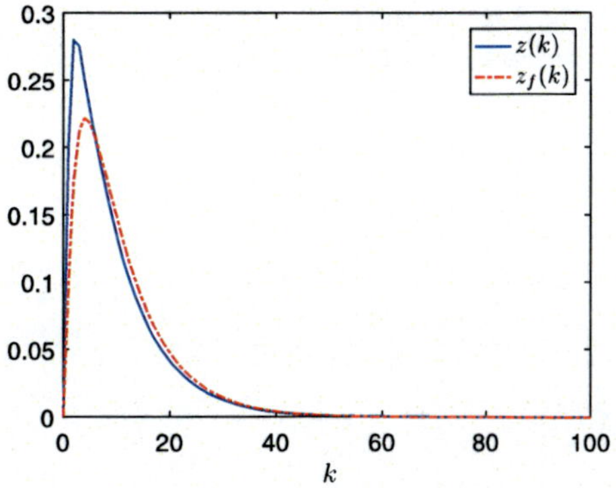

Fig. 3.13 Responses of $z(k)$ and $z_f(k)$

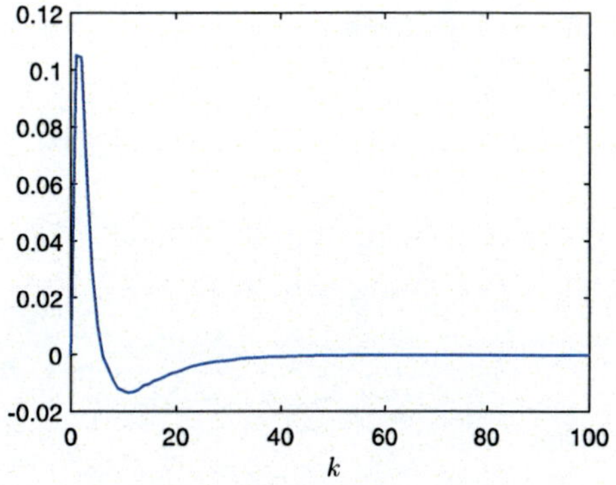

Fig. 3.14 Filtering error $e(k)$

Figure 3.16 plots the response of quantizer's parameter μ_z. Figure 3.17 plots the ratio of $\sqrt{\sum_{k=0}^{\infty} e^T(k)e(k)/\sum_{k=0}^{\infty} w^T(k)w(k)}$. It can be seen that the convergence value of the trajectory is $0.5079 < 3$ from Fig. 3.17.

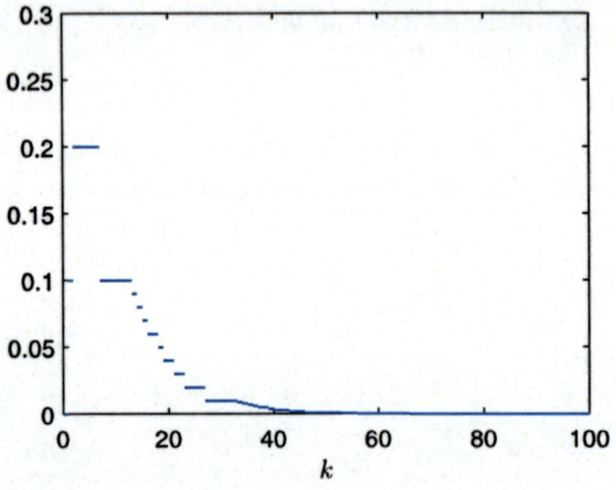

Fig. 3.15 Dynamic quantizer's parameter μ_y

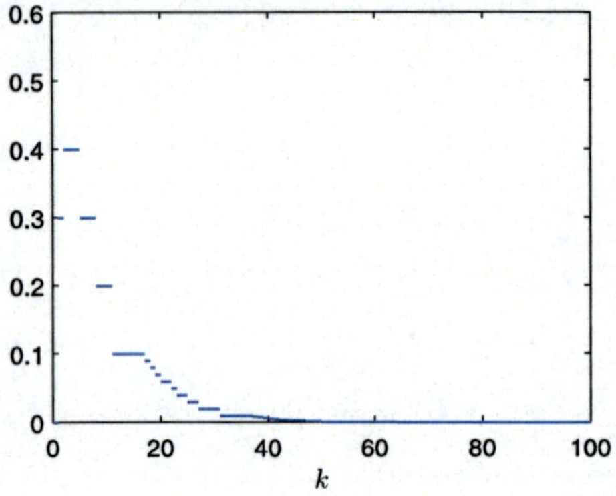

Fig. 3.16 Dynamic quantizer's parameter μ_z

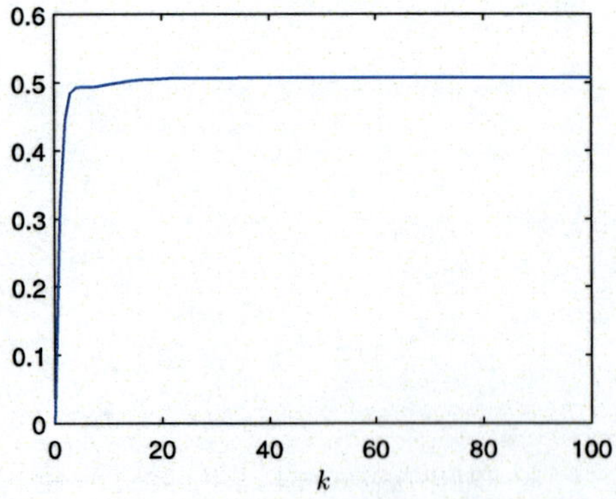

Fig. 3.17 History of $\sqrt{\sum_{k=0}^{\infty} e^T(k)e(k) / \sum_{k=0}^{\infty} w^T(k)w(k)}$

3.4 Conclusion

In this chapter, the problem of filtering for T–S fuzzy systems with the various induced phenomena, which occur during the measurement output signal and the performance output signal are transmitted through the communication channel, has been studied. The conditions have been given to make the filtering error systems satisfy the ℓ_2–ℓ_∞, induced ℓ_∞ and $\mathcal{H}_\infty$ performance, respectively, while the filtering error systems are stable. First of all, the ℓ_2–ℓ_∞ filter has been designed, when the measurement output signal and the performance output signal have been quantized by static quantizers. In the second place, the induced ℓ_∞ filter has been designed, when the measurement output signal and the performance output signal have been considered to occur data packet dropout. In addition, the $\mathcal{H}_\infty$ filter has been designed, when the measurement output signal and the performance output signal have been quantized by dynamic quantizers. At last, some simulation results have been provided to verify the effectiveness of these ℓ_2–ℓ_∞, induced ℓ_∞ and $\mathcal{H}_\infty$ filtering strategies for T–S fuzzy systems.

References

1. Fu M, Xie L (2005) The sector bound approach to quantized feedback control. IEEE Trans Autom Control 50:1698–1711
2. Chang XH, Wang YM (2018) Peak-to-peak filtering for networked nonlinear DC motor systems with quantization. IEEE Trans Industr Inf 14:5378–5388
3. Ahn CK, Shi P, Basin MV (2018) Two-dimensional peak-to-peak filtering for stochastic Fornasini-Marchesini systems. IEEE Trans Autom Control 63:1472–1479
4. Fan C, Lam J, Xie X (2018) Peak-to-peak filtering for periodic piecewise linear polytopic systems. Int J Syst Sci 49:1997–2011
5. Burchett B (2005) Parametric time domain system identification of a mass spring damper system. In 2005 Annual Conference
6. Liberzon D (2003) Hybrid feedback stabilization of systems with quantized signals. Automatica 39:1543–1554
7. Chang XH, Yang C, Xiong J (2019) Quantized fuzzy output feedback $\mathcal{H}_\infty$ control for nonlinear systems with adjustment of dynamic parameters. IEEE Trans Syst, Man, Cybern: Syst 40:2005–2015
8. Gao H, Chen T (2008) A new approach to quantized feedback control systems. Automatica 44:534–542
9. Zhang C, Feng G, Gao H, Qiu J (2011) $\mathcal{H}_\infty$ filtering for nonlinear discrete-time systems subject to quantization and packet dropouts. IEEE Trans Fuzzy Syst 19:353–365
10. Li ZM, Chang XH, Mathiyalagan K, Xiong J (2017) Robust energy-to-peak filtering for discrete-time nonlinear systems with measurement quantization. Signal Process 139:102–109

Chapter 4
Fuzzy Output Feedback Control with Communication Constraints

Abstract In this chapter, the problem of output feedback control is studied for T–S fuzzy systems with communication constraints. Firstly, considering the effect of the quantization error caused from multi-signal quantization, a novel quantization output feedback control scheme is proposed to ensure that the closed-loop fuzzy system is asymptotically stable and satisfies $\mathscr{H}_\infty$ performance. Secondly, by applying input delay method to describe the control input signal affected by data dropouts and time delays, a static output feedback control strategy is provided to guarantee the asymptotic stability and $\mathscr{H}_\infty$ performance of closed-loop fuzzy system. Thirdly, by using a Markov chain to describe the communication protocol scheduling behavior over the communication channel, a dynamic output feedback controller is designed to guarantee the stochastic stability and $\mathscr{H}_\infty$ performance of the closed-loop system. Finally, some simulation examples are given to prove the feasibility and the effectiveness of the developed design methods.

Keywords Multi-signal quantization · $\mathscr{H}_\infty$ output feedback controller · Time delays · Data dropouts · Stochastic communication protocol

4.1 Feedback Control with Quantization

In this section, the output feedback control problem is considered for T–S fuzzy system with the effects of multi-signal quantization. By using the concept of PDC, a new design method of output feedback controller under the quantized output signals and control signals is proposed to ensure that the closed-loop fuzzy control system is asymptotically stable and meets $\mathscr{H}_\infty$ performance.

X. Chang et al., *Control and Filtering of Fuzzy Systems Under Communication Channels*,
https://doi.org/10.1007/978-981-99-4346-3_4

4.1.1 Problem Formulation

Consider the following discrete-time T–S fuzzy model:

Plant Rule i_{th}: **IF** $\zeta_1(k)$ is J_{1i} and $\ldots\zeta_o(k)$ is J_{oi}, **THEN**

$$\begin{aligned} x(k+1) &= A_i x(k) + B_i \bar{u}(k), \\ y(k) &= C_i x(k), \end{aligned} \tag{4.1}$$

where $x(k) \in \mathbb{R}^{n_x}$, $\bar{u}(k) \in \mathbb{R}^{n_u}$, and $y(k) \in \mathbb{R}^{n_y}$ are the state vector, control input vector, and output vector, respectively; $\zeta(k) = \begin{bmatrix} \zeta_1(k) & \ldots & \zeta_o(k) \end{bmatrix}$, $\zeta_d(k), d = 1, \ldots, o$ are measurable premise variables; $J_{di}, i = 1, \ldots, r$, represent the fuzzy sets and the parameter r means the number of fuzzy rules; The matrices $A_i \in \mathbb{R}^{n_x \times n_x}$, $B_i \in \mathbb{R}^{n_x \times n_u}$ and $C_i \in \mathbb{R}^{n_y \times n_x}$ are known system matrices.

It is assumed that fuzzy basis function is given by

$$h_i(\zeta(k)) = \frac{\prod_{d=1}^{o} J_{di}(\zeta_d(k))}{\sum_{i=1}^{r} \prod_{d=1}^{o} J_{di}(\zeta_d(k))}, \tag{4.2}$$

where $J_{di}(\zeta_d(k))$ is the grade of membership function of $\zeta_d(k)$ in J_{di}.

Considering the properties of membership function, one has

$$h_i(\zeta(k)) \geq 0, \quad \sum_{i=1}^{r} h_i(\zeta(k)) = 1, \quad i = 1, \ldots, r. \tag{4.3}$$

Further, a more compact presentation of T–S fuzzy model (4.1) can be derived as follows:

$$\begin{aligned} x(k+1) &= A(h)x(k) + B(h)\bar{u}(k), \\ y(k) &= C(h)x(k), \end{aligned} \tag{4.4}$$

where

$$A(h) = \sum_{i=1}^{r} h_i(\zeta(k))A_i, \quad B(h) = \sum_{i=1}^{r} h_i(\zeta(k))B_i, \quad C(h) = \sum_{i=1}^{r} h_i(\zeta(k))C_i.$$

Next, assume that the measurement output $y(k)$ and control signal $u(k)$ should be quantized before the signals feedback to controller and system. Then, a dynamic quantizer is given as the following form:

$$\bar{\iota}(k) = q_{\mu_\iota}(\iota(k)) = \mu_\iota(k) q \left(\frac{\iota(k)}{\mu_\iota(k)} \right), \quad \mu_\iota(k) > 0, \tag{4.5}$$

where $\mu_\iota(k)$ is the quantization parameters of corresponding signals with $\iota = y, u$.

Set $\mu_\iota(k)M_\iota$ being the range of the quantizer and $\mu_\iota(k)\Delta_\iota$ being the quantization error bound of the quantizer. Then the quantizer $q_{\mu_\iota}(\cdot)$ satisfies the following conditions:

$$\begin{aligned} \|q_{\mu_\iota}(\iota(k)) - \iota(k)\| &\le \mu_\iota(k)\Delta_\iota, \quad \text{if } \|\iota(k)\| \le \mu_\iota(k)M_\iota, \\ \|q_{\mu_\iota}(\iota(k))\| &> \mu_\iota(k)M_\iota - \mu_\iota(k)\Delta_\iota, \quad \text{if } \|\iota(k)\| > \mu_\iota(k)M_\iota. \end{aligned} \tag{4.6}$$

Considering the effects of the quantization and data packet loss, the following fuzzy controller can be obtained by using the same membership function and premise variables

Plant Rule i_{th}: **IF** $\zeta_1(k)$ is J_{1i} and $\ldots\zeta_o(k)$ is J_{oi}, **THEN**

$$u(k) = K_i\bar{y}(k), \tag{4.7}$$

where $K_i \in \mathbb{R}^{n_u \times n_y}, i = 1, \ldots, r$, is controller gain matrix to be determined, $u(k)$ represents the control signal output by the controller, and $\bar{y}(k)$ denotes the signal passed into the controller after quantization of the system measurement output signal.

The following terse form is given to represent the fuzzy controller

$$u(k) = K(h)\bar{y}(k), \tag{4.8}$$

where

$$K(h) = \sum_{i=1}^{r} h_i(\zeta(k))K_i.$$

Considering the quantization effect of dynamic quantizer on the control signal input to fuzzy system (4.4), it yields

$$\begin{aligned} x(k+1) &= A(h)x(k) + B(h)u(k) + B(h)\left(q_{\mu_u}(u(k)) - u(k)\right), \\ y(k) &= C(h)x(k). \end{aligned} \tag{4.9}$$

Combining the above formula with the static fuzzy controller (4.8), one can obtain

$$\begin{aligned} x(k+1) &= A(h)x(k) + B(h)K(h)\bar{y}(k) + B(h)\left(q_{\mu_u}(u(k)) - u(k)\right), \\ y(k) &= C(h)x(k). \end{aligned} \tag{4.10}$$

The system output signal will also be quantized when entering the static controller, so the above equation can be rearranged as

$$\begin{aligned} x(k+1) &= A(h)x(k) + B(h)K(h)y(k) + B(h)\left(q_{\mu_u}(u(k)) - u(k)\right) \\ &\quad + B(h)K(h)\left(q_{\mu_y}(y(k)) - y(k)\right), \\ y(k) &= C(h)x(k). \end{aligned} \tag{4.11}$$

Substituting the equation (4.5) into the above system can be reduced to

$$
\begin{aligned}
x(k+1) &= A(h)x(k) + B(h)K(h)y(k) + B(h)\mu_u(k)\left(q\left(\frac{u(k)}{\mu_u(k)}\right) - \frac{u(k)}{\mu_u(k)}\right) \\
&\quad + B(h)K(h)\mu_y(k)\left(q\left(\frac{y(k)}{\mu_y(k)}\right) - \frac{y(k)}{\mu_y(k)}\right), \\
y(k) &= C(h)x(k). \qquad (4.12)
\end{aligned}
$$

Set $\vartheta_u(k) = \mu_u(k)\left(q\left(\frac{u(k)}{\mu_u(k)}\right) - \frac{u(k)}{\mu_u(k)}\right)$ and $\vartheta_y(k) = \mu_y(k)\left(q\left(\frac{y(k)}{\mu_y(k)}\right) - \frac{y(k)}{\mu_y(k)}\right)$. Then the discrete-time output feedback closed-loop control system is obtained

$$
x(k+1) = \mathscr{A}(h)x(k) + \mathscr{B}(h)\vartheta_y(k) + \mathscr{C}(h)\vartheta_u(k), \qquad (4.13)
$$

where

$$
\begin{aligned}
\mathscr{A}(h) &= A(h) + B(h)K(h)C(h), \\
\mathscr{B}(h) &= B(h)K(h), \ \mathscr{C}(h) = B(h).
\end{aligned} \qquad (4.14)
$$

Next, the quantized static output feedback control problem of fuzzy discrete-time systems can be formulated as:

Quantized static output feedback control problem. For the nonlinear discrete-time system, taking full account of the multiple quantization of the output signal $y(k)$ of the system and the output signal $u(k)$ of the controller, the goal of this section is to design the PDC controllers and the appropriate dynamic quantizers such that the closed-loop control system is stable and meets certain control performance.

4.1.2 Main Results

4.1.2.1 Output Feedback Stabilization Control Design

Based on the above construction of the discrete-time output feedback closed-loop control system (4.13) with multi-signal quantization, the problem of the stability analysis for the closed-loop system can be given by the following theorem.

Theorem 4.1 *For the T–S fuzzy system (4.13). Given quantization ranges M_y, M_u and quantization errors Δ_y, Δ_u. If there exist matrix $P > 0$, scalars σ_y, σ_u, ρ_y, and ρ_u satisfying the following matrix inequalities*

$$
M_y > \frac{1}{\rho_y}, \ M_u > \frac{1}{\rho_u}, \qquad (4.15)
$$

$$\begin{bmatrix} -P & * & * & * \\ 0 & -\sigma_y I & * & * \\ 0 & 0 & -\sigma_u I & * \\ \mathscr{A}(h) & \mathscr{B}(h) & \mathscr{C}(h) & -P^{-1} \end{bmatrix} + \sigma_y(Ha\{\Upsilon_1(h)\}) + \sigma_u(Ha\{\Upsilon_2(h)\}) < 0, \tag{4.16}$$

where

$$\Upsilon_1(h) = \begin{bmatrix} 2\rho_y \Delta_y C(h) & 0 & 0 \end{bmatrix}, \ \Upsilon_2(h) = \begin{bmatrix} \Upsilon_{21}(h) & \Upsilon_{22}(h) & 0 \end{bmatrix},$$
$$\Upsilon_{21}(h) = 2\rho_u \Delta_u K(h) C(h), \ \Upsilon_{22}(h) = 2\rho_u \Delta_u K(h),$$

and the online adjusting strategy for the parameters $\mu_y(k)$ and $\mu_u(k)$ are given as:

$$\begin{aligned} \rho_y \|y(k)\| \le \mu_y(k) \le 2\rho_y \|y(k)\|, \\ \rho_u \|u(k)\| \le \mu_u(k) \le 2\rho_u \|u(k)\|. \end{aligned} \tag{4.17}$$

Then the closed-loop control system is asymptotically stable.

Proof From the inequality (4.6), it is clear that $\left\| \frac{y(k)}{\mu_y(k)} \right\| \le M_y$ and one has

$$\left\| q\left(\frac{y(k)}{\mu_y(k)}\right) - \frac{y(k)}{\mu_y(k)} \right\| \le \Delta_y. \tag{4.18}$$

Consider the homogeneity property of Euclidean norm and combine the equation $\vartheta_y(k) = \mu_y(k)\left(q\left(\frac{y(k)}{\mu_y(k)}\right) - \frac{y(k)}{\mu_y(k)}\right)$. Then one gets with the inequalities (4.15) and (4.17)

$$\frac{1}{2\rho_y}\mu_y(k) \le \|y(k)\| \le \frac{1}{\rho_y}\mu_y(k) \le M_y \mu_y(k). \tag{4.19}$$

Further, one can obtain

$$\|\vartheta_y(k)\| \le 2\rho_y \Delta_y \|y(k)\|. \tag{4.20}$$

Square both sides of above inequality gives

$$\vartheta_y^T(k)\vartheta_y(k) \le 4\rho_y^2 \Delta_y^2 y^T(k) y(k). \tag{4.21}$$

Defining $\delta(k) = \begin{bmatrix} x^T(k) & \vartheta_y^T(k) & \vartheta_u^T(k) \end{bmatrix}^T$, one has

$$\delta^T(k)\Omega_1(h)\delta(k) \ge 0, \tag{4.22}$$

where $\Omega_1(h) = Ha\left\{\begin{bmatrix} 2\rho_y \Delta_y C(h) & 0 & 0 \end{bmatrix}\right\} - diag\{0, \ I, \ 0\}$.

Similarly, when the control signal is quantized there have $\left\| \frac{u(k)}{\mu_u(k)} \right\| \leq M_u$ and

$$\left\| q\left(\frac{u(k)}{\mu_u(k)}\right) - \frac{u(k)}{\mu_u(k)} \right\| \leq \Delta_u. \tag{4.23}$$

Consider the homogeneity property of Euclidean norm and combine the equation $\vartheta_u(k) = \mu_u(k)\left(q\left(\frac{u(k)}{\mu_u(k)}\right) - \frac{u(k)}{\mu_u(k)}\right)$. Then adjusting strategy $\frac{1}{2\rho_u}\mu_u(k) \leq \|u(k)\| \leq \frac{1}{\rho_u}\mu_u(k) \leq M_u\mu_u(k)$ with inequality (4.15) yields

$$\|\vartheta_u(k)\| \leq 2\rho_u\Delta_u\|u(k)\|. \tag{4.24}$$

Since the input signal $u(k)$ of the controller is the measurement output signal of the system after quantization, i.e., $u(k) = K(h)C(h)x(k) + K(h)\vartheta_y(k)$, one has

$$\delta^T(k)\Omega_2(h)\delta(k) \geq 0, \tag{4.25}$$

where $\Omega_2(h) = Ha\left\{\left[2\rho_u\Delta_u K(h)C(h)\ \ 2\rho_u\Delta_u K(h)\ \ 0\right]\right\} - diag\{0,\ 0,\ I\}$.

Next, a Lyapunov function is selected to analyze the stability of closed-loop control system

$$V(k, x(k)) = x^T(k)Px(k),\quad P = P^T > 0, \tag{4.26}$$

and combine system (4.13), it can be obtained that

$$\begin{aligned}
\Delta V(k, (x(k))) =& V(k+1, x(k+1)) - V(k, x(k)) \\
=& \left(\mathcal{A}(h)x(k) + \mathcal{B}(h)\vartheta_y(k) + \mathcal{C}(h)\vartheta_u(k)\right)^T P \\
& \times \left(\mathcal{A}(h)x(k) + \mathcal{B}(h)\vartheta_y(k) + \mathcal{C}(h)\vartheta_u(k)\right) - x^T(k)Px(k) \\
=& \delta^T(k)\Big(\left[\mathcal{A}(h)\ \ \mathcal{B}(h)\ \ \mathcal{C}(h)\right]^T P \left[\mathcal{A}(h)\ \ \mathcal{B}(h)\ \ \mathcal{C}(h)\right] \\
& - diag\{P,\ 0,\ 0\}\Big)\delta(k) \\
=& \delta^T(k)\Omega_0(h)\delta(k),
\end{aligned} \tag{4.27}$$

where $\Omega_0(h) = \left[\mathcal{A}(h)\ \ \mathcal{B}(h)\ \ \mathcal{C}(h)\right]^T P \left[\mathcal{A}(h)\ \ \mathcal{B}(h)\ \ \mathcal{C}(h)\right] - diag\{P,\ 0,\ 0\}$.

Using Lemma 1.3, the following inequality can be divided

$$\Omega_0(h) + \sigma_y\Omega_1(h) + \sigma_u\Omega_2(h) < 0. \tag{4.28}$$

Finally, the matrix inequality (4.16) can be obtained.

In Theorem 4.1, the stability analysis conditions of fuzzy system with quantization are given. However, because of the nonlinearity of the system and the controller and the non-convex terms in the matrix inequality, the inequality conditions in Theorem 4.1 cannot be directly used in controller design. Next, the main purpose is to decouple the nonlinearity terms so as to obtain controller design conditions in terms of the LMI form.

Theorem 4.2 *For the T–S fuzzy system (4.13). Given quantization ranges M_y, M_u and quantization errors Δ_y, Δ_u. If there exist matrices $\mathcal{P} > 0$, N, N_1, N_2, and U_l, $l = 1, \dots, r$, scalars η_y, and η_u satisfying the following matrix inequalities*

$$M_y > \eta_y, \quad M_u > \eta_u, \tag{4.29}$$

$$\Gamma_{ii} < 0, \quad i = 1, \dots, r, \tag{4.30}$$

$$\Gamma_{il} + \Gamma_{li} < 0, \quad i < l, \quad i, l = 1, \dots, r, \tag{4.31}$$

where

$$\Gamma_{il} = \begin{bmatrix} -\mathcal{P} & * & * & * & * & * \\ 0 & -\eta_y I & * & * & * & * \\ 0 & 0 & -\eta_u I & * & * & * \\ \tilde{\mathcal{A}}_{il} & \tilde{\mathcal{B}}_l & \hat{\mathcal{C}} & -\mathcal{N} - \mathcal{N}^T + \mathcal{P} & * & * \\ 2\Delta_y C_i Y_i^{-1} & 0 & 0 & 0 & -\eta_y I & * \\ \bar{\Upsilon}_{21il} Y_i^{-1} & \bar{\Upsilon}_{22l} & 0 & 0 & 0 & -\eta_u I \end{bmatrix},$$

$$\mathcal{U}_l = \begin{bmatrix} U_l \\ 0 \end{bmatrix}, \quad \bar{E} = \begin{bmatrix} I \\ 0 \end{bmatrix}, \quad \mathcal{N} = \begin{bmatrix} N & N_1 \\ 0 & N_2 \end{bmatrix},$$

$$\tilde{\mathcal{A}}_{il} = \mathcal{N} Y_i A_i Y_i^{-1} + \mathcal{U}_l C_i Y_i^{-1}, \quad \tilde{\mathcal{B}}_l = \mathcal{U}_l, \quad \hat{\mathcal{C}} = \mathcal{N} \bar{E},$$

$$\bar{\Upsilon}_{21il} = 2\Delta_u K_l C_i, \quad \bar{\Upsilon}_{22l} = 2\Delta_u K_l.$$

Then the closed-loop control system (4.13) is asymptotically stable. The corresponding dynamic quantization parameters $\mu_y(k)$ and $\mu_u(k)$ adjustment strategy in Theorem 4.1 can be rewritten as

$$\begin{aligned} \frac{1}{\eta_y}\|y(k)\| \le \mu_y(k) \le \frac{2}{\eta_y}\|y(k)\|, \\ \frac{1}{\eta_u}\|u(k)\| \le \mu_u(k) \le \frac{2}{\eta_u}\|u(k)\|. \end{aligned} \tag{4.32}$$

In addition, the controller gain matrix can be determined

$$K_l = N^{-1} U_l, \quad l = 1, \dots, r. \tag{4.33}$$

Proof Defining scalars $\eta_\iota = \sigma_\iota = \frac{1}{\rho_\iota}$, $\iota = y, u$ and using Lemma 1.1 to (4.16), it can be obtained that

$$\begin{bmatrix} -P & * & * & * & * & * \\ 0 & -\eta_y I & * & * & * & * \\ 0 & 0 & -\eta_u I & * & * & * \\ \mathscr{A}(h) & \mathscr{B}(h) & \mathscr{C}(h) & -P^{-1} & * & * \\ 2\Delta_y C(h) & 0 & 0 & 0 & -\eta_y I & * \\ 2\Delta_u K(h)C(h) & 2\Delta_u K(h) & 0 & 0 & 0 & -\eta_u I \end{bmatrix} < 0. \quad (4.34)$$

Choose a nonsingular state coordinate transformation matrix $Y(h)$, which satisfies $P = Y^T(h)\mathscr{P}Y(h)$ with $\mathscr{P} > 0$. Then, the inverse matrix of a matrix P is $P^{-1} = Y^{-1}(h)\mathscr{P}^{-1}Y^{-T}(h)$. Substituting matrices P and P^{-1} into the above inequality, and pre- and post-multiplying it by diag$\{Y^{-T}(h), I, I, Y(h), I, I\}$ and its transpose matrix, respectively, the following inequality can be obtained

$$\begin{bmatrix} -\mathscr{P} & * & * & * & * & * \\ 0 & -\eta_y I & * & * & * & * \\ 0 & 0 & -\eta_u I & * & * & * \\ Y(h)\mathscr{A}(h)Y^{-1}(h) & Y(h)\mathscr{B}(h) & Y(h)\mathscr{C}(h) & -\mathscr{P}^{-1} & * & * \\ 2\Delta_y C(h)Y^{-1}(h) & 0 & 0 & 0 & -\eta_y I & * \\ \bar{\Upsilon}_{21}(h)Y^{-1}(h) & \bar{\Upsilon}_{22}(h) & 0 & 0 & 0 & -\eta_u I \end{bmatrix} < 0, \quad (4.35)$$

where $\bar{\Upsilon}_{21}(h) = 2\Delta_u K(h)C(h)$, $\bar{\Upsilon}_{22}(h) = 2\Delta_u K(h)$.

Define $\bar{E} = Y(h)B(h)$. Then the matrix $Y(h)$ is used as the state coordinate transformation matrix. It yields

$$\begin{bmatrix} -\mathscr{P} & * & * & * & * & * \\ 0 & -\eta_y I & * & * & * & * \\ 0 & 0 & -\eta_u I & * & * & * \\ \bar{\mathscr{A}}(h) & \bar{\mathscr{B}}(h) & \bar{\mathscr{C}} & -\mathscr{P}^{-1} & * & * \\ 2\Delta_y C(h)Y^{-1}(h) & 0 & 0 & 0 & -\eta_y I & * \\ \bar{\Upsilon}_{21}(h)Y^{-1}(h) & \bar{\Upsilon}_{22}(h) & 0 & 0 & 0 & -\eta_u I \end{bmatrix} < 0, \quad (4.36)$$

where

$$\bar{\mathscr{A}}(h) = Y(h)A(h)Y^{-1}(h) + \bar{E}K(h)C(h)Y^{-1}(h),\ \bar{\mathscr{B}}(h) = \bar{E}K(h),\ \bar{\mathscr{C}} = \bar{E}.$$

Pre- and post-multiplying (4.36) by diag$\{I, I, I, \mathscr{N}, I, I, I, I\}$ and its transpose, respectively, one has

$$\begin{bmatrix} -\mathscr{P} & * & * & * & * & * \\ 0 & -\eta_y I & * & * & * & * \\ 0 & 0 & -\eta_u I & * & * & * \\ \hat{\mathscr{A}}(h) & \hat{\mathscr{B}}(h) & \hat{\mathscr{C}} & -\mathscr{N}\mathscr{P}^{-1}\mathscr{N}^T & * & * \\ 2\Delta_y C(h)Y^{-1}(h) & 0 & 0 & 0 & -\eta_y I & * \\ \bar{\Upsilon}_{21}(h)Y^{-1}(h) & \bar{\Upsilon}_{22}(h) & 0 & 0 & 0 & -\eta_u I \end{bmatrix} < 0, \quad (4.37)$$

where

$$\hat{\mathscr{A}}(h) = \mathscr{N} Y(h)A(h)Y^{-1}(h) + \mathscr{N}\bar{E}K(h)C(h)Y^{-1}(h),$$
$$\hat{\mathscr{B}}(h) = \mathscr{N}\bar{E}K(h),\ \hat{\mathscr{C}} = \mathscr{N}\bar{E}.$$

Based on the constructed matrices $\mathscr{N}$, $\mathscr{U}(h)$, and the new definition matrix $U(h) = NK(h)$, it's easy to get

$$\mathscr{N}\bar{E}K(h) = \begin{bmatrix} N & N_1 \\ 0 & N_2 \end{bmatrix}\begin{bmatrix} I \\ 0 \end{bmatrix} K(h) = \begin{bmatrix} NK(h) \\ 0 \end{bmatrix} = \begin{bmatrix} U(h) \\ 0 \end{bmatrix} = \mathscr{U}(h). \tag{4.38}$$

So the inequality (4.37) can be rearranged as

$$\begin{bmatrix} -\mathscr{P} & * & * & * & * & * \\ 0 & -\eta_y I & * & * & * & * \\ 0 & 0 & -\eta_u I & * & * & * \\ \tilde{\mathscr{A}}(h) & \tilde{\mathscr{B}}(h) & \hat{\mathscr{C}} & -\mathscr{N}\mathscr{P}^{-1}\mathscr{N}^T & * & * \\ 2\Delta_y C(h)Y^{-1}(h) & 0 & 0 & 0 & -\eta_y I & * \\ \bar{\Upsilon}_{21}(h)Y^{-1}(h) & \bar{\Upsilon}_{22}(h) & 0 & 0 & 0 & -\eta_u I \end{bmatrix} < 0, \tag{4.39}$$

where

$$\tilde{\mathscr{A}}(h) = \mathscr{N} Y(h)A(h)Y^{-1}(h) + \mathscr{U}(h)C(h)Y^{-1}(h),\ \ \tilde{\mathscr{B}}(h) = \mathscr{U}(h).$$

Considering the fact that $-\mathscr{N}\mathscr{P}^{-1}\mathscr{N}^T \le -\mathscr{N} - \mathscr{N}^T + \mathscr{P}$, the above inequality can be reorganized as

$$\begin{bmatrix} -\mathscr{P} & * & * & * & * & * \\ 0 & -\eta_y I & * & * & * & * \\ 0 & 0 & -\eta_u I & * & * & * \\ \tilde{\mathscr{A}}(h) & \tilde{\mathscr{B}}(h) & \hat{\mathscr{C}} & -\mathscr{N} - \mathscr{N}^T + \mathscr{P} & * & * \\ 2\Delta_y C(h)Y^{-1}(h) & 0 & 0 & 0 & -\eta_y I & * \\ \bar{\Upsilon}_{21}(h)Y^{-1}(h) & \bar{\Upsilon}_{22}(h) & 0 & 0 & 0 & -\eta_u I \end{bmatrix} < 0. \tag{4.40}$$

According to the T–S fuzzy model and membership function, the above inequality can be rewritten as

$$\begin{aligned} &\sum_{i=1}^{r}\sum_{l=1}^{r} h_i(\zeta(k))h_l(\zeta(k))\Gamma_{il} \\ &= \sum_{i=1}^{r} h_i^2(\zeta(k))\Gamma_{ii} + \sum_{i=1}^{r}\sum_{i<l}^{r} h_i(\zeta(k))h_l(\zeta(k))(\Gamma_{il} + \Gamma_{li}) < 0. \end{aligned} \tag{4.41}$$

Based on (4.3) and [1], the inequality (4.41) yields the feasibility conditions (4.30) and (4.31).

Remark 4.1 In Theorem 4.2, for the full column rank matrix B_i, the state coordinate transformation matrix Y_i can be selected as $Y_i = \begin{bmatrix} (B_i^T B_i)^{-1} B_i^T \\ \left((B_i^T)^{\perp}\right)^T \end{bmatrix}$, where $(B_i^T)^{\perp}$ means an orthogonal basis for the null space of B_i^T and satisfies the condition that $B_i^T (B_i^T)^{\perp} = 0$.

Remark 4.2 The dynamic parameters $\mu_\iota(k)$, $\iota = y, u$, are adopted the same adjusting rule as [2] to update the quantizer's parameter $\mu_\iota(k)$ on both sides of encoder/decoder. And the adjusting rule is

$$\mu_\iota(k) = \begin{cases} floor(2\beta(k) \times 10^s) \times 10^{-s}, & 0 \le \beta(k) < \frac{1}{2}, \\ 1, & \frac{1}{2} \le \beta(k) < 1, \\ floor(2\beta(k)), & 1 \le \beta(k), \end{cases}$$

where $\beta(k) = \frac{1}{\eta_\iota} \|\iota(k)\|$, $s = min\{s \in \mathcal{N}^+ | (2\beta(k) \times 10^s) > 1\}$, and the function $floor(\varepsilon)$ means the maximum integer which does not exceed ε.

4.1.2.2 Output Feedback $\mathcal{H}_\infty$ Control Design

The above content is the analysis and proof of the stability of discrete-time output feedback control problem with quantization. When considering the $\mathcal{H}_\infty$ performance of the system, based on the same fuzzy rules, the fuzzy control system can be described as

$$\begin{aligned} x(k+1) &= A(h)x(k) + B(h)\bar{u}(k) + E(h)w(k), \\ z(k) &= L(h)x(k) + D(h)\bar{u}(k) + H(h)w(k), \\ y(k) &= C(h)x(k) + F(h)w(k), \end{aligned} \tag{4.42}$$

where $w(k) \in \mathbb{R}^{n_w}$ is the external disturbance signal of the system in $\ell_2[0, \infty)$; $z(k) \in \mathbb{R}^{n_z}$ is the controlled output variable; $E(h) \in \mathbb{R}^{n_x \times n_w}$, $L(h) \in \mathbb{R}^{n_z \times n_x}$, $D(h) \in \mathbb{R}^{n_z \times n_u}$, $H(h) \in \mathbb{R}^{n_z \times n_w}$ and $F(h) \in \mathbb{R}^{n_y \times n_w}$ are all known system matrices with appropriate dimensions.

Using the same steps, the corresponding closed-loop control system can be obtained as

$$\begin{aligned} x(k+1) &= \mathbb{A}(h)x(k) + \mathbb{B}(h)w(k) + \mathbb{C}(h)\vartheta_y(k) + \mathbb{D}(h)\vartheta_u(k), \\ z(k) &= \mathbb{J}(h)x(k) + \mathbb{F}(h)w(k) + \mathbb{G}(h)\vartheta_y(k) + \mathbb{H}(h)\vartheta_u(k), \end{aligned} \tag{4.43}$$

where

$$\begin{aligned} &\mathbb{A}(h) = A(h) + B(h)K(h)C(h), \ \mathbb{B}(h) = E(h) + B(h)K(h)F(h), \\ &\mathbb{C}(h) = B(h)K(h), \ \mathbb{D}(h) = B(h), \\ &\mathbb{J}(h) = L(h) + D(h)K(h)C(h), \ \mathbb{F}(h) = H(h) + D(h)K(h)F(h), \\ &\mathbb{G}(h) = D(h)K(h), \ \mathbb{H}(h) = D(h). \end{aligned} \tag{4.44}$$

Similarly, we first analyze the stability and $\mathscr{H}_\infty$ performance of the closed-loop control system to facilitate the subsequent design conditions.

Theorem 4.3 *For the T–S fuzzy system (4.43). Given the performance index $\gamma > 0$, quantization ranges M_y, M_u and quantization errors Δ_y, Δ_u. If there exist matrix $P > 0$, scalars σ_y, σ_u, ρ_y, and ρ_u satisfying the following matrix inequalities*

$$M_y > \frac{1}{\rho_y}, \quad M_u > \frac{1}{\rho_u}, \tag{4.45}$$

$$\begin{bmatrix} -P & * & * & * & * & * \\ 0 & -\gamma^2 I & * & * & * & * \\ 0 & 0 & -\sigma_y I & * & * & * \\ 0 & 0 & 0 & -\sigma_u I & * & * \\ \mathbb{A}(h) & \mathbb{B}(h) & \mathbb{C}(h) & \mathbb{D}(h) & -P^{-1} & * \\ \mathbb{J}(h) & \mathbb{F}(h) & \mathbb{G}(h) & \mathbb{H}(h) & 0 & -I \end{bmatrix} + \sigma_y(Ha\{\Psi_1(h)\}) + \sigma_u(Ha\{\Psi_2(h)\}) < 0, \tag{4.46}$$

where

$$\begin{aligned} \Psi_1(h) &= \begin{bmatrix} 2\rho_y\Delta_y C(h) & 2\rho_y\Delta_y F(h) & 0 & 0 \end{bmatrix}, \\ \Psi_2(h) &= \begin{bmatrix} \Psi_{21}(h) & \Psi_{22}(h) & \Psi_{23}(h) & 0 \end{bmatrix}, \\ \Psi_{21}(h) &= 2\rho_u\Delta_u K(h)C(h), \ \Psi_{22}(h) = 2\rho_u\Delta_u K(h)F(h), \ \Psi_{23}(h) = 2\rho_u\Delta_u K(h), \end{aligned}$$

and use the same online adjusting strategy for the parameters $\mu_y(k)$ and $\mu_u(k)$ in Theorem 4.1. Then the closed-loop control system is asymptotically stable and satisfies $\mathscr{H}_\infty$ control performance.

Proof Similar to the derivation of the quantization process in Theorem 4.1, by redefinition the vector $\phi(k) = [\, x^T(k) \ \ w^T(k) \ \ \vartheta_y^T(k) \ \ \vartheta_u^T(k)\,]^T$, the conditions (4.22) and (4.25) can be adjusted as the following form

$$\phi^T(k)\Lambda_1(h)\phi(k) \geq 0, \tag{4.47}$$

$$\phi^T(k)\Lambda_2(h)\phi(k) \geq 0, \tag{4.48}$$

where

$$\begin{aligned} \Lambda_1(h) &= Ha\left\{\begin{bmatrix} 2\rho_y\Delta_y C(h) & 2\rho_y\Delta_y F(h) & 0 & 0 \end{bmatrix}\right\} - diag\{0,\ 0,\ I,\ 0\}, \\ \Lambda_2(h) &= Ha\left\{\begin{bmatrix} 2\rho_u\Delta_u K(h)C(h) & 2\rho_u\Delta_u K(h)F(h) & 2\rho_u\Delta_u K(h) & 0 \end{bmatrix}\right\} \\ &\quad - diag\{0,\ 0,\ 0,\ I\}. \end{aligned}$$

Taking the same Lyapunov function $V(k, (x(k)))$ and the resulting equation (4.27), one has

$$\begin{aligned}
&\Delta V(k, (x(k))) + z^T(k)z(k) - \gamma^2 w^T(k)w(k) \\
=&V(k+1, x(k+1)) - V(k, x(k)) + z^T(k)z(k) - \gamma^2 w^T(k)w(k) \\
=&\left[\mathbb{A}(h)x(k) + \mathbb{B}(h)w(k) + \mathbb{C}(h)\vartheta_y(k) + \mathbb{D}(h)\vartheta_u(k)\right]^T P \\
&\times \left[\mathbb{A}(h)x(k) + \mathbb{B}(h)w(k) + \mathbb{C}(h)\vartheta_y(k) + \mathbb{D}(h)\vartheta_u(k)\right] \\
&+ \left[\mathbb{J}(h)x(k) + \mathbb{F}(h)w(k) + \mathbb{G}(h)\vartheta_y(k) + \mathbb{H}(h)\vartheta_u(k)\right]^T \\
&\times \left[\mathbb{J}(h)x(k) + \mathbb{F}(h)w(k) + \mathbb{G}(h)\vartheta_y(k) + \mathbb{H}(h)\vartheta_u(k)\right] \\
&- x^T(k)Px(k) - \gamma^2 w^T(k)w(k) \\
=&\phi^T(k)\Lambda_0(h)\phi(k),
\end{aligned} \tag{4.49}$$

where

$$\begin{aligned}
\Lambda_0(h) =& \left[\mathbb{A}(h) \quad \mathbb{B}(h) \quad \mathbb{C}(h) \quad \mathbb{D}(h)\right]^T P \left[\mathbb{A}(h) \quad \mathbb{B}(h) \quad \mathbb{C}(h) \quad \mathbb{D}(h)\right] \\
&+ \left[\mathbb{J}(h) \quad \mathbb{F}(h) \quad \mathbb{G}(h) \quad \mathbb{H}(h)\right]^T \left[\mathbb{J}(h) \quad \mathbb{F}(h) \quad \mathbb{G}(h) \quad \mathbb{H}(h)\right] \\
&- diag\{P, \ \gamma^2 I, \ 0, \ 0\}.
\end{aligned}$$

Using Lemma 1.3 with (4.47) and (4.48), we get the following inequality

$$\Lambda_0(h) + \sigma_y \Lambda_1(h) + \sigma_u \Lambda_2(h) < 0. \tag{4.50}$$

From the result that $\Lambda_0(h) < 0$ and taking the sum from 0 to ∞ of (4.49), the following expression can be get

$$V(\infty, (x(\infty))) - V(0, (x(0))) + \sum_{k=0}^{\infty} z^T(k)z(k) - \gamma^2 \sum_{k=0}^{\infty} w^T(k)w(k) < 0. \tag{4.51}$$

Due to the fact that $V(k, (x(k))) \geq 0$ and $V(0, x(0)) = 0$, the following inequality is obtained under zero initial conditions

$$\sum_{k=0}^{\infty} z^T(k)z(k) \leq \gamma^2 \sum_{k=0}^{\infty} w^T(k)w(k), \tag{4.52}$$

which demonstrates that the designed performance conditions can ensure that the closed-loop control system is asymptotically stable and satisfies $\mathscr{H}_\infty$ control performance. The proof is completed.

Theorem 4.4 *For the T–S fuzzy system (4.43). Given the performance index $\gamma > 0$, quantization ranges M_y, M_u and quantization errors Δ_y, Δ_u. If there exist matrix $P > 0$, Q, W_l, α_i, and R, scalars β, η_y, and η_u satisfying the following matrix inequalities*

$$M_y > \eta_y, \quad M_u > \eta_u, \tag{4.53}$$

$$\Theta_{ii} < 0, \ \ i = 1, \ldots, r, \tag{4.54}$$

$$\Theta_{il} + \Theta_{li} < 0, \ \ i < l, \ \ i, l = 1, \ldots, r, \tag{4.55}$$

where

$$\Theta_{il} = \begin{bmatrix} -P & * & * & * & * & * & * & * & * \\ 0 & -\gamma^2 I & * & * & * & * & * & * & * \\ 0 & 0 & -\eta_y I & * & * & * & * & * & * \\ 0 & 0 & 0 & -\eta_u I & * & * & * & * & * \\ \Theta_{51il} & \Theta_{52il} & \alpha_i W_l & QB_i & \Theta_{55} & * & * & * & * \\ \Theta_{61il} & \Theta_{62il} & D_i W_l & D_i & 0 & -I & * & * & * \\ 2\Delta_y C_i & 2\Delta_y F_i & 0 & 0 & 0 & 0 & -\eta_y I & * & * \\ \bar{\Psi}_{21il} & \bar{\Psi}_{22il} & \bar{\Psi}_{23l} & 0 & 0 & 0 & 0 & -\eta_u I & * \\ W_l C_i & W_l F_i & W_l & 0 & \Theta_{95i} & \Theta_{96i} & 0 & 0 & -He\{\beta R\} \end{bmatrix},$$

$$\Theta_{51il} = QA_i + \alpha_i W_l C_i, \ \ \Theta_{52il} = QE_i + \alpha_i W_l F_i, \ \ \Theta_{55} = -Q - Q^T + P,$$
$$\Theta_{61il} = L_i + D_i W_l C_i, \ \ \Theta_{62il} = H_i + D_i W_l F_i,$$
$$\bar{\Psi}_{21il} = 2\Delta_u K_l C_i, \ \ \bar{\Psi}_{22il} = 2\Delta_u K_l F_i, \ \ \bar{\Psi}_{23l} = 2\Delta_u K_l,$$
$$\Theta_{95i} = \beta(QB_i - \alpha_i R)^T, \ \ \Theta_{96i} = \beta(D_i - D_i R)^T,$$

and use the same online adjusting strategy for the parameters $\mu_y(k)$ and $\mu_u(k)$ in Theorem 4.2. Then the closed-loop control system is asymptotically stable and satisfies $\mathcal{H}_\infty$ control performance.

In addition, the controller gain matrix can be determined

$$K_l = R^{-1} W_l, \ \ l = 1, \ldots, r. \tag{4.56}$$

Proof Similarly, define scalars $\eta_\iota = \sigma_\iota = \frac{1}{\rho_\iota}, \iota = y, u$ and adopt Lemma 1.1 to (4.46), it can be obtained that

$$\begin{bmatrix} -P & * & * & * & * & * & * & * \\ 0 & -\gamma^2 I & * & * & * & * & * & * \\ 0 & 0 & -\eta_y I & * & * & * & * & * \\ 0 & 0 & 0 & -\eta_u I & * & * & * & * \\ \mathbb{A}(h) & \mathbb{B}(h) & \mathbb{C}(h) & \mathbb{D}(h) & -P^{-1} & * & * & * \\ \mathbb{J}(h) & \mathbb{F}(h) & \mathbb{G}(h) & \mathbb{H}(h) & 0 & -I & * & * \\ 2\Delta_y C(h) & 2\Delta_y F(h) & 0 & 0 & 0 & 0 & -\eta_y I & * \\ \bar{\Psi}_{21}(h) & \bar{\Psi}_{22}(h) & \bar{\Psi}_{23}(h) & 0 & 0 & 0 & 0 & -\eta_u I \end{bmatrix} < 0, \tag{4.57}$$

where $\bar{\Psi}_{21}(h) = 2\Delta_u K(h)C(h)$, $\bar{\Psi}_{22}(h) = 2\Delta_u K(h)F(h)$, and $\bar{\Psi}_{23}(h) = 2\Delta_u K(h)$.

Performing congruence transformation with diag$\{I, \ I, \ I, \ I, \ Q, \ I, \ I, \ I\}$ to above inequality has

$$\begin{bmatrix} -P & * & * & * & * & * & * & * \\ 0 & -\gamma^2 I & * & * & * & * & * & * \\ 0 & 0 & -\eta_y I & * & * & * & * & * \\ 0 & 0 & 0 & -\eta_u I & * & * & * & * \\ Q\mathbb{A}(h) & Q\mathbb{B}(h) & Q\mathbb{C}(h) & Q\mathbb{D}(h) & -QP^{-1}Q^T & * & * & * \\ \mathbb{J}(h) & \mathbb{F}(h) & \mathbb{G}(h) & \mathbb{H}(h) & 0 & -I & * & * \\ 2\Delta_y C(h) & 2\Delta_y F(h) & 0 & 0 & 0 & 0 & -\eta_y I & * \\ \bar{\Psi}_{21}(h) & \bar{\Psi}_{22}(h) & \bar{\Psi}_{23}(h) & 0 & 0 & 0 & 0 & -\eta_u I \end{bmatrix} < 0. \tag{4.58}$$

Considering the fact that $-QP^{-1}Q^T \leq -Q - Q^T + P$, one can obtain

$$\begin{bmatrix} -P & * & * & * & * & * & * & * \\ 0 & -\gamma^2 I & * & * & * & * & * & * \\ 0 & 0 & -\eta_y I & * & * & * & * & * \\ 0 & 0 & 0 & -\eta_u I & * & * & * & * \\ Q\mathbb{A}(h) & Q\mathbb{B}(h) & Q\mathbb{C}(h) & Q\mathbb{D}(h) & -Q - Q^T + P & * & * & * \\ \mathbb{J}(h) & \mathbb{F}(h) & \mathbb{G}(h) & \mathbb{H}(h) & 0 & -I & * & * \\ 2\Delta_y C(h) & 2\Delta_y F(h) & 0 & 0 & 0 & 0 & -\eta_y I & * \\ \bar{\Psi}_{21}(h) & \bar{\Psi}_{22}(h) & \bar{\Psi}_{23}(h) & 0 & 0 & 0 & 0 & -\eta_u I \end{bmatrix} < 0. \tag{4.59}$$

The above inequality can be rewritten as follows

$$\begin{aligned} &\begin{bmatrix} -P & * & * & * & * & * & * & * \\ 0 & -\gamma^2 I & * & * & * & * & * & * \\ 0 & 0 & -\eta_y I & * & * & * & * & * \\ 0 & 0 & 0 & -\eta_u I & * & * & * & * \\ QA(h) & QE(h) & 0 & QB(h) & -Q - Q^T + P & * & * & * \\ L(h) & H(h) & 0 & D(h) & 0 & -I & * & * \\ 2\Delta_y C(h) & 2\Delta_y F(h) & 0 & 0 & 0 & 0 & -\eta_y I & * \\ \bar{\Psi}_{21}(h) & \bar{\Psi}_{22}(h) & \bar{\Psi}_{23}(h) & 0 & 0 & 0 & 0 & -\eta_u I \end{bmatrix} \\ &+ He\left\{\begin{bmatrix} 0 \\ 0 \\ 0 \\ 0 \\ QB(h) \\ D(h) \\ 0 \\ 0 \end{bmatrix} K(h)\begin{bmatrix} C(h) & F(h) & I & 0 & 0 & 0 & 0 & 0 \end{bmatrix}\right\} < 0. \end{aligned} \tag{4.60}$$

Define a nonsingular matrix R, which satisfies $K(h) = R^{-1}W(h)$. Then (4.60) can be rewritten as

$$\begin{bmatrix} -P & * & * & * & * & * & * & * \\ 0 & -\gamma^2 I & * & * & * & * & * & * \\ 0 & 0 & -\eta_y I & * & * & * & * & * \\ 0 & 0 & 0 & -\eta_u I & * & * & * & * \\ \Theta_{51}(h) & \Theta_{52}(h) & \alpha(h)W(h) & QB(h) & \Theta_{52} & * & * & * \\ \Theta_{61}(h) & \Theta_{62}(h) & D(h)W(h) & D(h) & 0 & -I & * & * \\ 2\Delta_y C(h) & 2\Delta_y F(h) & 0 & 0 & 0 & 0 & -\eta_y I & * \\ \bar{\Psi}_{21}(h) & \bar{\Psi}_{22}(h) & \bar{\Psi}_{23}(h) & 0 & 0 & 0 & 0 & -\eta_u I \end{bmatrix}$$
$$+ He\left\{\begin{bmatrix} 0 \\ 0 \\ 0 \\ 0 \\ QB(h)-\alpha(h)R \\ D(h)-D(h)R \\ 0 \\ 0 \end{bmatrix} R^{-1}W(h)\begin{bmatrix} C(h) & F(h) & I & 0 & 0 & 0 & 0 & 0 \end{bmatrix}\right\} < 0, \tag{4.61}$$

where

$$\Theta_{51}(h) = QA(h) + \alpha(h)W(h)C(h), \quad \Theta_{52} = QE(h) + \alpha(h)W(h)F(h),$$
$$\Theta_{61}(h) = L(h) + D(h)W(h)C(h), \quad \Theta_{62} = H(h) + D(h)W(h)F(h).$$

By Lemma 1.4, set $M_4 = R$, $\lambda = \beta$, and the following matrices

$$M_1 = \begin{bmatrix} -P & * & * & * & * & * & * & * \\ 0 & -\gamma^2 I & * & * & * & * & * & * \\ 0 & 0 & -\eta_y I & * & * & * & * & * \\ 0 & 0 & 0 & -\eta_u I & * & * & * & * \\ \Theta_{51}(h) & \Theta_{52}(h) & \alpha(h)W(h) & QB(h) & \Theta_{52} & * & * & * \\ \Theta_{61}(h) & \Theta_{62}(h) & D(h)W(h) & D(h) & 0 & -I & * & * \\ 2\Delta_y C(h) & 2\Delta_y F(h) & 0 & 0 & 0 & 0 & -\eta_y I & * \\ \bar{\Psi}_{21}(h) & \bar{\Psi}_{22}(h) & \bar{\Psi}_{23}(h) & 0 & 0 & 0 & 0 & -\eta_u I \end{bmatrix},$$

$$M_2 = \begin{bmatrix} 0 \\ 0 \\ 0 \\ 0 \\ QB(h)-\alpha(h)R \\ D(h)-D(h)R \\ 0 \\ 0 \end{bmatrix}, \quad M_3 = R^{-1}W(h)\begin{bmatrix} C(h) & F(h) & I & 0 & 0 & 0 & 0 & 0 \end{bmatrix}.$$

Then according to the T–S fuzzy model and membership function, inequality (4.61) can be rewritten as

$$\begin{aligned}&\sum_{i=1}^{r}\sum_{l=1}^{r}h_i(\zeta(k))h_l(\zeta(k))\Theta_{il}\\&=\sum_{i=1}^{r}h_i^2(\zeta(k))\Theta_{ii}+\sum_{i=1}^{r}\sum_{i<l}^{r}h_i(\zeta(k))h_l(\zeta(k))(\Theta_{il}+\Theta_{li})<0.\end{aligned}\tag{4.62}$$

Therefore, based on (4.3) and [1], the inequality (4.62) yields the feasibility conditions (4.54) and (4.55).

Remark 4.3 Different from Theorem 4.2, the system matrix B_i in the design condition of PDC controllers in Theorem 4.4 is not required to be full column rank. We assume that it can be divided into two cases. i.e., if B_i is of full column rank, the dimension adjustment matrix α_i can be chosen as $\alpha_i = B_i(B_i^T B_i)^{-1}$ or $\alpha_i = Y_i^T\left[I \in \mathbb{R}^{n_u\times n_u}; 0 \in \mathbb{R}^{(n_x-n_u)\times n_u}\right]$; if not, $\alpha_i = B_i$ or $\alpha_i = \left[I \in \mathbb{R}^{n_u\times n_u}; 0 \in \mathbb{R}^{(n_x-n_u)\times n_u}\right]$ for $n_x > n_u$.

4.1.3 Simulation Example

In order to verify the feasibility and effectiveness of the algorithm in the stability design of closed-loop control system with quantization and the controller design with quantization satisfying $\mathcal{H}_\infty$ performance, consider the following second-order discrete-time chaotic system [5, 6]

$$\begin{aligned}x_1(k+1)&=ax_1(k)-x_1^3(k)+x_2(k)+\bar{u}(k)+e_1w(k),\\x_2(k+1)&=bx_1(k)+e_2w(k),\\z(k)&=Lx(k)+D\bar{u}(k)+Hw(k),\\y(k)&=Cx(k)+Fw(k),\end{aligned}\tag{4.63}$$

where $x(k) = [x_1^T(k)\ \ x_2^T(k)]^T$; the parameters a and b are uncertain satisfying $a_0 - \delta_a \le a \le a_0 + \delta_a$ and $b_0 - \delta_b \le b \le b_0 + \delta_b$, in which the $a_0, b_0, \delta_a \ge 0$, and $\delta_b \ge 0$ are the nominal values of a and b; $\bar{u}(k)$ is the quantized signal value.

Suppose that $x_1(k) \in [-d, d]$ and $d > 0$ so that $0 \le x_1^2(k) \le d^2$. Then using the sector nonlinearity approach [7] in this example, two T–S fuzzy rules are selected to approximate the state-space equation and corresponding fuzzy system parameter matrices are

$$\begin{aligned}&A_1=\begin{bmatrix}a_0&1\\b_0&0\end{bmatrix},\quad A_2=\begin{bmatrix}a_0-d^2&1\\b_0&0\end{bmatrix},\quad B_1=B_2=\begin{bmatrix}1\\0\end{bmatrix},\\&E_1=E_2=\begin{bmatrix}0.2\\0.1\end{bmatrix},\quad L_1=L_2=\begin{bmatrix}-0.7&0.2\\0.1&1\end{bmatrix},\quad D_1=D_2=\begin{bmatrix}-0.2\\-0.1\end{bmatrix},\\&H_1=H_2=\begin{bmatrix}-0.3\\-0.3\end{bmatrix},\quad F_1=F_2=0.2,\quad C_1=C_2=\begin{bmatrix}-4&0.1\end{bmatrix}.\end{aligned}\tag{4.64}$$

Considering the range of the variable $x_1(k)$ and the fuzzy basis functions $h_1(\zeta_1(k))$ and $h_2(\zeta_1(k))$ corresponding to the fuzzy rules, and defining the premise variable $\zeta_1(k) = x_1(k)$, the fuzzy basis functions need to satisfy

$$x_1^2(k) = h_1(x_1(k)) \times 0 + h_2(x_1(k)) \times d^2,$$

$$h_1(x_1(k)) + h_2(x_1(k)) = 1.$$

After equality transformation, the mathematical expression of fuzzy basis functions can be obtained

$$\begin{aligned} h_1(x_1(k)) &= 1 - \frac{x_1^2(k)}{d^2}, \\ h_2(x_1(k)) &= 1 - h_1(x_1(k)). \end{aligned} \tag{4.65}$$

In this section, it is assumed that some parameters in the system are $a_0 = 1$, $b_0 = 0.3$, $\delta_a = 0.5$, $\delta_b = 0.1$, and $d = 1$, respectively.

Afterwards, select the quantization errors bounds $\Delta_y = \Delta_u = 0.01$, the quantizers ranges $M_y = M_u = 50$, and initial condition $x(0) = [\,0.5 \;\; 0\,]^T$. Then, LMI toolbox of MATLAB is used to solve the formulas (4.29), (4.30), and (4.31). When there is no external disturbance, i.e., $w(k) = 0$, the feasible matrices satisfying asymptotic stability can be obtained

$$N = 6.4448, \;\; U_1 = 1.5358, \;\; U_2 = -0.0111. \tag{4.66}$$

According to (4.33), the controller gain matrices gives

$$K_1 = 0.2383, \;\; K_2 = -0.0017. \tag{4.67}$$

In the design of $\mathcal{H}_\infty$ controller for the closed-loop control system, adopting the same quantization ranges and quantization errors and selecting the initial condition $x(0) = [\,0 \;\; 0\,]^T$ and the external interference signal $w(k) = 2\cos(2k)e^{-0.25k}$ with the performance index $\gamma = 1$, the LMIs (4.53), (4.54), and (4.55) in Theorem 4.4 can be substituted into the feasibility matrices satisfying the stability and $\mathcal{H}_\infty$ performance of the system are

$$R = 3.5217, \;\; W_1 = 0.8178, \;\; W_2 = 0.0832. \tag{4.68}$$

According to (4.56), the controller gain matrices are

$$K_1 = 0.2322, \;\; K_2 = 0.0236. \tag{4.69}$$

In the discrete-time output feedback closed-loop control system (4.13) with multi-signal quantization, the corresponding simulation results are shown in the Figs. 4.1, 4.2, and 4.3. When the external interference is zero, but is not in a stable state, i.e.,

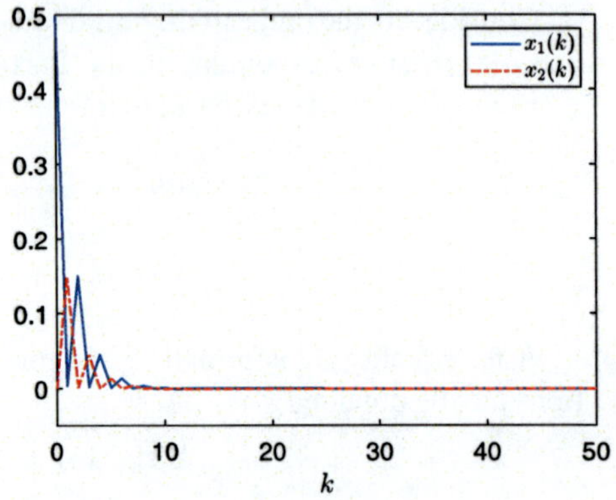

Fig. 4.1 Response of the system state $x(k)$ with $w(k) = 0$

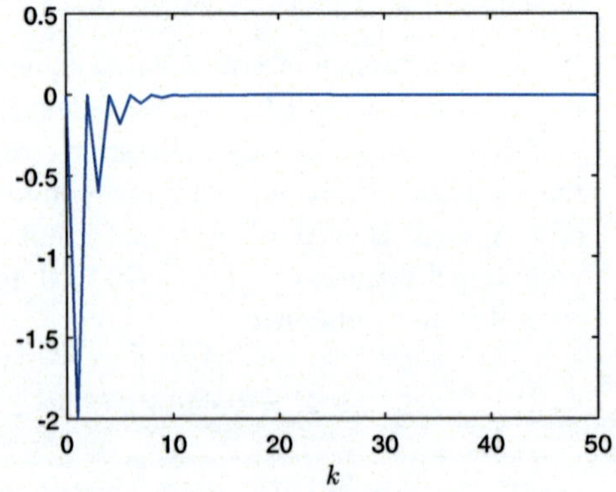

Fig. 4.2 Response of the measurement output $y(k)$ with $w(k) = 0$

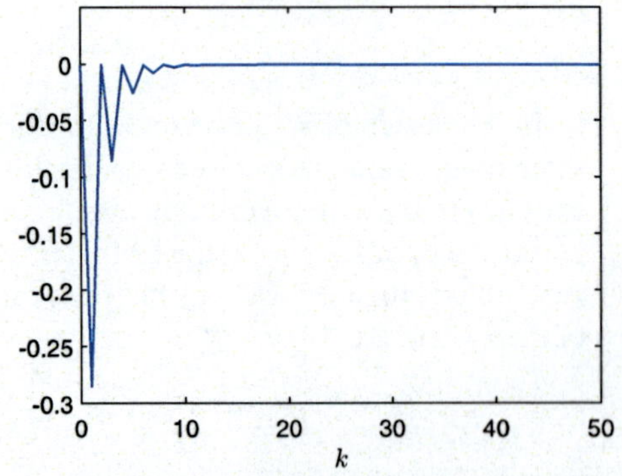

Fig. 4.3 Response of the control input $u(k)$ with $w(k) = 0$

$x(0) = [0.5\ \ 0]^T$, under the control of the controller Fig. 4.3, the system can quickly return to the stable state (Fig. 4.1).

For the closed-loop control system (4.43), when there has external interference, the corresponding simulation results are shown in the Figs. 4.4, 4.5, 4.6, 4.7, and 4.8. In the zero initial state, the system state, output and measurement output can return to the stable state (Figs. 4.4, 4.5, 4.6, and 4.7) and obtain the ratio of $\mathscr{H}_\infty$ performance

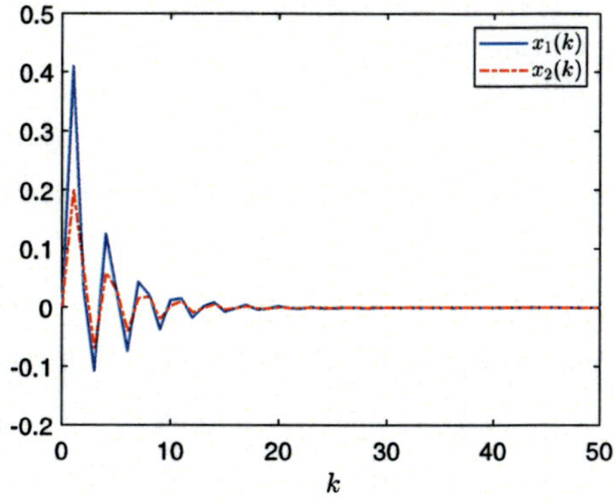

Fig. 4.4 Response of $x(k)$ with the quantized $\mathscr{H}_\infty$ controller

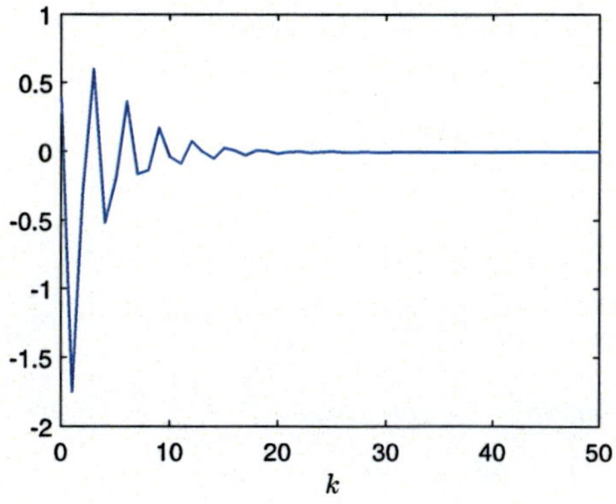

Fig. 4.5 Response of $y(k)$ with the quantized $\mathscr{H}_\infty$ controller

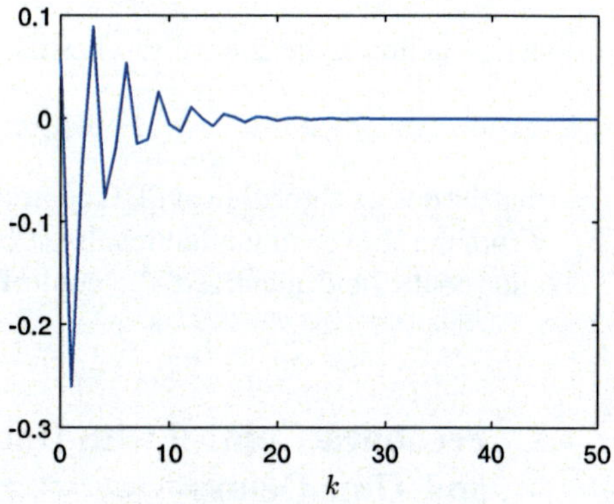

Fig. 4.6 Response of $u(k)$ with the quantized $\mathscr{H}_\infty$ controller

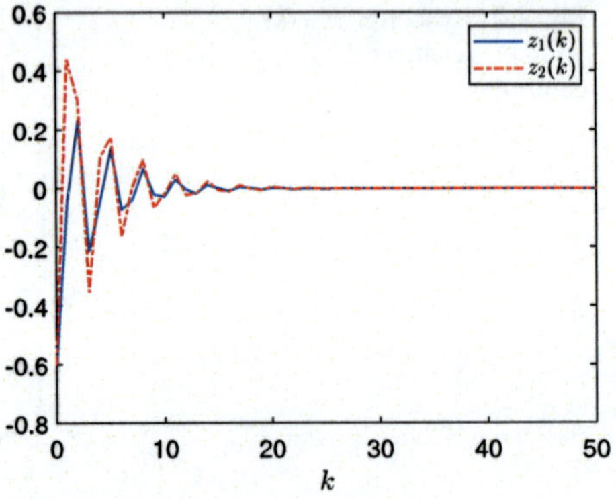

Fig. 4.7 Response of $z(k)$ with the quantized $\mathscr{H}_\infty$ controller

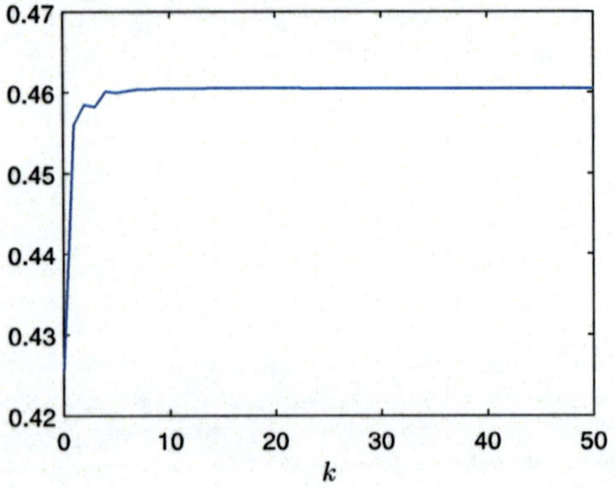

Fig. 4.8 History of $\sqrt{\sum_{\phi=0}^{k} z^T(\phi)z(\phi)/\sum_{\phi=0}^{k} w^T(\phi)w(\phi)}$ with the quantized $\mathscr{H}_\infty$ controller

under the action of the controller when the system is affected by external interference. The convergence value of $\sqrt{\sum_{\phi=0}^{k} z^T(\phi)z(\phi)/\sum_{\phi=0}^{k} w^T(\phi)w(\phi)}$ is $0.4606 < 1$ (Fig. 4.8), so that the design algorithm of PDC controller satisfying $\mathscr{H}_\infty$ performance is feasible.

From the above simulation results, it can be shown that both the quantized stabilization control and quantized $\mathscr{H}_\infty$ control design approaches developed are effective.

4.2 Feedback Control with Data Packet Dropouts and Time Delays

This section addresses the output feedback control problem for the T–S fuzzy system subject to time delays and data dropouts. An input delay method is proposed to describe the control input signal affected by packet losses and time delays, and the output feedback controller design strategy is derived in the form of LMIs to ensure that the closed-loop system is asymptotically stable and meets $\mathscr{H}_\infty$ performance.

4.2.1 Problem Formulation

The following fuzzy system is considered:

Plant Rule i_{th}: **IF** $f_1(t)$ is M_{1i} and ... $f_o(t)$ is M_{oi}, **THEN**

$$\begin{aligned} \dot{x}(t) &= A_i x(t) + B_i u(t) + E_i w(t), \\ y(t) &= C_1 x(t), \\ z(t) &= C_2 x(t), \end{aligned} \tag{4.70}$$

where $M_{ji}, j = 1, 2, ..., o, i = 1, \ldots, r,$ are the fuzzy sets; $f(t) = [f_1(t), f_2(t), \ldots, f_o(t)]$ is the vector of premise variables; $x(t) \in \mathbb{R}^{n_x}$ and $u(t) \in \mathbb{R}^{n_u}$ stand for the state variable and the control input, respectively; $w(t) \in \mathbb{R}^{n_w}$ refers to the noise input, which belongs to $\mathcal{L}_2[0, \infty)$; $y(t) \in \mathbb{R}^{n_y}$ and $z(t) \in \mathbb{R}^{n_z}$ represent the measurement output and the controlled output; $A_i \in \mathbb{R}^{n_x \times n_x}$, $B_i \in \mathbb{R}^{n_x \times n_u}$, $E_i \in \mathbb{R}^{n_x \times n_w}$, $C_1 \in \mathbb{R}^{n_y \times n_x}$, and $C_2 \in \mathbb{R}^{n_z \times n_x}$ are given system matrices. It is assumed that fuzzy basis function is given by

$$\varpi_i(f(t)) = \frac{\prod_{j=1}^{o} M_{ji}\left(f_j(t)\right)}{\sum_{i=1}^{r} \prod_{j=1}^{o} M_{ji}\left(f_j(t)\right)}, \tag{4.71}$$

where $M_{ji}\left(f_j(t)\right)$ is the grade of membership function of $f_j(t)$ in M_{ji}.

Due to the nature of the member function, it needs to satisfy

$$\varpi_i(f(t)) \geq 0, \quad \sum_{i=1}^{r} \varpi_i(f(t)) = 1, \quad i = 1, \ldots, r. \tag{4.72}$$

Then, the following fuzzy system can be expressed as:

$$\begin{aligned} \dot{x}(t) &= A(\varpi)x(t) + B(\varpi)u(t) + E(\varpi)w(t), \\ y(t) &= C_1 x(t), \\ z(t) &= C_2 x(t), \end{aligned} \tag{4.73}$$

where

$$A(\varpi) = \sum_{i=1}^{r} \varpi_i(f(t))A_i, \quad B(\varpi) = \sum_{i=1}^{r} \varpi_i(f(t))B_i, \quad E(\varpi) = \sum_{i=1}^{r} \varpi_i(f(t))E_i.$$

In the fuzzy system (4.73), the system stability is improved by considering a static output feedback controller. Hence, the considered controller can be represented by the following equation:

Plant Rule i_{th}: **IF** $f_1(t)$ is M_{1i} and ... $f_o(t)$ is M_{oi}, **THEN**

$$u(t) = K_i \hat{y}(t_k), \quad i = 1, \ldots, r. \tag{4.74}$$

The following terse form is given to represent the fuzzy control

$$u(t) = K(\varpi)\hat{y}(t_k), \tag{4.75}$$

where

$$K(\varpi) = \sum_{i=1}^{r} \varpi_i(f(t))K_i.$$

In practical applications, signals transmitted through data channels will be affected by a series of practical factors, such as unreliable communication lines and bandwidth delays, which may lead to data dropouts and transmission delays between sensors, controllers, and actuators. In order to fully consider and reduce the impact of this phenomenon, we assume that t_k is the sampling instant and $\bar{h}$ is the sampling time. The transmission delays in communication channel at each sampling period is denoted as d_k with $d_k = d_{sc} + d_{ca}$, where d_{sc} is the transmission delays between measurement output and controller and d_{ca} denotes the transmission delays between controller and actuator, respectively. Now, the packets transmitted to the controller are represented as:

$$\hat{y}(t_k) = y(t_k - p_{ck}\bar{h}), \tag{4.76}$$

where $p_{ck} = p_{1k} + p_{2k}$ represents the number of data dropouts at time t_k. Moreover, p_{1k} and p_{2k} denote the data dropouts in sensor-to-controller and controller-to-actuator, respectively. In this way, the output signal including transmission delays and data dropouts is given as:

$$\hat{y}(t_k) = y(t_k - p_{ck}\bar{h} - d_k). \tag{4.77}$$

Based on the input delay method, define a new variable $d(t) = t - (t_k - p_{ck}\bar{h} - d_k)$. Then the controller can be described as:

$$u(t) = K(\varpi)y(t - d(t)), \tag{4.78}$$

where $d(t)$ satisfies $0 < d(t) < d_{max}$ and d_{max} is the upper bound of the delay.

Combining (4.73) and the control strategy (4.78), the closed-loop system is described as follows:

$$\begin{aligned}\dot{x}(t) &= \sum_{i=1}^{r}\sum_{l=1}^{r}\varpi_i(f(t))\varpi_l(f(t))\Big\{A_i x(t) + B_i K_l C_1 x(t-d(t)) + E_i w(t)\Big\} \\ &= A(\varpi)x(t) + B(\varpi)K(\varpi)C_1 x(t-d(t)) + E(\varpi)w(t), \\ y(t) &= C_1 x(t), \\ z(t) &= C_2 x(t).\end{aligned} \tag{4.79}$$

Next, we can formulate the $\mathcal{H}_\infty$ feedback controller design problem:

$\mathcal{H}_\infty$ feedback controller design problem. The purpose of this section is to design a static output feedback controller given in (4.77) with transmission delays and data dropouts such that the closed-loop system (4.79) satisfies the following conditions:

1) When $w(t) = 0$, the closed-loop system (4.79) is asymptotically stable;

2) Under zero initial conditions, for $w(t) \neq 0$ and a given $\gamma > 0$, the controlled output $z(t)$ satisfies

$$\int_0^\infty z^T(t)z(t)dt < \gamma^2 \int_0^\infty w^T(t)w(t)dt. \tag{4.80}$$

4.2.2 Main Results

In this subsection, the output feedback control design problem for the T–S fuzzy system (4.79) with time delays and data packet dropouts will be solved. To begin, we will provide a meaningful stability analysis criterion to ensure that the closed-loop system (4.79) is asymptotically stable with $\mathcal{H}_\infty$ performance level. It is presented by the following theorem.

Theorem 4.5 *For given scalar $\gamma > 0$, the closed-loop system (4.79) is asymptotically stable with $\mathcal{H}_\infty$ performance level γ, if there exist positive definite matrices P, Q, and R such that the following matrix inequality holds:*

$$\begin{bmatrix}\Omega_{11}(\varpi) & * \\ \Omega_{21}(\varpi) & \Omega_{22}\end{bmatrix} + d_{max}\Phi^T(\varpi)R\Phi(\varpi) < 0, \tag{4.81}$$

where

$$\begin{aligned}\Omega_{11}(\varpi) &= He\{P(A(\varpi) + B(\varpi)K(\varpi)C_1)\} + Q + C_2^T C_2, \\ \Omega_{21}(\varpi) &= \begin{bmatrix}PB(\varpi)K(\varpi)C_1 & 0 & PE(\varpi)\end{bmatrix}^T, \\ \Phi(\varpi) &= \begin{bmatrix}A(\varpi) & B(\varpi)K(\varpi)C_1 & 0 & E(\varpi)\end{bmatrix}, \\ \Omega_{22} &= diag\{-d_{max}^{-1}R, \ -Q, \ -\gamma^2 I\}.\end{aligned}$$

Proof Select the following Lyapunov function as

$$V(x(t)) = x^T(t)Px(t) + \int_{t-d_{max}}^{t} x^T(s)Qx(s)ds + \int_{-d_{max}}^{0}\int_{t+\theta}^{t} \dot{x}^T(s)R\dot{x}(s)dsd\theta. \tag{4.82}$$

Then, one can obtain

$$\begin{aligned}\dot{V}(x(t)) = &\ \dot{x}^T(t)Px(t) + x^T(t)P\dot{x}(t) + d_{max}\dot{x}^T(t)R\dot{x}(t)\\ &-x^T(t-d_{max})Qx(t-d_{max}) + x^T(t)Qx(t)\\ &-\int_{t-d_{max}}^{t} \dot{x}^T(s)R\dot{x}(s)ds.\end{aligned} \tag{4.83}$$

Consider the fact that

$$x(t) - x(t-d(t)) - \int_{t-d(t)}^{t} \dot{x}(s)ds = 0. \tag{4.84}$$

Then according to the closed-loop system (4.79), one has

$$\begin{aligned}\dot{x}(t) = &\ (A(\varpi) + B(\varpi)K(\varpi)C_1)x(t)\\ &+E(\varpi)w(t) - B(\varpi)K(\varpi)C_1\int_{t-d(t)}^{t} \dot{x}(s)ds.\end{aligned} \tag{4.85}$$

The following equation can be obtained by calculation

$$\begin{aligned}&\dot{x}^T(t)Px(t) + x^T(t)P\dot{x}(t)\\ &= \Big((A(\varpi) + B(\varpi)K(\varpi)C_1)x(t) + E(\varpi)w(t) - B(\varpi)K(\varpi)C_1\int_{t-d(t)}^{t} \dot{x}(s)ds\Big)^T Px(t)\\ &\quad +x^T(t)P\Big((A(\varpi) + B(\varpi)K(\varpi)C_1)x(t) + E(\varpi)w(t) - B(\varpi)K(\varpi)C_1\int_{t-d(t)}^{t} \dot{x}(s)ds\Big)\\ &= x^T(t)\Big(P(A(\varpi) + B(\varpi)K(\varpi)C_1) + (A(\varpi) + B(\varpi)K(\varpi)C_1)^T P\Big)x(t) + x^T(t)P\\ &\quad \times E(\varpi)w(t) + (x^T(t)PE(\varpi)w(t))^T - 2x^T(t)PB(\varpi)K(\varpi)C_1\int_{t-d(t)}^{t} \dot{x}(s)ds.\end{aligned} \tag{4.86}$$

By using Lemma 1.5, one can obtain

$$-2x^T(t)PB(\varpi)K(\varpi)C_1\int_{t-d(t)}^{t}\dot{x}(s)ds \le d_{max}x^T(t)PB(\varpi)K(\varpi)C_1R^{-1}\big(PB(\varpi)K(\varpi)C_1\big)^Tx(t) + \int_{t-d_{max}}^{t}\dot{x}^T(s)R\dot{x}(s)ds. \quad (4.87)$$

According to (4.86) and (4.87), (4.83) can be rewritten as

$$\begin{aligned}\dot{V}(x(t)) \le\; & x^T(t)\phi(\varpi)x(t) - x^T(t-d_{max})Qx(t-d_{max}) + d_{max}\dot{x}^T(t)R\dot{x}(t) \\ & + x^T(t)PE(\varpi)w(t) + \big(x^T(t)PE(\varpi)w(t)\big)^T,\end{aligned} \quad (4.88)$$

where $\phi(\varpi) = P\big(A(\varpi)+B(\varpi)K(\varpi)C_1\big) + \Big(P\big(A(\varpi)+B(\varpi)K(\varpi)C_1\big)\Big)^T + Q + d_{max}\big(PB(\varpi)K(\varpi)C_1\big)R^{-1}\Big(PB(\varpi)K(\varpi)C_1\Big)^T$.

Define

$$\zeta_1^T(t) = \big[x^T(t)\;\; x^T(t-d(t))\;\; x^T(t-d_{max})\big].$$

Then one has

$$\begin{aligned}d_{max}\dot{x}^T(t)R\dot{x}(t) = d_{max}\zeta_1^T(t)&\big[A(\varpi)\quad B(\varpi)K(\varpi)C_1\quad 0\big]^T R \\ &\times\big[A(\varpi)\quad B(\varpi)K(\varpi)C_1\quad 0\big]\zeta_1(t).\end{aligned} \quad (4.89)$$

According to (4.81), it's easy to know that

$$\dot{V}(x(t)) \le 0, \quad (4.90)$$

which means that the system (4.79) is asymptotically stable for $w(t) = 0$.

When $w(t) \ne 0$, defining $\zeta^T(t) = [x^T(t)\;\; x^T(t-d(t))\;\; x^T(t-d_{max})\;\; w^T(t)]$, one has

$$\begin{aligned}d_{max}\dot{x}^T(t)R\dot{x}(t) = d_{max}\zeta^T(t)&\big[A(\varpi)\quad B(\varpi)K(\varpi)C_1\quad 0\quad E(\varpi)\big]^T R \\ &\times\big[A(\varpi)\quad B(\varpi)K(\varpi)C_1\quad 0\quad E(\varpi)\big]\zeta(t).\end{aligned} \quad (4.91)$$

Once the condition in (4.81) can be guaranteed, based on (4.83), (4.91), Lemma 1.1, and congruence transformation, the following inequality can be easily obtained

$$\dot{V}(x(t)) + z^T(t)z(t) - \gamma^2 w^T(t)w(t) < 0, \quad (4.92)$$

which implies that

$$V(\infty) - V(0) + \int_0^\infty z^T(t)z(t)dt - \gamma^2 \int_0^\infty w^T(t)w(t)dt < 0. \tag{4.93}$$

Since zero initial condition $x(0) = 0$ and $V(\infty) > 0$, the following inequality can be obtained

$$\int_0^\infty z^T(t)z(t)dt < \gamma^2 \int_0^\infty w^T(t)w(t)dt, \tag{4.94}$$

which means that the system (4.79) is asymptotically stable with $w(t) = 0$ and satisfies the prescribed $\mathcal{H}_\infty$ performance index γ with $w(t) \neq 0$. The proof is completed.

Theorem 4.6 *For given scalars $\epsilon, \gamma > 0$, the T–S fuzzy system (4.79) is asymptotically stable with $\mathcal{H}_\infty$ performance level γ, if there exist positive definite matrices P, Q, R, matrices U, G_l such that the following matrix inequalities hold:*

$$\Psi_{ii} < 0, \quad i = 1, 2, \cdots, r, \tag{4.95}$$

$$\Psi_{il} + \Psi_{li} < 0, \quad i < l = 1, 2, \cdots, r, \tag{4.96}$$

where

$$\Psi_{il} = \begin{bmatrix} \Upsilon_{11} & * & * & * & * & * & * & * \\ (B_i G_l C_1)^T & -d_{max}^{-1} R & * & * & * & * & * & * \\ 0 & 0 & -Q & * & * & * & * & * \\ (PE_i)^T & 0 & 0 & -\gamma^2 I & * & * & * & * \\ PA_i & B_i G_l C_1 & 0 & PE_i & -d_{max}^{-1}(2P - R) & * & * & * \\ C_2 & 0 & 0 & 0 & 0 & -I & * & * \\ \Upsilon_{71} & G_l C_1 & 0 & 0 & 0 & 0 & \Upsilon_{77} & * \\ 0 & G_l C_1 & 0 & 0 & \Upsilon_{85} & 0 & 0 & \Upsilon_{88} \end{bmatrix},$$

$$\begin{aligned} \Upsilon_{11} &= PA_i + B_i G_l C_1 + \left(PA_i + B_i G_l C_1\right)^T + Q, \\ \Upsilon_{71} &= \epsilon B_i^T P^T - \epsilon U^T B_i^T + G_l C_1, \\ \Upsilon_{77} &= -\epsilon U - \epsilon U^T, \\ \Upsilon_{85} &= \epsilon B_i^T P^T - \epsilon U^T B_i^T, \\ \Upsilon_{88} &= -\epsilon U - \epsilon U^T. \end{aligned}$$

Moreover, the controller gain can be obtained by $K_l = U^{-1} G_l$.

Proof It should be emphasized that the condition in Theorem 4.5 cannot be employed directly for controller design due to many couple terms. So, in the following, these terms from Theorem 4.5 will be removed, and the condition to design controller will be given in the form of LMIs. One can use Schur complement (Lemma 1.1) to Theorem 4.5, which is equivalent to the following inequality

$$\begin{bmatrix} \hat{\Omega}_{11}+Q & * & * & * & * & * \\ (PB(\varpi)K(\varpi)C_1)^T & -d_{max}^{-1}R & * & * & * & * \\ 0 & 0 & -Q & * & * & * \\ (PE(\varpi))^T & 0 & 0 & -\gamma^2 I & * & * \\ A(\varpi) & B(\varpi)K(\varpi)C_1 & 0 & E(\varpi) & -d_{max}^{-1}R^{-1} & * \\ C_2 & 0 & 0 & 0 & 0 & -I \end{bmatrix} < 0, \tag{4.97}$$

where $\hat{\Omega}_{11}(\varpi) = P(A(\varpi) + B(\varpi)K(\varpi)C_1) + \big(P\big(A(\varpi) + B(\varpi)K(\varpi)C_1\big)\big)^T$.

Perform congruence transformation with $\mathfrak{P} = diag\{I,\ I,\ I,\ I,\ P,\ I\}$ to (4.97). Then, due to the fact that $-d_{max}^{-1}PR^{-1}P \leq -d_{max}^{-1}(2P - R)$, one obtains

$$\begin{bmatrix} \hat{\Omega}_{11}+Q & * & * & * & * & * \\ (PB(\varpi)K(\varpi)C_1)^T & -d_{max}^{-1}R & * & * & * & * \\ 0 & 0 & -Q & * & * & * \\ (PE(\varpi))^T & 0 & 0 & -\gamma^2 I & * & * \\ PA(\varpi) & PB(\varpi)K(\varpi)C_1 & 0 & PE(\varpi) & -d_{max}^{-1}(2P-R) & * \\ C_2 & 0 & 0 & 0 & 0 & -I \end{bmatrix} < 0. \tag{4.98}$$

It's worth noting that there is a coupling term $PB(\varpi)K(\varpi)C_1$ in (4.98), which needs to be solved. Now, define $K(\varpi) = U^{-1}G(\varpi)$ with $G(\varpi) = \sum_{l=1}^{r} \varpi_l(f(t))G_l$. Then (4.98) can be rewritten as follows

$$\begin{bmatrix} \tilde{\Omega}_{11}(\varpi) & * & * & * & * & * \\ (PB(\varpi)U^{-1}G(\varpi)C_1)^T & -d_{max}^{-1}R & * & * & * & * \\ 0 & 0 & -Q & * & * & * \\ (PE(\varpi))^T & 0 & 0 & -\gamma^2 I & * & * \\ PA(\varpi) & PB(\varpi)U^{-1}G(\varpi)C_1 & 0 & PE(\varpi) & -d_{max}^{-1}(2P-R) & * \\ C_2 & 0 & 0 & 0 & 0 & -I \end{bmatrix} < 0, \tag{4.99}$$

where

$$\begin{aligned}\tilde{\Omega}_{11}(\varpi) = {} & P(A(\varpi) + B(\varpi)U^{-1}G(\varpi)C_1) \\ & + \big(P\big(A(\varpi) + B(\varpi)U^{-1}G(\varpi)C_1\big)\big)^T + Q.\end{aligned}$$

The following inequality can be obtained by matrix decomposition on (4.99)

$$
\begin{bmatrix}
\bar{\Omega}_{11}(\varpi) & * & * & * & * & * \\
(B(\varpi)G(\varpi)C_1)^T & -d_{max}^{-1}R & * & * & * & * \\
0 & 0 & -Q & * & * & * \\
(PE(\varpi))^T & 0 & 0 & -\gamma^2 I & * & * \\
PA(\varpi) & B(\varpi)G(\varpi)C_1 & 0 & PE(\varpi) & -d_{max}^{-1}(2P-R) & * \\
C_2 & 0 & 0 & 0 & 0 & -I
\end{bmatrix}
+ \begin{bmatrix}
C_1^T & 0 \\
C_1^T & C_1^T \\
0 & 0 \\
0 & 0 \\
0 & 0 \\
0 & 0
\end{bmatrix}
\times \begin{bmatrix}
U^{-1}G(\varpi) & 0 \\
0 & U^{-1}G(\varpi)
\end{bmatrix}^T
\times \begin{bmatrix}
PB(\varpi) - B(\varpi)U & 0 \\
0 & 0 \\
0 & 0 \\
0 & 0 \\
0 & PB(\varpi) - B(\varpi)U \\
0 & 0
\end{bmatrix}^T
+ \begin{bmatrix}
PB(\varpi) - B(\varpi)U & 0 \\
0 & 0 \\
0 & 0 \\
0 & 0 \\
0 & PB(\varpi) - B(\varpi)U \\
0 & 0
\end{bmatrix}
\times \begin{bmatrix}
U^{-1}G(\varpi) & 0 \\
0 & U^{-1}G(\varpi)
\end{bmatrix}
\times \begin{bmatrix}
C_1^T & 0 \\
C_1^T & C_1^T \\
0 & 0 \\
0 & 0 \\
0 & 0 \\
0 & 0
\end{bmatrix}^T < 0,
$$

where $\bar{\Omega}_{11}(\varpi) = PA(\varpi) + B(\varpi)G(\varpi)C_1 + \left(PA(\varpi) + B(\varpi)G(\varpi)C_1\right)^T + Q$.

To make it easier to deal with the previous inequality, let's define the following equation

$$\mathbb{Z} = \begin{bmatrix} PB(\varpi) - B(\varpi)U & 0 \\ 0 & 0 \\ 0 & 0 \\ 0 & 0 \\ 0 & PB(\varpi) - B(\varpi)U \\ 0 & 0 \end{bmatrix}, \quad \mathbb{U} = \begin{bmatrix} U & 0 \\ 0 & U \end{bmatrix},$$
$$\mathbb{Y} = \begin{bmatrix} U^{-1}G(\varpi) & 0 \\ 0 & U^{-1}G(\varpi) \end{bmatrix} \begin{bmatrix} C_1 & C_1 & 0 & 0 & 0 & 0 \\ 0 & C_1 & 0 & 0 & 0 & 0 \end{bmatrix}. \tag{4.100}$$

Then by using Lemma 1.4, it can be obtained that

$$\sum_{i=1}^{r}\sum_{l=1}^{r} \varpi_i(f(t))\varpi_l(f(t))\Xi_{il} < 0, \tag{4.101}$$

where

$$\Xi_{il} = \begin{bmatrix} \Xi_{11} & * & * & * & * & * & * & * \\ (B_iG_lC_1)^T & -d_{max}^{-1}R & * & * & * & * & * & * \\ 0 & 0 & -Q & * & * & * & * & * \\ (PE_i)^T & 0 & 0 & -\gamma^2 I & * & * & * & * \\ PA_i & B_iG_lC_1 & 0 & PE_i & -d_{max}^{-1}(2P-R) & * & * & * \\ C_2 & 0 & 0 & 0 & 0 & -I & * & * \\ \Xi_{71} & G_lC_1 & 0 & 0 & 0 & 0 & \Xi_{77} & * \\ 0 & G_lC_1 & 0 & 0 & \Xi_{85} & 0 & 0 & \Xi_{88} \end{bmatrix},$$
$$\Xi_{11} = PA_i + B_iG_lC_1 + (PA_i + B_iG_lC_1)^T + Q,$$
$$\Xi_{71} = \epsilon B_i^T P^T - \epsilon U^T B_i^T + G_lC_1,$$
$$\Xi_{77} = -\epsilon U - \epsilon U^T,$$
$$\Xi_{85} = \epsilon B_i^T P^T - \epsilon U^T B_i^T,$$
$$\Xi_{88} = -\epsilon U - \epsilon U^T.$$

Further, the following inequality can be obtained

$$\sum_{i=1}^{r} \varpi_i^2(f(t))\Xi_{ii} + \sum_{i=1}^{r}\sum_{i<l}^{r} \varpi_i(f(t))\varpi_l(f(t))\Xi_{il} + \sum_{i=1}^{r}\sum_{i<l}^{r} \varpi_i(f(t))\varpi_l(f(t))\Xi_{li} < 0. \tag{4.102}$$

If LMIs (4.95) and (4.96) are satisfied, then the inequality (4.102) holds.

Remark 4.4 For Theorem 4.6, in order to obtain LMI-based design condition, we split K_l into the form of $K_l = U^{-1}G_l$ to decouple the term $PB_iK_lC_1$. In fact, the controller gain K_l can also be define as $K_l = U^{-1}G_l$ for further reducing design conservativeness.

Remark 4.5 The proposed conditions in Theorem 4.6 are strict LMIs. Therefore, the controller gain matrices can be obtained by solving (4.95) and (4.96) in MATLAB

LMI toolbox. In order to achieve the required performance as much as possible, the following optimization problem is utilized:

min γ subject to LMIs (4.95) and (4.96).

Then the optimal value of $\mathcal{H}_\infty$ performance level γ can be obtained.

4.2.3 Simulation Example

In this subsection, in order to verify the effectiveness of the feedback controller design, a mass-spring-damping system is introduced whose the dynamics equation is defined as [8]

$$m\ddot{\theta}(t) + c\dot{\theta}(t) + k\left(1 + a^2\theta^2(t)\right)\theta(t) = u(t), \tag{4.103}$$

where $\theta(t)$ stands for the displacement from a reference point; m stands for the mass; $u(t)$ denotes the external control input; k, c and a are constants. Let $m = 1, k = 8, c = 2, a = 0.3$, and $\theta(t) \in [-2, 2]$. Then, defining $\boldsymbol{x}(t) = \begin{bmatrix} \theta(t) \\ \dot{\theta}(t) \end{bmatrix}$, the original nonlinear system can be represented as the following T–S fuzzy model

Plant Rule 1: **IF** $x_1(t)$ is M_{11}, **THEN**

$$\dot{\boldsymbol{x}}(t) = A_1\boldsymbol{x}(t) + B_1\boldsymbol{u}(t)$$

Plant Rule 2: **IF** $x_1(t)$ is M_{12}, **THEN**

$$\dot{\boldsymbol{x}}(t) = A_2\boldsymbol{x}(t) + B_2\boldsymbol{u}(t)$$

The matrices for the T–S fuzzy system can be obtained

$$A_1 = \begin{bmatrix} 0 & 1 \\ \frac{-k-4ka^2}{m} & -\frac{c}{m} \end{bmatrix},\ A_2 = \begin{bmatrix} 0 & 1 \\ \frac{-k}{m} & -\frac{c}{m} \end{bmatrix},\ B_1 = \begin{bmatrix} 0 \\ \frac{1}{m} \end{bmatrix},\ B_2 = \begin{bmatrix} 0 \\ \frac{1}{m} \end{bmatrix}.$$

The other system matrices are defined as

$$E_1 = E_2 = \begin{bmatrix} -0.03 \\ -0.02 \end{bmatrix},\ C_1 = [1 \quad 1],\ C_2 = \begin{bmatrix} 1 & 0 \\ 0 & 1 \end{bmatrix}.$$

The normalized grades of membership for each rule are as follows

$$\varpi_1(x_1(t)) = \frac{x_1^2(t)}{4},\ \varpi_2(x_1(t)) = 1 - \frac{x_1^2(t)}{4}.$$

Now, we suppose that the maximum value of transmission delay d_k and the maximum number of data packet dropouts p_{ck} in sensor-to-controller and controller-to-actuator are 15ms and 5ms, respectively. Since the sampling time is set to 2.5ms, we can get $d_{max} = 15 + 2.5 \times 5 = 27.5$ms. Choose the scalar $\epsilon = 0.02$ and the $\mathscr{H}_\infty$ performance $\gamma = 1$. Then by solving the LMIs in (4.95) and (4.96), the controller gain matrices can be calculated as

$$U = 675.3543, \quad G_1 = -16.5941, \quad G_2 = -10.5851, \tag{4.104}$$

and

$$K_1 = -0.0246, \quad K_2 = -0.0157. \tag{4.105}$$

Moreover, assume that $x(0) = [\,0 \;\; 0\,]^T$ and $w(t) = 5\sin(2t)e^{-0.15t}$. Then the simulation responses are given in Figs. 4.9, 4.10, 4.11, 4.12, and 4.13. Figure 4.9 plots the trajectories of $w(t)$. Figure 4.10 plots the trajectories of $x(t)$. Figure 4.11 plots the trajectories of $z(t)$. Figure 4.12 plots the trajectory of $u(t)$. Figure 4.13 plots the trajectories of ratio of $\sqrt{\int_0^t z^T(\tau)z(\tau)d\tau / \int_0^t w^T(\tau)w(\tau)d\tau}$. This is clearly rising and converging to a fixed value at the macro level. We can see that the convergence value 0.0447 is less than the given performance index 1.0.

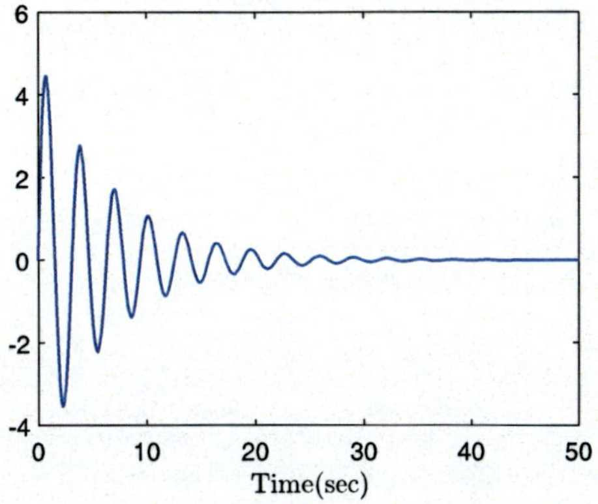

Fig. 4.9 History of $w(t)$

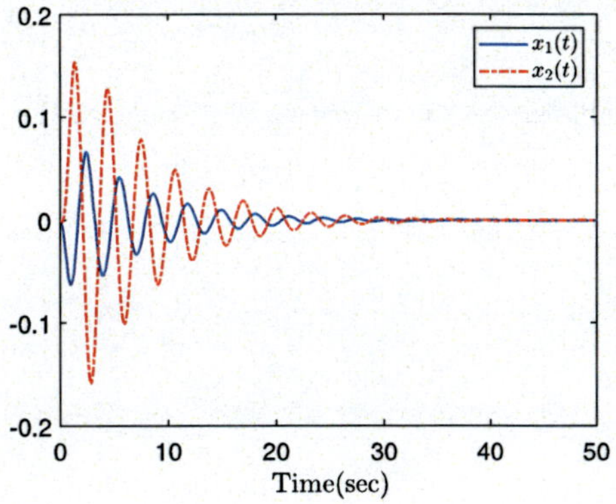

Fig. 4.10 Response of $x(t)$

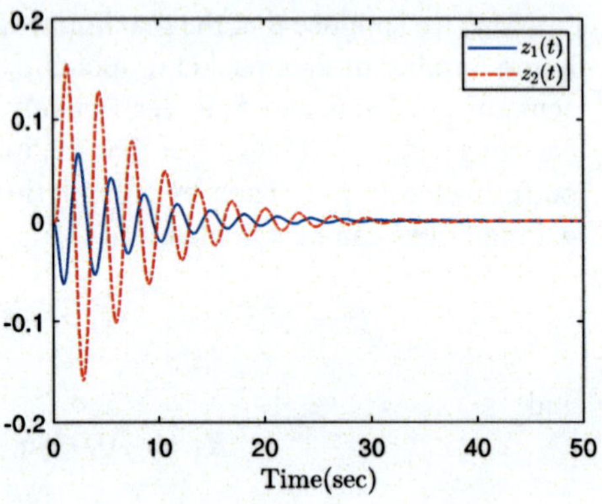

Fig. 4.11 Response of $z(t)$

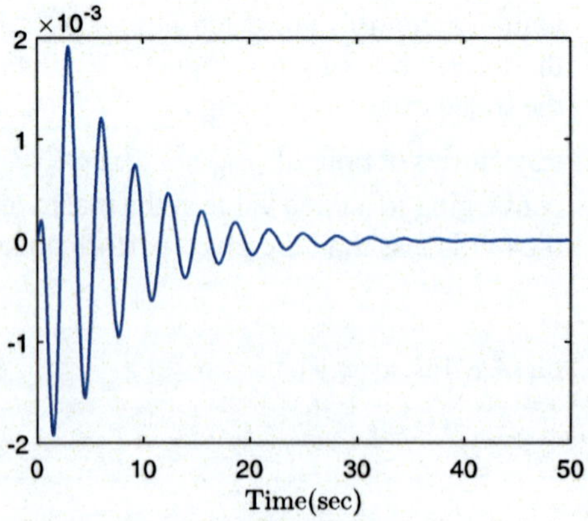

Fig. 4.12 Response of $u(t)$

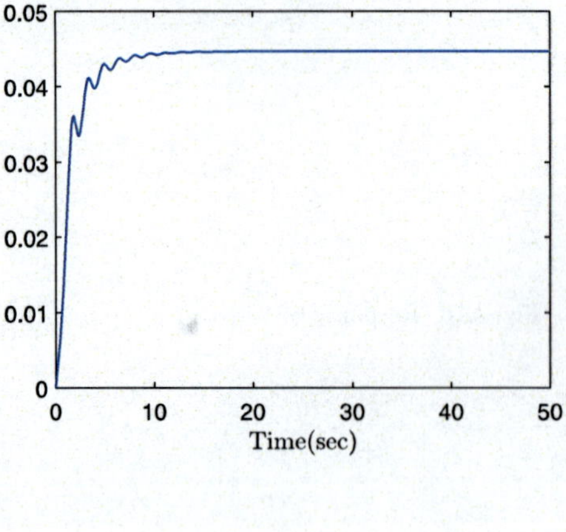

Fig. 4.13 History of $\sqrt{\int_0^t z^T(\tau)z(\tau)d\tau / \int_0^t w^T(\tau)w(\tau)d\tau}$

4.3 Feedback Control with Stochastic Communication Protocol

In this section, the problem of dynamic output feedback (DOF) control is addressed for T–S fuzzy system with the effect of stochastic communication protocol (SCP) protocol. A Markov chain is introduced to describe the SCP scheduling behavior over the communication channel. By using the fuzzy Markov jump model method, the DOF controller design scheme is proposed to ensure the stochastic stability and the required $\mathscr{H}_\infty$ performance of the closed-loop system.

4.3.1 Problem Formulation

Consider the following discrete-time nonlinear plant based on T–S fuzzy model:

Plant Rule i_{th}: **IF** $\zeta_1(k)$ is J_{1i} $,\cdots,$ and $\zeta_o(k)$ is J_{oi}, **THEN**

$$\begin{aligned} x(k+1) &= A_i x(k) + B_i u(k) + E_i \omega(k), \\ y(k) &= C_i x(k) + F_i \omega(k), \\ z(k) &= D_i x(k), \end{aligned} \tag{4.106}$$

where $\zeta_d(k)$, $d = 1, \ldots, o$, are represented the premise variables; J_{di}, $i = 1, \ldots, r$, are used to represent the fuzzy sets, which r is the number of fuzzy rules; $x(k) \in \mathbb{R}^{n_x}$, $y(k) \in \mathbb{R}^{n_y}$, and $z(k) \in \mathbb{R}^{n_z}$ denote, respectively, the state, the measurement output, and the control output; $u(k) \in \mathbb{R}^{n_u}$ denotes the control input; $\omega(k) \in \mathbb{R}^{n_w}$ is the disturbance input belongs to $\ell_2[0, \infty)$; $A_i \in \mathbb{R}^{n_x \times n_x}$, $B_i \in \mathbb{R}^{n_x \times n_u}$, $C_i \in \mathbb{R}^{n_y \times n_x}$, $D_i \in \mathbb{R}^{n_z \times n_x}$, $E_i \in \mathbb{R}^{n_x \times n_w}$, and $F_i \in \mathbb{R}^{n_y \times n_w}$ are known matrices.

According to (4.2) in Sect. 4.1, the T–S fuzzy system in (4.106) can be rewritten as follows:

$$\begin{aligned} x(k+1) &= A(h)x(k) + B(h)u(k) + E(h)\omega(k), \\ y(k) &= C(h)x(k) + F(h)\omega(k), \\ z(k) &= D(h)x(k), \end{aligned} \tag{4.107}$$

where

$$A(h) = \sum_{i=1}^{r} h_i(\zeta(k))A_i, \quad B(h) = \sum_{i=1}^{r} h_i(\zeta(k))B_i, \quad C(h) = \sum_{i=1}^{r} h_i(\zeta(k))C_i,$$

$$D(h) = \sum_{i=1}^{r} h_i(\zeta(k))D_i, \quad E(h) = \sum_{i=1}^{r} h_i(\zeta(k))E_i, \quad F(h) = \sum_{i=1}^{r} h_i(\zeta(k))F_i.$$

In order to alleviate the negative influence from the limited communication capacity of the communication channel, the SCP is adopted to determine which sensor can obtain the access to the communication channel at time instant k.

Let $\sigma(k) \in \mathbb{N} = \{1, 2, \cdots, n_y\}$ denote the select sensor. Then as shown in [9], $\sigma(k)$ can be governed by a Markov chain with the transition probability matrix $\mathcal{P} = [\pi_{st}] \in \mathbb{R}^{n_y \times n_y}$, which the transition probabilities π_{st} are defined as follows:

$$\pi_{st} \triangleq Prob(\sigma(k+1) = t | \sigma(k) = s), \quad s, t \in \mathbb{N}. \tag{4.108}$$

Let $\hat{y}(k) = \left[\hat{y}_1(k), \hat{y}_2(k), \cdots, \hat{y}_{n_y}(k)\right]^T$ denote the measurement output signal after being transmitted through the SCP scheduling. Then the updating rules for $\hat{y}_{\sigma(k)}(k)$ subject to the SCP are set to be

$$\hat{y}_s(k) = \begin{cases} y_s(k), & if\ \sigma(k) = s, \\ \hat{y}_s(k-1), & otherwise. \end{cases} \tag{4.109}$$

By defining the Kronecker delta function $\bar{\delta}_a^b = \delta(a-b)$, one has

$$\hat{y}(k) = \Phi_{\sigma(k)} y(k) + (I - \Phi_{\sigma(k)}) \hat{y}(k-1), \tag{4.110}$$

where $\Phi_{\sigma(k)} = diag\{\bar{\delta}^1_{\sigma(k)}, \bar{\delta}^2_{\sigma(k)}, \cdots, \bar{\delta}^{n_y}_{\sigma(k)}\}$. It should be noted that, due to the effect of the SCP, the output signal $\hat{y}(k)$ includes a time-delay term $\hat{y}(k-1)$ in equation (4.110), which will cause great difficulty for designing the DOF controller. Therefore, a method based on augmented vector is adopted to solve this issue.

By denoting $\bar{x}(k) \triangleq [x^T(k)\hat{y}^T(k-1)]^T$, the T–S fuzzy system (4.107) can be reformulated as follows:

$$\begin{aligned} \bar{x}(k+1) &= \bar{A}_{\sigma(k)h}\bar{x}(k) + \bar{B}_{\sigma(k)h}u(k) + \bar{E}_{\sigma(k)h}\omega(k), \\ z(k) &= \bar{D}_{\sigma(k)h}\bar{x}(k), \end{aligned} \tag{4.111}$$

where

$$\bar{A}_{\sigma(k)h} = \begin{bmatrix} A(h) & 0 \\ \Phi_{\sigma(k)}C(h) & I - \Phi_{\sigma(k)}, \end{bmatrix}, \quad \bar{B}_{\sigma(k)h} = \begin{bmatrix} B(h) \\ 0 \end{bmatrix}, \quad \bar{E}_{\sigma(k)h} = \begin{bmatrix} E(h) \\ \Phi_{\sigma(k)}F(h) \end{bmatrix},$$
$$\bar{D}_{\sigma(k)h} = \begin{bmatrix} D(h) & 0 \end{bmatrix}.$$

In this section, the main objective is to design an appropriate controller such that the discrete-time T–S fuzzy system in (4.106) is stable with prescribed $\mathcal{H}_\infty$ performance. Thus, we consider the following mode-dependent fuzzy DOF controller:

Plant Rule i_{th}: **IF** $\zeta_1(k)$ is J_{1i} ,$\cdots$, and $\zeta_o(k)$ is J_{oi}, **THEN**

$$\begin{aligned} \hat{x}(k+1) &= \mathcal{A}_{\sigma(k)i}\hat{x}(k) + \mathcal{B}_{\sigma(k)i}\hat{y}(k), \\ u(k) &= \mathcal{C}_{\sigma(k)i}\hat{x}(k) + \mathcal{D}_{\sigma(k)i}\hat{y}(k), \end{aligned} \tag{4.112}$$

where $\hat{x}(k) \in \mathbb{R}^{n_{\hat{x}}}$ is the state vector of the mode-dependent fuzzy DOF controller with $n_{\hat{x}} = n_x + n_y$; $\mathscr{A}_{\sigma(k)i}$, $\mathscr{B}_{\sigma(k)i}$, $\mathscr{C}_{\sigma(k)i}$, and $\mathscr{D}_{\sigma(k)i}$, $i = 1, \ldots, r$, are controller gain matrices with appropriate dimensions.

Remark 4.6 It should be noted that, due to the Markov chain applied to stand for the SCP scheduling, the synchronous Markov mode is considered for the system (4.111) in the mode-dependent fuzzy DOF controller (4.112).

Before the analysis and synthesis of the DOF controller design problem, the corresponding closed-loop system should be formulated firstly.

For the sake of analysis, let $\sigma(k)=s \in \mathbb{N}$. Then by defining $\tilde{x}(k)=\begin{bmatrix}\bar{x}^T(k) & \hat{x}^T(k)\end{bmatrix}^T$, the closed-loop system is achieved by combining system (4.111) and controller (4.112) as follows:

$$\begin{aligned} \tilde{x}(k+1) &= \tilde{A}_{sh}\tilde{x}(k) + \tilde{E}_{sh}\omega(k), \\ z(k) &= \tilde{D}_{sh}\tilde{x}(k), \end{aligned} \tag{4.113}$$

where

$$\begin{aligned} &\tilde{A}_{sh} = \sum_{i=1}^{r}\sum_{l=1}^{r} h_i(\zeta(k))h_l(\zeta(k))\tilde{A}_{sil}, \quad \tilde{E}_{sh} = \sum_{i=1}^{r}\sum_{l=1}^{r} h_i(\zeta(k))h_l(\zeta(k))\tilde{E}_{sil}, \\ &\tilde{D}_{sh} = \sum_{i=1}^{r} \tilde{D}_{si}, \quad \tilde{A}_{sil} = \begin{bmatrix} \bar{A}_{si} + \bar{B}_{si}\mathscr{D}_{sl}\bar{C}_{si} & \bar{B}_{si}\mathscr{C}_{sl} \\ \mathscr{B}_{sl}\bar{C}_{si} & \mathscr{A}_{sl} \end{bmatrix}, \quad \tilde{E}_{sil} = \begin{bmatrix} \bar{B}_{si}\mathscr{D}_{sl}\Phi_s F_i + \bar{E}_{si} \\ \mathscr{B}_{sl}\Phi_s F_i \end{bmatrix}, \\ &\tilde{D}_{si} = \begin{bmatrix}\bar{D}_{si} & 0\end{bmatrix}, \quad \bar{C}_{si} = \begin{bmatrix}\Phi_s C_i & I - \Phi_s\end{bmatrix}. \end{aligned}$$

$\mathscr{H}_\infty$ dynamic output feedback control problem. For the discrete-time T–S fuzzy system with the effect of the SCP scheduling, the main purpose of this section is to design a suitable controller to guarantee the stability and prescribed $\mathscr{H}_\infty$ performance of the closed-loop system.

Before proceeding further, we first recall the following definitions.

Definition 4.1 The closed-loop system (4.113) is said to be stochastically stable, if when $\omega(k) = 0$ for any initial condition $\tilde{x}(0)$ and $\sigma(0)$, the following condition holds

$$\mathbb{E}\left\{\sum_{k=0}^{\infty} \|\tilde{x}(k)\|^2 | \tilde{x}(0), \sigma(0)\right\} < \infty. \tag{4.114}$$

Definition 4.2 For a given scalar $\gamma > 0$, under zero initial condition, if the following condition holds

$$\mathbb{E}\left\{\sum_{k=0}^{\infty} \left(\|z(k)\|^2 - \gamma^2\|\omega(k)\|^2\right)\right\} \le 0, \tag{4.115}$$

then the closed-loop system (4.113) is said to satisfy $\mathscr{H}_\infty$ performance γ.

4.3.2 Main Results

Sufficient conditions will be presented to prove the stochastic stability and $\mathscr{H}_\infty$ performance of the closed-loop system (4.113) with the control law (4.112) on the assumption that the controller gains have been given.

Theorem 4.7 *For the T–S fuzzy system (4.113), given the controller in the form of (4.112) with the gain matrices (i.e., $\mathscr{A}_{sl}, \mathscr{B}_{sl}, \mathscr{C}_{sl}, \mathscr{D}_{sl}$), transition probabilities π_{st}, $i, l, \ell \in \mathbb{M}$ and $s, t \in \mathbb{N}$, if there exist matrices $P_{si} > 0$ and G_{sl} satisfying*

$$\begin{aligned} &\Sigma_{s\ell ii} < 0, \\ &\Sigma_{s\ell il} + \Sigma_{s\ell li} < 0,\ i < l, \end{aligned} \tag{4.116}$$

where

$$\Sigma_{s\ell il} = \begin{bmatrix} -P_{si} & * & * & * \\ 0 & -\gamma^2 I & * & * \\ G_{sl}^T \tilde{A}_{sil} & G_{sl}^T \tilde{E}_{sil} & Q_{s\ell l} & * \\ \tilde{D}_{si} & 0 & 0 & -I \end{bmatrix},$$

$$Q_{s\ell l} = -He\{G_{sl}\} + \sum_{t\in\mathbb{N}} \pi_{st} P_{t\ell},$$

then the stochastic stability and $\mathscr{H}_\infty$ performance $\gamma > 0$ for the closed-loop system (4.113) can be guaranteed by the controller (4.112) under the SCP scheduling governed by the Markov chain σ_k whose transition probabilities are defined in (4.108).

Proof Construct a mode-dependent fuzzy Lyapunov function as follows:

$$V(\tilde{x}(k)) = \tilde{x}^T(k) P_{sh} \tilde{x}(k), \tag{4.117}$$

where $P_{sh} = \sum_{i=1}^{r} h_i(\zeta(k)) P_{si}$. Then, we assume that $P_{t\hat{h}} = \sum_{\ell=1}^{r} \hat{h}_\ell P_{t\ell}$ with $t = \sigma(k+1)$ for any $\sigma(k) = s$, where $\hat{h}_\ell$ denotes the normalized membership function at time $k+1$. Along the trajectories of the closed-loop system (4.113), the difference of the Lyapunov function $V(\tilde{x}(k))$ can be calculated as follows:

$$\begin{aligned} \mathbb{E}\{\Delta V(\tilde{x}(k))\} &= \mathbb{E}\{\tilde{x}^T(k+1) P_{t\hat{h}} \tilde{x}(k+1) - \tilde{x}^T(k) P_{sh} \tilde{x}(k)\} \\ &= \mathbb{E}\left\{\tilde{x}^T(k+1)\left(\sum_{t=1}^{n_y} \pi_{st} P_{t\hat{h}}\right)\tilde{x}(k+1)\right\} - \tilde{x}^T(k) P_{sh} \tilde{x}(k) \\ &= \eta^T(k)\left\{\left[\tilde{A}_{sh} \ \ \tilde{E}_{sh}\right]^T \bar{P}_{s\hat{h}} \left[\tilde{A}_{sh} \ \ \tilde{E}_{sh}\right] - diag\{P_{sh}, 0\}\right\}\eta(k) \\ &= \eta^T(k) \Psi_{sh} \eta(k), \end{aligned} \tag{4.118}$$

where $\eta(k) = \begin{bmatrix} \tilde{x}(k) \\ \omega(k) \end{bmatrix}$ and $\bar{P}_{s\hat{h}} = \sum_{t=1}^{n_y} \pi_{st} P_{t\hat{h}}$.

On the other hand, it can be inferred from condition (4.116) that

$$\begin{aligned} \Sigma_{sh} &= \sum_{\ell=1}^{r}\sum_{i=1}^{r}\sum_{l=1}^{r} \hat{h}_\ell h_i(\zeta(k))h_l(\zeta(k))\Sigma_{s\ell il} \\ &= \sum_{\ell=1}^{r} \hat{h}_\ell \left(\sum_{i=1}^{r} h_i^2(\zeta(k))\Xi_{s\ell ii} + \sum_{i=1}^{r-1}\sum_{l=i+1}^{r} h_i(\zeta(k))h_l(\zeta(k))(\Sigma_{s\ell il} + \Sigma_{s\ell li}) \right) < 0, \end{aligned} \tag{4.119}$$

where

$$\Sigma_{sh} = \begin{bmatrix} -P_{sh} & * & * & * \\ 0 & -\gamma^2 I & * & * \\ G_{sh}^T \tilde{A}_{sh} & G_{sh}^T \tilde{E}_{sh} & Q_{sh} & * \\ \tilde{D}_{sh} & 0 & 0 & -I \end{bmatrix},$$

$$G_{sh} = \sum_{l=1}^{r} h_l(\zeta(k))G_{sl}, \quad Q_{sh} = \sum_{\ell=1}^{r}\sum_{l=1}^{r} \hat{h}_\ell h_l(\zeta(k))Q_{s\ell l}.$$

According to that the transition probabilities $\pi_{st} > 0$ and the Lyapunov matrices $P_{si} > 0$, we obtain $\bar{P}_{s\hat{h}} > 0$. And it is easy to know that $(\bar{P}_{s\hat{h}} - G_{sh})^T \bar{P}_{s\hat{h}}^{-1}(\bar{P}_{s\hat{h}} - G_{sh}) \geq 0$, which implies that

$$-He\{G_{sh}\} + \bar{P}_{s\hat{h}} \geq -G_{sh}^T \bar{P}_{s\hat{h}}^{-1} G_{sh}. \tag{4.120}$$

Based on (4.119) and (4.120), one has

$$\begin{bmatrix} -P_{sh} & * & * & * \\ 0 & -\gamma^2 I & * & * \\ G_{sh}^T \tilde{A}_{sh} & G_{sh}^T \tilde{E}_{sh} & -G_{sh}^T \bar{P}_{s\hat{h}}^{-1} G_{sh} & * \\ \tilde{D}_{sh} & 0 & 0 & -I \end{bmatrix} < 0. \tag{4.121}$$

By pre- and post-multiplying $diag\{I, I, G_{sh}^{-T}, I\}$ and its transpose to (4.121), one can obtain that

$$\begin{bmatrix} -P_{sh} & * & * & * \\ 0 & -\gamma^2 I & * & * \\ \tilde{A}_{sh} & \tilde{E}_{sh} & -\bar{P}_{s\hat{h}}^{-1} & * \\ \tilde{D}_{sh} & 0 & 0 & -I \end{bmatrix} < 0. \tag{4.122}$$

By using Lemma 1.5 to (4.122), one gets

$$diag\{-P_{sh}, -\gamma^2 I\} + \Gamma_1^T \bar{P}_{s\hat{h}} \Gamma_1 + \Gamma_2^T \Gamma_2 < 0, \tag{4.123}$$

where $\Gamma_1 = \begin{bmatrix} \tilde{A}_{sh} & \tilde{E}_{sh} \end{bmatrix}$ and $\Gamma_2 = \begin{bmatrix} \tilde{D}_{sh} & 0 \end{bmatrix}$.

By pre-multiplying $\eta^T(k)$ and post-multiplying $\eta(k)$ to (4.123), it can be inferred from (4.118) that

$$\eta^T(k)\Psi_{sh}\eta(k) + z^T(k)z(k) - \gamma^2\omega^T(k)\omega(k) < 0. \tag{4.124}$$

For (4.124), when $\omega(k) = 0$, one has $\Psi_{sh} < 0$, which implies that

$$\mathbb{E}\{\Delta V(\tilde{x}(k))\} < -\lambda_{min}(-\Psi_{sh})\eta^T(k)\eta(k) < 0. \tag{4.125}$$

Summing both sides of (4.125) from 0 to ∞ and taking mathematical expectation, it is easily obtained that

$$\mathbb{E}\left\{\sum_{k=0}^{\infty}\|\eta(k)\|^2|\eta(0), \sigma(0)\right\} < \lambda_{min}^{-1}(-\Psi_{sh})V(\tilde{x}(0)) < \infty. \tag{4.126}$$

Obviously, $\mathbb{E}\left\{\sum_{k=0}^{\infty}\|\eta(k)\|^2|\eta(0), \sigma(0)\right\} \geq \mathbb{E}\left\{\sum_{k=0}^{\infty}\|\tilde{x}(k)\|^2|\tilde{x}(0), \sigma(0)\right\}$. According to Definition 4.1, it can be concluded that the closed-loop system (4.113) is stochastically stable.

Next, for any non-zero $\omega(k)$, according to (4.118) and (4.125), one has

$$\mathbb{E}\{\Delta V(\tilde{x}(k))\} + z^T(k)z(k) - \gamma^2\omega^T(k)\omega(k) < 0. \tag{4.127}$$

Under zero initial condition $V(\tilde{x}(0)) = 0$, integrating both sides of (4.127) from 0 to ∞ and taking mathematical expectation, one can obtain

$$\mathbb{E}\left\{\sum_{k=0}^{\infty}(\|z(k)\|^2 - \gamma^2\|\omega(k)\|^2)\right\} < -V(\tilde{x}(\infty)) < 0. \tag{4.128}$$

From the above analysis and Definition 4.2, the closed-loop system (4.113) is stochastically stable with $\mathcal{H}_\infty$ performance $\gamma > 0$. The proof is completed.

Next, we will give design conditions to determine the gain matrices.

Theorem 4.8 *The T–S fuzzy system (4.113) is stochastically stable with $\mathcal{H}_\infty$ performance $\gamma > 0$, given a scalar λ, transition probabilities π_{st}, $i, l, \ell \in \mathbb{M}$ and $s, t \in \mathbb{N}$, if there exist matrices $P_{si} = \begin{bmatrix} P_{1si} & * \\ P_{2si} & P_{3si} \end{bmatrix} > 0$, $G_{sl} = \begin{bmatrix} G_{1sl} & G_{2sl} \\ \lambda G_{3sl} & G_{3sl} \end{bmatrix}$, R_{sl}, and scalar ϵ such that the following inequalities hold:*

$$\begin{aligned} \Lambda_{s\ell ii} &< 0, \\ \Lambda_{s\ell il} + \Lambda_{s\ell li} &< 0, \quad i < l, \end{aligned} \tag{4.129}$$

where

$$\Lambda_{s\ell il} = \begin{bmatrix} -P_{1si} & * & * & * & * & * & * & * \\ -P_{2si} & -P_{3si} & * & * & * & * & * & * \\ 0 & 0 & -\gamma^2 I & * & * & * & * & * \\ \psi_{1sil} & \psi_{2sl} & \psi_{3sil} & \psi_{4s\ell l} & * & * & * & * \\ \psi_{5sil} & \psi_{6sl} & \psi_{7sil} & \psi_{8s\ell l} & \psi_{9s\ell l} & * & * & * \\ \bar{D}_{si} & 0 & 0 & 0 & 0 & -I & * & * \\ \bar{\mathscr{D}}_{sl}\bar{C}_{si} & \bar{\mathscr{C}}_{sl} & 0 & \epsilon\psi_{10sil}^T & \epsilon\psi_{11sil}^T & 0 & -\epsilon He\{R_{sl}\} & * \\ 0 & 0 & \bar{\mathscr{D}}_{sl}\Phi_s F_i & \epsilon\bar{B}_{si}^T G_{1sl} & \epsilon\bar{B}_{si}^T G_{2sl} & 0 & 0 & -\epsilon He\{R_{sl}\} \end{bmatrix},$$

$\psi_{1sil} = G_{1sl}^T\bar{A}_{si} + \lambda\bar{\mathscr{B}}_{sl}\bar{C}_{si} + U\bar{\mathscr{D}}_{sl}\bar{C}_{si}$, $\psi_{2sl} = \lambda\bar{\mathscr{A}}_{sl} + U\bar{\mathscr{C}}_{sl}$,

$\psi_{3sil} = G_{1sl}^T\bar{E}_{si} + \lambda\bar{\mathscr{B}}_{sl}\Phi_s F_i$, $\psi_{4s\ell l} = -He\{G_{1sl}\} + \sum_{\iota\in\mathbb{N}}\pi_{s\iota}P_{1\iota\ell}$,

$\psi_{5sil} = G_{2sl}^T\bar{A}_{si} + \bar{\mathscr{B}}_{sl}\bar{C}_{si} + U\bar{\mathscr{D}}_{sl}\bar{C}_{si}$, $\psi_{6sl} = \bar{\mathscr{A}}_{sl} + U\bar{\mathscr{C}}_{sl}$,

$\psi_{7sil} = G_{2sl}^T\bar{E}_{si} + \bar{\mathscr{B}}_{sl}\Phi_s F_i$, $\psi_{8s\ell l} = -\lambda G_{3sl} - G_{2sl}^T + \sum_{\iota\in\mathbb{N}}\pi_{s\iota}P_{2\iota\ell}$,

$\psi_{9s\ell l} = -He\{G_{3sl}\} + \sum_{\iota\in\mathbb{N}}\pi_{s\iota}P_{3\iota\ell}$, $\psi_{10sil} = G_{1sl}^T\bar{B}_{si} - UR_{sl}$, $\psi_{11sil} = G_{2sl}^T\bar{B}_{si} - UR_{sl}$,

and U is a dimension regulation matrix and defined as I, $\begin{bmatrix} I_{n_{\hat{x}}\times n_{\hat{x}}} & 0_{n_{\hat{x}}\times(n_u-n_{\hat{x}})}\end{bmatrix}$, and $\begin{bmatrix} I_{n_u\times n_u} & 0_{n_u\times(n_{\hat{x}}-n_u)}\end{bmatrix}^T$ when $n_{\hat{x}} = n_u$, $n_{\hat{x}} < n_u$, and $n_{\hat{x}} > n_u$, respectively. Then, the dynamic output feedback controller is given as

$$\mathscr{A}_{sl} = (G_{3sl}^T)^{-1}\bar{\mathscr{A}}_{sl},\ \mathscr{B}_{sl} = (G_{3sl}^T)^{-1}\bar{\mathscr{B}}_{sl},$$
$$\mathscr{C}_{sl} = R_{sl}^{-1}\bar{\mathscr{C}}_{sl},\ \mathscr{D}_{sl} = R_{sl}^{-1}\bar{\mathscr{D}}_{sl}.$$

Proof Firstly, by defining

$$P_{si} = \begin{bmatrix} P_{1si} & * \\ P_{2si} & P_{3si} \end{bmatrix},\ G_{sl} = \begin{bmatrix} G_{1sl} & G_{2sl} \\ \lambda G_{3sl} & G_{3sl} \end{bmatrix}, \tag{4.130}$$

and $\bar{\mathscr{A}}_{sl} = G_{3sl}^T\mathscr{A}_{sl}$, $\bar{\mathscr{B}}_{sl} = G_{3sl}^T\mathscr{B}_{sl}$, $\bar{\mathscr{C}}_{sl} = R_{sl}\mathscr{C}_{sl}$, and $\bar{\mathscr{D}}_{sl} = R_{sl}\mathscr{D}_{sl}$, the matrix $\Sigma_{s\ell il}$ in Theorem 4.7 can be rewritten as follows:

$$\Sigma_{s\ell il} = \Lambda_{1s\ell il} + He\{\Lambda_{2s\ell il}\bar{R}_{sl}^{-1}\Lambda_{3s\ell il}\}, \tag{4.131}$$

where

$$\Lambda_{1s\ell il} = \begin{bmatrix} -P_{1si} & * & * & * & * & * \\ -P_{2si} & -P_{3si} & * & * & * & * \\ 0 & 0 & -\gamma^2 I & * & * & * \\ \psi_{1sil} & \psi_{2sl} & \psi_{3sil} & \psi_{4s\ell l} & * & * \\ \psi_{5sil} & \psi_{6sl} & \psi_{7sil} & \psi_{8s\ell l} & \psi_{9s\ell l} & * \\ \bar{D}_{si} & 0 & 0 & 0 & 0 & -I \end{bmatrix}, \; \Lambda_{2s\ell il} = \begin{bmatrix} 0 & 0 \\ 0 & 0 \\ 0 & 0 \\ \psi_{10sil} & G_{1sl}^T \bar{B}_{si} \\ \psi_{11sil} & G_{2sl}^T \bar{B}_{si} \\ 0 & 0 \end{bmatrix},$$

$$\Lambda_{3s\ell il} = \begin{bmatrix} \mathfrak{D}_{sl}\bar{C}_{si} & \mathscr{C}_{sl} & 0 & 0 & 0 & 0 \\ 0 & 0 & \bar{\mathfrak{D}}_{sl}\Phi_s F_i & 0 & 0 & 0 \end{bmatrix}, \; \bar{R}_{sl} = diag\{R_{sl}, R_{sl}\}.$$

Further, (4.116) in Theorem 4.7 can be rewritten as follows:

$$\begin{aligned} &\Lambda_{1s\ell ii} + He\left\{\Lambda_{2s\ell ii}\bar{R}_{sl}^{-1}\Lambda_{3s\ell ii}\right\} < 0, \\ &(\Lambda_{1s\ell il} + \Lambda_{1s\ell li}) + He\left\{\Lambda_{2s\ell il}\bar{R}_{sl}^{-1}\Lambda_{3s\ell il}\right\} \\ &\qquad + He\left\{\Lambda_{2s\ell li}\bar{R}_{sl}^{-1}\Lambda_{3s\ell li}\right\} < 0, \quad i < l. \end{aligned} \tag{4.132}$$

It is easy to know that (4.132) can be guaranteed if the condition (4.129) is satisfied by applying Lemma 1.4. The proof is completed.

4.3.3 Simulation Example

Let us consider the discrete-time T–S fuzzy model in the form of (4.106) with $x(k) = \left[x_1^T(k) \;\; x_2^T(k)\right]^T$. Then the rules are given as follows:

Plant Rule 1: **IF** $x_1(k)$ is J_{11}, **THEN**

$$\begin{aligned} x(k+1) &= A_1 x(k) + B_1 u(k) + E_1 \omega(k), \\ y(k) &= C_1 x(k) + F_1 \omega(k), \\ z(k) &= D_1 x(k), \end{aligned}$$

Plant Rule 2: **IF** $x_1(k)$ is J_{12}, **THEN**

$$\begin{aligned} x(k+1) &= A_2 x(k) + B_2 u(k) + E_2 \omega(k), \\ y(k) &= C_2 x(k) + F_2 \omega(k), \\ z(k) &= D_2 x(k), \end{aligned}$$

where

$$A_1 = \begin{bmatrix} 0.7238 & 0.4972 \\ 0.5148 & 0.0877 \end{bmatrix}, \quad B_1 = \begin{bmatrix} 0.1176 \\ 0.5759 \end{bmatrix}, \quad E_1 = \begin{bmatrix} -0.1250 \\ 0.0148 \end{bmatrix},$$

$$A_2 = \begin{bmatrix} 0.9652 & 0.0969 \\ 0.4813 & 0.0027 \end{bmatrix}, \quad B_2 = \begin{bmatrix} 0.1470 \\ 0.6748 \end{bmatrix}, \quad E_2 = \begin{bmatrix} -0.0530 \\ -0.2478 \end{bmatrix},$$

$$C_1 = \begin{bmatrix} 0.8282 & 0.4565 \\ -0.5649 & 0.1428 \end{bmatrix}, \quad F_1 = \begin{bmatrix} -0.0169 \\ 0.0842 \end{bmatrix}, \quad D_1 = \begin{bmatrix} 0.1426 & -0.4029 \end{bmatrix},$$

$$C_2 = \begin{bmatrix} 0.1408 & 0.5053 \\ -0.6831 & 0.2858 \end{bmatrix}, \quad F_2 = \begin{bmatrix} -0.3267 \\ -0.0144 \end{bmatrix}, \quad D_2 = \begin{bmatrix} -0.4803 & -0.3073 \end{bmatrix}.$$

The disturbance signal is given as $\omega(k) = 10e^{-0.4k}\cos(0.1k)$. In addition, the membership functions are set to be $h_1(x_1(k)) = \sin^2(x_1(k))$ and $h_2(x_1(k)) = 1 - h_1(x_1(k))$. Then, under the initial condition $x(0) = \begin{bmatrix} 0 & 0 \end{bmatrix}^T$ and $u(k) = 0$, the state response of the open-loop system is depicted in Fig. 4.14, which shows that this example is instable without considering the controller.

Next, the effectiveness of the controller designed in this paper is demonstrated. Due to the SCP scheduling, we assume the transition probability matrix

$$\mathcal{P} = \begin{bmatrix} 0.45 & 0.55 \\ 0.65 & 0.35 \end{bmatrix}.$$

By solving LMI-based inequalities conditions in Theorem 4.8 with $\mathcal{H}_\infty$ index $\gamma = 1$, we obtain the designed controller gain matrices as follows:

$$\mathcal{A}_{11} = \begin{bmatrix} 0.0010 & 0.0013 & 0.0002 & -0.0001 \\ -0.0010 & -0.0010 & -0.0003 & 0.0001 \\ 0.0006 & 0.0008 & 0.0001 & -0.0001 \\ -0.0002 & -0.0003 & -0.0001 & 0 \end{bmatrix}, \quad \mathcal{B}_{11} = \begin{bmatrix} -0.0773 & 0.0097 \\ 0.0193 & 0.0147 \\ -0.0929 & -0.0042 \\ 0.0194 & -0.0510 \end{bmatrix},$$

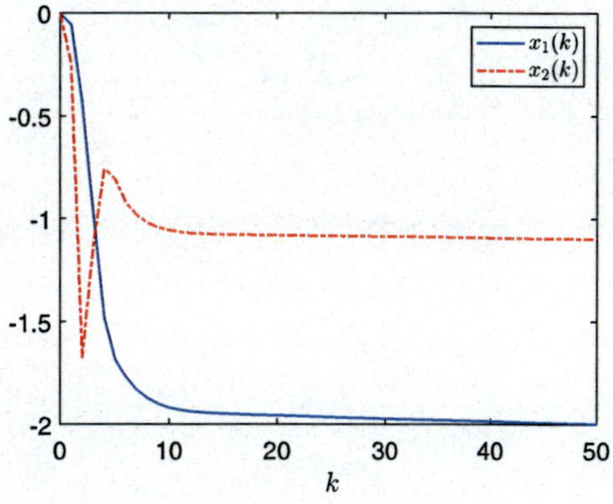

Fig. 4.14 Response of the state of open-loop system

$$\mathscr{A}_{12}=\begin{bmatrix}0.0003 & -0.0026 & -0.0026 & 0.0011\\ -0.0025 & -0.0044 & -0.0110 & 0.0041\\ 0.0006 & -0.0000 & 0.0012 & -0.0004\\ -0.0002 & 0.0002 & -0.0001 & 0\end{bmatrix},\ \mathscr{B}_{12}=\begin{bmatrix}-0.0351 & 0.0354\\ 0.0133 & -0.0266\\ -0.0762 & 0.0111\\ 0.0089 & -0.0637\end{bmatrix},$$

$$\mathscr{A}_{21}=\begin{bmatrix}0.0057 & 0.0056 & -0.0010 & 0.0014\\ -0.0067 & -0.0061 & -0.0004 & 0.0002\\ 0.0038 & 0.0037 & -0.0003 & 0.0005\\ 0.0004 & 0.0004 & 0 & 0.0001\end{bmatrix},\ \mathscr{B}_{21}=\begin{bmatrix}-0.0215 & 0.0845\\ -0.0226 & -0.0855\\ -0.0746 & 0.0241\\ 0.0198 & -0.0676\end{bmatrix},$$

$$\mathscr{A}_{22}=\begin{bmatrix}0.0026 & 0.0019 & 0.0013 & -0.0006\\ -0.0017 & -0.0037 & 0.0016 & 0.0003\\ 0 & 0 & 0 & 0\\ -0.0008 & -0.0009 & -0.0001 & 0.0002\end{bmatrix},\ \mathscr{B}_{22}=\begin{bmatrix}-0.0223 & 0.0673\\ -0.0205 & -0.0656\\ -0.0904 & 0.0164\\ -0.0143 & -0.1102\end{bmatrix},$$

$$\mathscr{C}_{11}=\begin{bmatrix}-0.0147 & -0.0188 & -0.0111 & 0.0042\end{bmatrix},\ \mathscr{D}_{11}=\begin{bmatrix}-0.9218 & -0.0810\end{bmatrix},$$
$$\mathscr{C}_{12}=\begin{bmatrix}0.0267 & 0.0514 & 0.1456 & -0.0555\end{bmatrix},\ \mathscr{D}_{12}=\begin{bmatrix}-1.1533 & 0.4405\end{bmatrix},$$
$$\mathscr{C}_{21}=\begin{bmatrix}0.1169 & 0.1106 & -0.0029 & 0.0084\end{bmatrix},\ \mathscr{D}_{21}=\begin{bmatrix}0.0467 & 1.8981\end{bmatrix},$$
$$\mathscr{C}_{22}=\begin{bmatrix}0.0387 & 0.0517 & -0.0048 & -0.0076\end{bmatrix},\ \mathscr{D}_{22}=\begin{bmatrix}-0.0097 & 1.2577\end{bmatrix}.$$

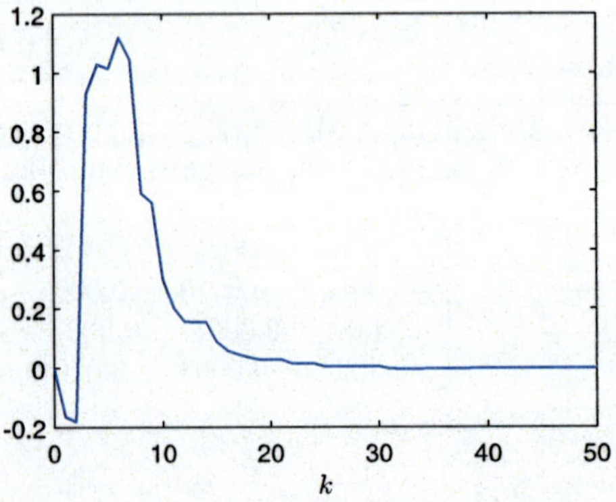

Fig. 4.15 Response of the control input $u(k)$

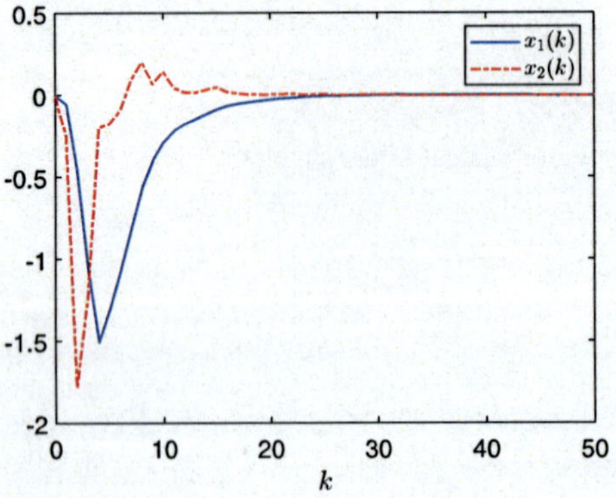

Fig. 4.16 Response of the state of closed-loop system

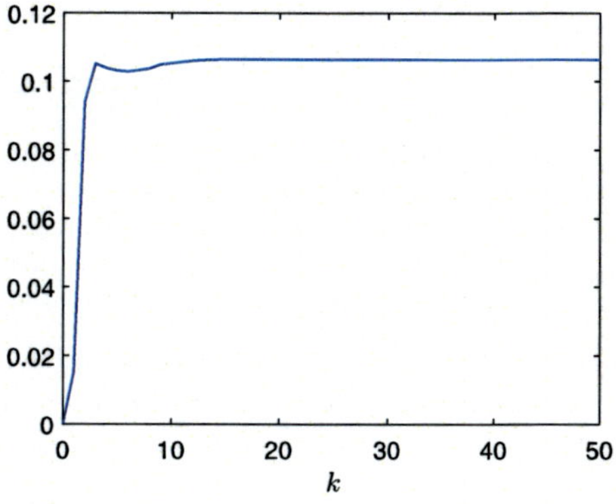

Fig. 4.17 History of $\sqrt{\sum_{\tau=0}^{k} z^T(\tau)z(\tau)/\sum_{\tau=0}^{k}\omega^T(\tau)\omega(\tau)}$

We apply the above controller to the simulation example. The control input is shown in Fig. 4.15. The state response of the closed-loop system is shown in Fig. 4.16. It is obvious that the system is stabilized by the controller designed above. Figure 4.17 depicts the ratio of $\sqrt{\sum_{\tau=0}^{k} z^T(\tau)z(\tau)/\sum_{\tau=0}^{k}\omega^T(\tau)\omega(\tau)}$, which is less than $\gamma = 1$.

4.4 Conclusion

In this chapter, the output feedback control problem of T–S fuzzy system with several typical communication constraints has been addressed. Firstly, the quantization effects from measurement output of the system and the output of the controller have been fully considered. By using the concept of PDC, the novel design strategy for both the output feedback controller and the quantizers' dynamic parameters has been proposed. Secondly, based on sampling data, the input delay approach has been introduced to the control synthesis for the fuzzy system with time delays and data dropouts. By using the Lyapunov function approach, the desired output feedback controller gain has been obtained. Thirdly, the Markov chain has been adopted to describe the SCP scheduling behavior. By using the fuzzy Markov jump model method, the dynamic output feedback control scheme has been provided. Finally, some examples have been given to demonstrate the effectiveness of the proposed methods.

References

1. Tanaka K, Ikeda T, Wang HO (1998) Fuzzy regulator and fuzzy observer: Relaxed stability conditions and LMI-based designs. IEEE Trans Fuzzy Syst 6:250-265
2. Chang XH, Huang R, Wang H, Liu L (2020) Robust design strategy of quantized feedback control. IEEE Trans Circuits Syst II Exp Briefs 67:730–734
3. Chang XH, Xiong J, Li ZM, Park JH (2018) Quantized static output feedback control for discrete-time systems, IEEE Trans Industr Inf 14:3426–3435
4. Chang XH, Park JH, Zhou J (2015) Robust static output feedback $\mathscr{H}_\infty$ control design for linear systems with polytopic uncertainties. Syst Control Lett 85:23–32

5. Ushio T (1995) Chaotic synchronization and controlling chaos based on contraction mappings. Phys Lett A 198:14–22
6. Wu HN (2007) Robust $\mathscr{H}_2$ fuzzy output feedback control for discrete-time nonlinear systems with parametric uncertainties. Int J Approx Reason 46:151–165
7. Collins EG (2003) Fuzzy control systems design and analysis: a linear matrix inequality approach. Automatica 39:2011–2013
8. Zhao T, Dian SY (2017) Fuzzy dynamic output feedback $\mathscr{H}_\infty$ control for continuous-time T-S fuzzy systems under imperfect premise matching. ISA Trans 70:248–259
9. Zou L, Wang Z, Gao H (2016) Observer-based $\mathscr{H}_\infty$ control of networked systems with stochastic communication protocol: The finite-horizon case. Automatica 63:366–373

Chapter 5
Event-Triggered Fuzzy Control and Filtering with Communication Constraints

Abstract This chapter investigates the problem of control and filtering under the event-triggered mechanism for T–S fuzzy systems with different communication constraints. Firstly, the event-triggered output feedback tracking control is studied for discrete-time T–S fuzzy systems with static quantization, and sufficient conditions to design tracking controller are given by LMIs. Secondly, a piecewise Lyapunov–Krasovskii functional method is applied to the event-triggered $\mathscr{L}_2$–$\mathscr{L}_\infty$ filtering issue for T–S fuzzy systems with time-delay and external interference under the limited communication resources and DoS attacks, so that the filtering error system is exponentially stable and maintains the prescribed $\mathscr{L}_2$–$\mathscr{L}_\infty$ performance. Thirdly, the problem of event-triggered filtering for discrete-time T–S fuzzy systems with dynamic quantization and stochastic deception attacks is investigated. The sufficient design conditions for the full- and reduced-order $\mathscr{H}_\infty$ filters are presented as two strict LMIs. Finally, some examples are provided to show the feasibility and effectiveness of the proposed design methods, respectively.

Keywords T–S fuzzy systems · Event-triggered mechanism · Quantization · Denial-of-service attacks · Stochastic deception attacks

5.1 Tracking Control with Static Quantization

In this section, an event-triggered static output feedback tracking controller is designed. In the presence of input and output static quantization, both the static logarithmic quantizers and the asynchronous event-triggered communication scheme are considered such that the closed-loop system is asymptotically stable with the prescribed $\mathscr{H}_\infty$ tracking performance.

X. Chang et al., *Control and Filtering of Fuzzy Systems Under Communication Channels*,
https://doi.org/10.1007/978-981-99-4346-3_5

5.1.1 Problem Formulation

Consider a nonlinear system represented by T–S fuzzy dynamic model, in which the ith rule is described as follows:

Plant Rule i_{th} : **IF** $\epsilon_1(k)\, is\, M_{1i},\ \epsilon_2(k)\, is M_{2i},\ and,\ \cdots,\ and\ \epsilon_d(k)\, is\, M_{di}$, **THEN**

$$
\begin{aligned}
x(k+1) &= A_i x(k) + B_i u_g(k) + D_i w(k), \\
y(k) &= C_i x(k),
\end{aligned}
\tag{5.1}
$$

where $x(k) \in \mathbb{R}^{n_x}$ stands for the state variable, $u_g(k) \in \mathbb{R}^{n_u}$ represents the control input, $y(k) \in \mathbb{R}^{n_y}$ means the measured output, and $w(k) \in \mathbb{R}^{n_w}$ denotes the noise signal in $\ell_2[0,\ \infty)$. The fuzzy sets are described by M_{pi}, $i = 1, 2, \ldots, r$, $p = 1, 2, \ldots, d$, where r is the number of fuzzy rules. The premise variables are denoted by $\epsilon_1(k), \epsilon_2(k), \ldots, \epsilon_d(k)$. The matrices $A_i \in \mathbb{R}^{n_x \times n_x}$, $B_i \in \mathbb{R}^{n_x \times n_u}$, $C_i \in \mathbb{R}^{n_y \times n_x}$, and $D_i \in \mathbb{R}^{n_x \times n_w}$ with $i = 1, 2, \ldots, r$ stand for given system matrices.

By introducing the following fuzzy basis functions:

$$
\rho_i(\epsilon(k)) = \frac{\prod_{p=1}^{d} M_{pi}(\epsilon_p(k))}{\sum_{i=1}^{r} \prod_{p=1}^{d} M_{pi}(\epsilon_p(k))},
\tag{5.2}
$$

where $M_{pi}(\epsilon_p(k))$ is the grade of membership function of $\epsilon_p(k)$ in M_{pi}.

According to the basis functions defined in (5.2), it can be obtained that

$$
\rho_i(\epsilon(k)) \geq 0, \quad \sum_{i=1}^{r} \rho_i(\epsilon(k)) = 1, \quad i = 1, 2, \ldots, r.
\tag{5.3}
$$

Besides, the discrete-time T–S fuzzy model in (5.1) can be expressed as

$$
\begin{aligned}
x(k+1) &= A(\rho)x(k) + B(\rho)u_g(k) + D(\rho)w(k), \\
y(k) &= C(\rho)x(k),
\end{aligned}
\tag{5.4}
$$

with

$$
\begin{aligned}
A(\rho) &= \sum_{i=1}^{r} \rho_i(\epsilon(k))A_i, \quad B(\rho) = \sum_{i=1}^{r} \rho_i(\epsilon(k))B_i, \\
C(\rho) &= \sum_{i=1}^{r} \rho_i(\epsilon(k))C_i, \quad D(\rho) = \sum_{i=1}^{r} \rho_i(\epsilon(k))D_i.
\end{aligned}
$$

In this section, the reference model is presented as follows:

$$\begin{aligned} \widehat{x}(k+1) &= A_\iota \widehat{x}(k) + B_\iota \iota(k), \\ \widehat{y}(k) &= C_\iota \widehat{x}(k), \end{aligned} \tag{5.5}$$

where $\widehat{x}(k) \in \mathbb{R}^{n_{\widehat{x}}}$, $\iota \in \mathbb{R}^{n_\iota}$, and $\widehat{y}(k) \in \mathbb{R}^{n_y}$ are used to denote the state, the energy bounded reference input, and the measurement output of the reference model, respectively. $A_\iota \in \mathbb{R}^{n_{\widehat{x}} \times n_{\widehat{x}}}$ is a known Hurwitz matrix, $B_\iota \in \mathbb{R}^{n_{\widehat{x}} \times n_\iota}$ and $C_\iota \in \mathbb{R}^{n_y \times n_{\widehat{x}}}$ stand for known matrices.

In order to reduce the amount of data in the communication of network and realize the effective utilization of the limited network communication resources, the following improved event-triggered mechanism will be employed.

The output of the fuzzy model (5.4), i.e., $y(k)$ will be transmitted to the quantizer $g_y(\star)$ only if

$$\xi_y^T(k)\mathbf{J}_y\xi_y(k) - b_y(y(k) - \xi_y(k))^T\mathbf{J}_y(y(k) - \xi_y(k)) \geq 0 \tag{5.6}$$

is satisfied. Here, $\xi_y(k) = y(k) - y(k_s)$, $y(k)$ is the present data for output, and $y(k_s)$ represents the last triggering instant. $b_y \in [0,\ 1)$ is a predetermined scalar, $\mathbf{J}_y > 0$ stands for a weighting matrix to be determined.

The output of the reference model (5.5), i.e., $\widehat{y}(k)$ will be transmitted to the quantizer $g_{\widehat{y}}(\star)$ only when

$$\xi_{\widehat{y}}^T(k)\mathbf{J}_{\widehat{y}}\xi_{\widehat{y}}(k) - b_{\widehat{y}}(\widehat{y}(k) - \xi_{\widehat{y}}(k))^T\mathbf{J}_{\widehat{y}}(\widehat{y}(k) - \xi_{\widehat{y}}(k)) \geq 0 \tag{5.7}$$

is satisfied. Here, $\xi_{\widehat{y}}(k) = \widehat{y}(k) - \widehat{y}(k_r)$, $\widehat{y}(k)$ represents the present data for reference output, and $\widehat{y}(k_r)$ is the last triggering instant. $b_{\widehat{y}} \in [0,\ 1)$ means a predetermined scalar, $\mathbf{J}_{\widehat{y}} > 0$ is a weighting matrix to be determined.

Remark 5.1 It is worth noting that the event-triggered mechanism used in this section is asynchronous, i.e., different event-triggered conditions (5.6) and (5.7) are predefined to independently determine whether the plant's output $y(k)$ and the reference model's output $\widehat{y}(k)$ should be released to the tracking controller or not. In contrast with the synchronous event-triggered mechanism used in [1–3], i.e., a specific event-triggered condition is predefined to judge whether the augmented variable of the states for the controlled system and the reference model should be released to the tracking controller at the same certain time-point, the asynchronous event-triggered mechanism used in this section is more general. This is mainly because it is difficult for various parts of the system to be located at the same place in networked based tracking control and the asynchronous event-triggered mechanism can effectively reduce the network communication burdens in both communication channels from the plant to the controller and from the reference model to the controller.

To further reduce the amount of data in the communication of network and realize the effective utilization of the limited network communication resources, three static logarithmic quantizers will be utilized to quantize $y(k_s)$, $\widehat{y}(k_r)$, and $u(k)$ before they are transmitted to the tracking controller and the plant, respectively. The static logarithmic quantizers considered in this section are given as

$$\begin{aligned}
y_g(k_s) &= g_y(y(k_s)) = \left[g_y^1(y^1(k_s))\ g_y^2(y^2(k_s))\ \ \dots\ g_y^{n_y}(y^{n_y}(k_s))\right]^T, \\
g_y^{j_y}(-y^{j_y}(k_s)) &= -g_y^{j_y}(y^{j_y}(k_s)),\ j_y = 1, 2, \dots, n_y, \\
\widehat{y}_g(k_r) &= g_{\widehat{y}}(\widehat{y}(k_r)) = \left[g_{\widehat{y}}^1(\widehat{y}^1(k_r))\ g_{\widehat{y}}^2(\widehat{y}^2(k_r))\ \ \dots\ g_{\widehat{y}}^{n_y}(\widehat{y}^{n_y}(k_r))\right]^T, \\
g_{\widehat{y}}^{j_{\widehat{y}}}(-\widehat{y}^{j_{\widehat{y}}}(k_r)) &= -g_{\widehat{y}}^{j_{\widehat{y}}}(\widehat{y}^{j_{\widehat{y}}}(k_r)),\ j_{\widehat{y}} = 1, 2, \dots, n_y, \\
u_g(k) &= g_u(u(k)) = \left[g_u^1(u^1(k))\ g_u^2(u^2(k))\ \ \dots\ g_u^{n_u}(u^{n_u}(k))\right]^T, \\
g_u^{j_u}(-u^{j_u}(k)) &= -g_u^{j_u}(u^{j_u}(k)),\ j_u = 1, 2, \dots, n_u.
\end{aligned} \tag{5.8}$$

As in [4–6], the set of quantization levels for quantizer $g_u^{j_u}(u^{j_u}(k))$ is assumed to be

$$\mathbf{U}_{j_u} = \{\pm\eta_i^{j_u}, \eta_i^{j_u} =. (\theta_u^{j_u})^i \eta_0^{j_u}, i = \pm 1, \pm 2, \cdots\} \cup \{\pm\eta_0^{j_u}\} \cup \{0\}, \tag{5.9}$$

where $\theta_u^{j_u} \in (0,\ 1)$ stands for the quantization density of $g_u^{j_u}(u^{j_u}(k))$, $\eta_0^{j_u} > 0$. The definition of $g_u^{j_u}(u^{j_u}(k))$ is formulated as

$$g_u^{j_u}(u^{j_u}(k)) = \begin{cases} \eta_i^{j_u}, & \text{if } \frac{\eta_i^{j_u}}{1+\beta_u^{j_u}} < u^{j_u}(k) \le \frac{\eta_i^{j_u}}{1-\beta_u^{j_u}}, \\ 0, & \text{if } u^{j_u}(k) = 0, \\ -g_u^{j_u}(-u^{j_u}(k)), & \text{if } u^{j_u}(k) < 0, \end{cases} \tag{5.10}$$

where

$$\beta_u^{j_u} = \frac{1 - \theta_u^{j_u}}{1 + \theta_u^{j_u}}. \tag{5.11}$$

Besides, assuming that $\theta_y^{j_y} \in (0,\ 1)$ and $\theta_{\widehat{y}}^{j_{\widehat{y}}} \in (0,\ 1)$ are quantization densities for $g_y^{j_y}(y^{j_y}(k_s))$ and $g_{\widehat{y}}^{j_{\widehat{y}}}(\widehat{y}^{j_{\widehat{y}}}(k_r))$, respectively. Then, there are

$$\beta_y^{j_y} = \frac{1 - \theta_y^{j_y}}{1 + \theta_y^{j_y}},\ \ \beta_{\widehat{y}}^{j_{\widehat{y}}} = \frac{1 - \theta_{\widehat{y}}^{j_{\widehat{y}}}}{1 + \theta_{\widehat{y}}^{j_{\widehat{y}}}}. \tag{5.12}$$

Based on the important conclusions developed in [4], it can be obtained that

$$\begin{aligned}
y_g(k_s) - y(k_s) &= \varXi_y(k)y(k_s),\ \varXi_y(k) \in [-\beta_y,\ \beta_y], \\
\widehat{y}_g(k_r) - \widehat{y}(k_r) &= \varXi_{\widehat{y}}(k)\widehat{y}(k_r),\ \varXi_{\widehat{y}}(k) \in [-\beta_{\widehat{y}},\ \beta_{\widehat{y}}], \\
u_g(k) - u(k) &= \varXi_u(k)u(k),\ \varXi_u(k) \in [-\beta_u,\ \beta_u],
\end{aligned} \tag{5.13}$$

where

$$\begin{aligned}
\Xi_y(k) &= diag\{\Xi_y^1(k),\ \Xi_y^2(k),\ \ldots,\ \Xi_y^{n_y}(k)\},\\
\Xi_{\widehat{y}}(k) &= diag\{\Xi_{\widehat{y}}^1(k),\ \Xi_{\widehat{y}}^2(k),\ \ldots,\ \Xi_{\widehat{y}}^{n_y}(k)\},\\
\Xi_u(k) &= diag\{\Xi_u^1(k),\ \Xi_u^2(k),\ \ldots,\ \Xi_u^{n_u}(k)\},\\
\beta_y &= diag\{\beta_y^1,\ \beta_y^2,\ \ldots,\ \beta_y^{n_y}\},\\
\beta_{\widehat{y}} &= diag\{\beta_{\widehat{y}}^1,\ \beta_{\widehat{y}}^2,\ \ldots,\ \beta_{\widehat{y}}^{n_y}\},\\
\beta_u &= diag\{\beta_u^1,\ \beta_u^2,\ \ldots,\ \beta_u^{n_u}\}.
\end{aligned}$$

Remark 5.2 In this section, three static logarithmic quantizers are used to reduce the number of data transmission in the communication channels from the plant and the reference model to the controller and from the controller to the plant, which is a more general assumption in practical situations. It is a significant contribution relative to the existing results on quantized tracking control where only quantized measurement output (see [7] for more details) or quantized control input (see [8] for more details) was considered.

The output feedback tracking controller discussed in this section is presented as

$$u(k) = G_1 y_g(k_s) + G_2 \widehat{y}_g(k_r), \tag{5.14}$$

where G_1 and G_2 denote the parameters for the tracking controller.

By considering the equation in (5.13), $u(k)$ defined in (5.14) can be expressed as

$$u(k) = G_1(I + \Xi_y(k))y(k_s) + G_2(I + \Xi_{\widehat{y}}(k))\widehat{y}(k_r). \tag{5.15}$$

Moreover, by taking $y(k_s) = y(k) - \xi_y(k)$ and $\widehat{y}(k_r) = \widehat{y}(k) - \xi_{\widehat{y}}(k)$ into consideration, one can be obtained that

$$\begin{aligned}
u(k) &= G_1(I + \Xi_y(k))(C(\rho)x(k) - \xi_y(k))\\
&\quad + G_2(I + \Xi_{\widehat{y}}(k))(C_\iota\widehat{x}(k) - \xi_{\widehat{y}}(k))\\
&= G_1(I + \Xi_y(k))C(\rho)x(k) - G_1(I + \Xi_y(k))\xi_y(k))\\
&\quad + G_2(I + \Xi_{\widehat{y}}(k))C_\iota\widehat{x}(k) - G_2(I + \Xi_{\widehat{y}}(k))\xi_{\widehat{y}}(k)).
\end{aligned} \tag{5.16}$$

Based on the discussions above, the closed-loop system can be established as

$$\begin{aligned}
\delta(k+1) &= A_c(\rho)\sigma(k) + B_c(\rho)w_\iota(k),\\
e_c(k) &= C_c(\rho)\sigma(k),
\end{aligned} \tag{5.17}$$

where

$$\delta^T(k) = \begin{bmatrix} x^T(k) & \widehat{x}^T(k) \end{bmatrix}, \ \sigma^T(k) = \begin{bmatrix} x^T(k) & \widehat{x}^T(k) & \xi_y^T(k) & \xi_{\widehat{y}}^T(k) \end{bmatrix},$$
$$w_\iota^T(k) = \begin{bmatrix} w^T(k) & \iota^T(k) \end{bmatrix}, \ e_c(k) = y(k) - \widehat{y}(k),$$
$$A_c(\rho) = \begin{bmatrix} \Delta_1(\rho) & \Delta_2(\rho) & \Delta_3(\rho) & \Delta_4(\rho) \\ 0 & A_\iota & 0 & 0 \end{bmatrix}, \ B_c(\rho) = \begin{bmatrix} D(\rho) & 0 \\ 0 & B_\iota \end{bmatrix},$$
$$C_c(\rho) = \begin{bmatrix} C(\rho) & -C_\iota & 0 & 0 \end{bmatrix},$$
$$\Delta_1(\rho) = A(\rho) + B(\rho)G_1(I + \Xi_y(k))C(\rho) + B(\rho)\Xi_u(k)G_1(I + \Xi_y(k))C(\rho),$$
$$\Delta_2(\rho) = B(\rho)G_2(I + \Xi_{\widehat{y}}(k))C_\iota + B(\rho)\Xi_u(k)G_2(I + \Xi_{\widehat{y}}(k))C_\iota,$$
$$\Delta_3(\rho) = -B(\rho)G_1(I + \Xi_y(k)) - B(\rho)\Xi_u(k)G_1(I + \Xi_y(k)),$$
$$\Delta_4(\rho) = -B(\rho)G_2(I + \Xi_{\widehat{y}}(k)) - B(\rho)\Xi_u(k)G_2(I + \Xi_{\widehat{y}}(k)).$$

The event-triggered static output feedback tracking control problem considered in this section is to design an event-based output feedback tracking controller in (5.14) such that

(1) The closed-loop system (5.17) is asymptotically stable when $w_\iota(k) = 0$.
(2) The closed-loop system (5.17) has a prescribed $\mathcal{H}_\infty$ tracking performance $\gamma > 0$, i.e., under the zero-initial condition,

$$\sum_{k=0}^{\infty} e_c^T(k)e_c(k) < \gamma^2 \sum_{k=0}^{\infty} w_\iota{}^T(k)w_\iota(k)$$

is satisfied for any $w_\iota(k) \neq 0$.

5.1.2 *Main Results*

In this subsection, firstly, the sufficient conditions will be given, such that the closed-loop system (5.17) is asymptotically stable with the specified $\mathcal{H}_\infty$ tracking performance $\gamma > 0$ in the presence of event-triggered communication scheme and quantization. Based on the Lyapunov theory, a useful $\mathcal{H}_\infty$ tracking performance analysis criterion will be given in the following theorem, which can ensure the closed-loop system (5.17) is asymptotically stable and has the specified $\mathcal{H}_\infty$ tracking performance.

Theorem 5.1 *Consider the fuzzy system (5.1), the reference model (5.5), and the output feedback tracking controller (5.14). Suppose that the scalars $0 \le b_y < 1$ and $0 \le b_{\widehat{y}} < 1$ are known. The closed-loop system (5.17) is asymptotically stable with a specified $\mathcal{H}_\infty$ tracking performance $\gamma > 0$, if there exist matrices $P > 0$, $\mathbf{J}_y > 0$, and $\mathbf{J}_{\widehat{y}} > 0$ satisfying*

$$\begin{bmatrix} -\widehat{\mathbf{J}}_y - \widehat{\mathbf{J}}_{\widehat{y}} - \widehat{P} + b_y \Sigma_y^T \mathbf{J}_y \Sigma_y + b_{\widehat{y}} \Sigma_{\widehat{y}}^T \mathbf{J}_{\widehat{y}} \Sigma_{\widehat{y}} & * & * & * \\ 0 & -\gamma^2 I & * & * \\ A_c(\rho) & B_c(\rho) & -P^{-1} & * \\ C_c(\rho) & 0 & 0 & -I \end{bmatrix} < 0, \tag{5.18}$$

where

$$\begin{aligned} \widehat{\mathbf{J}}_y &= diag\left\{0,\ 0,\ \mathbf{J}_y,\ 0\right\}, \\ \widehat{\mathbf{J}}_{\widehat{y}} &= diag\left\{0,\ 0,\ 0,\ \mathbf{J}_{\widehat{y}}\right\}, \\ \widehat{P} &= diag\left\{P,\ 0,\ 0\right\}, \\ \Sigma_y &= \left[C(\rho)\ 0\ -I\ 0\right], \\ \Sigma_{\widehat{y}} &= \left[0\ C_\iota\ 0\ -I\right]. \end{aligned}$$

Proof Firstly, we will show that if the condition (5.18) is satisfied, then, for $w_\iota(k) = 0$, the closed-loop system (5.17) is asymptotically stable.

To this end, a Lyapunov function is constructed as

$$V(\delta(k)) = \delta^T(k) P \delta(k),\ P > 0. \tag{5.19}$$

According to the event-triggered conditions given in (5.6) and (5.7), one can be obtained that

$$\begin{aligned} & b_y(y(k) - \xi_y(k))^T \mathbf{J}_y (y(k) - \xi_y(k)) - \xi_y^T(k) \mathbf{J}_y \xi_y(k) \geq 0, \\ & b_{\widehat{y}}(\widehat{y}(k) - \xi_{\widehat{y}}(k))^T \mathbf{J}_{\widehat{y}} (\widehat{y}(k) - \xi_{\widehat{y}}(k)) - \xi_{\widehat{y}}^T(k) \mathbf{J}_{\widehat{y}} \xi_{\widehat{y}}(k) \geq 0. \end{aligned}$$

Then, it is easily to have that

$$\begin{aligned} & V(\delta(k+1)) - V(\delta(k)) \\ & \leq \delta^T(k+1) P \delta(k+1) - \delta^T(k) P \delta(k) \\ & \quad - \xi_y^T(k) \mathbf{J}_y \xi_y(k) + b_y(y(k) - \xi_y(k))^T \mathbf{J}_y (y(k) - \xi_y(k)) \\ & \quad - \xi_{\widehat{y}}^T(k) \mathbf{J}_{\widehat{y}} \xi_{\widehat{y}}(k) + b_{\widehat{y}}(\widehat{y}(k) - \xi_{\widehat{y}}(k))^T \mathbf{J}_{\widehat{y}} (\widehat{y}(k) - \xi_{\widehat{y}}(k)) \\ & = \sigma^T(k) A_c^T(\rho) P A_c(\rho) \sigma(k) - \delta^T(k) P \delta(k) \\ & \quad - \xi_{\widehat{y}}^T(k) \mathbf{J}_{\widehat{y}} \xi_{\widehat{y}}(k) + b_{\widehat{y}}(C_\iota \widehat{x}(k) - \xi_{\widehat{y}}(k))^T \mathbf{J}_{\widehat{y}} (C_\iota \widehat{x}(k) - \xi_{\widehat{y}}(k)) \\ & \quad - \xi_y^T(k) \mathbf{J}_y \xi_y(k) + b_y(C(\rho) x(k) - \xi_y(k))^T \mathbf{J}_y (C(\rho) x(k) - \xi_y(k)) \\ & = \sigma^T(k) \left(A_c^T(\rho) P A_c(\rho) - \widehat{P} - \widehat{\mathbf{J}}_y - \widehat{\mathbf{J}}_{\widehat{y}} + b_y \Sigma_y^T \mathbf{J}_y \Sigma_y + b_{\widehat{y}} \Sigma_{\widehat{y}}^T \mathbf{J}_{\widehat{y}} \Sigma_{\widehat{y}}\right) \sigma(k). \end{aligned} \tag{5.20}$$

When $w_\iota(k) = 0$, the inequality in (5.18) reduces to

$$\begin{bmatrix} -\widehat{\mathbf{J}}_y - \widehat{\mathbf{J}}_{\widehat{y}} - \widehat{P} + b_y \Sigma_y^T \mathbf{J}_y \Sigma_y + b_{\widehat{y}} \Sigma_{\widehat{y}}^T \mathbf{J}_{\widehat{y}} \Sigma_{\widehat{y}} & * \\ A_c(\rho) & -P^{-1} \end{bmatrix} < 0. \tag{5.21}$$

By using Lemma 1.1 to (5.21), it can be obtained that

$$A_c^T(\rho)PA_c(\rho) - \widehat{\mathbf{J}}_y - \widehat{\mathbf{J}}_{\widehat{y}} - \widehat{P} + b_y\Sigma_y^T\mathbf{J}_y\Sigma_y + b_{\widehat{y}}\Sigma_{\widehat{y}}^T\mathbf{J}_{\widehat{y}}\Sigma_{\widehat{y}} < 0. \tag{5.22}$$

Then, for $\sigma(k) \neq 0$, it can be obtained that

$$V(\delta(k+1)) - V(\delta(k)) < 0. \tag{5.23}$$

Thus, it can be concluded that the closed-loop system (5.17) is asymptotically stable for $w_\iota(k) = 0$ if the condition (5.18) is satisfied.

Next, for $w_\iota(k) \neq 0$, the $\mathscr{H}_\infty$ tracking performance for the closed-loop system (5.17) under zero-initial conditions will be discussed.

When $w_\iota(k) \neq 0$, the following conclusion can be drawn

$$\begin{aligned}
&V(\delta(k+1)) - V(\delta(k)) + e_c^T(k)e_c(k) - \gamma^2 w_\iota^T(k)w_\iota(k)\\
&\le \delta^T(k+1)P\delta(k+1) - \delta^T(k)P\delta(k) + e_c^T(k)e_c(k) - \gamma^2 w_\iota^T(k)w_\iota(k)\\
&\quad + b_y(y(k) - \xi_y(k))^T\mathbf{J}_y(y(k) - \xi_y(k)) - \xi_y^T(k)\mathbf{J}_y\xi_y(k)\\
&\quad + b_{\widehat{y}}(\widehat{y}(k) - \xi_{\widehat{y}}(k))^T\mathbf{J}_{\widehat{y}}(\widehat{y}(k) - \xi_{\widehat{y}}(k)) - \xi_{\widehat{y}}^T(k)\mathbf{J}_{\widehat{y}}\xi_{\widehat{y}}(k)\\
&= (A_c(\rho)\sigma(k) + B_c(\rho)w_\iota(k))^T P(A_c(\rho)\sigma(k) + B_c(\rho)w_\iota(k))\\
&\quad - \delta^T(k)P\delta(k) + \sigma^T(k)C_c^T(\rho)C_c(\rho)\sigma(k) - \gamma^2 w_\iota^T(k)w_\iota(k)\\
&\quad - \xi_{\widehat{y}}^T(k)\mathbf{J}_{\widehat{y}}\xi_{\widehat{y}}(k) + b_{\widehat{y}}(C_\iota\widehat{x}(k) - \xi_{\widehat{y}}(k))^T\mathbf{J}_{\widehat{y}}(C_\iota\widehat{x}(k) - \xi_{\widehat{y}}(k))\\
&\quad - \xi_y^T(k)\mathbf{J}_y\xi_y(k) + b_y(C(\rho)x(k) - \xi_y(k))^T\mathbf{J}_y(C(\rho)x(k) - \xi_y(k))\\
&= \widehat{\sigma}^T(k)\mathbf{Z}(\rho)\widehat{\sigma}(k),
\end{aligned} \tag{5.24}$$

where $\widehat{\sigma}^T(k) = \begin{bmatrix}\sigma^T(k) & w_\iota^T(k)\end{bmatrix}$ and

$$\begin{aligned}
\mathbf{Z}(\rho) = &\begin{bmatrix}A_c(\rho) & B_c(\rho)\end{bmatrix}^T P\begin{bmatrix}A_c(\rho) & B_c(\rho)\end{bmatrix} + \begin{bmatrix}C_c(\rho) & 0\end{bmatrix}^T\begin{bmatrix}C_c(\rho) & 0\end{bmatrix}\\
&- diag\{\widehat{P}, \gamma^2 I\} + b_y\begin{bmatrix}\Sigma_y & 0\end{bmatrix}^T\mathbf{J}_y\begin{bmatrix}\Sigma_y & 0\end{bmatrix} - diag\{\widehat{\mathbf{J}}_y, 0\}\\
&+ b_{\widehat{y}}\begin{bmatrix}\Sigma_{\widehat{y}} & 0\end{bmatrix}^T\mathbf{J}_{\widehat{y}}\begin{bmatrix}\Sigma_{\widehat{y}} & 0\end{bmatrix} - diag\{\widehat{\mathbf{J}}_{\widehat{y}}, 0\}.
\end{aligned}$$

By applying Lemma 1.1 to (5.18), it can be obtained that

$$\mathbf{Z}(\rho) < 0. \tag{5.25}$$

Then, for $\widehat{\sigma}(k) \neq 0$, there is

$$V(\delta(k+1)) - V(\delta(k)) + e_c^T(k)e_c(k) - \gamma^2 w_\iota^T(k)w_\iota(k) < 0. \tag{5.26}$$

Moreover, by summing up (5.26) from $k = 0$ to $k = \infty$, it can be obtained that

$$V(\delta(\infty)) - V(\delta(0)) + \sum_{k=0}^{\infty} e_c^T(k)e_c(k) - \gamma^2\sum_{k=0}^{\infty} w_\iota^T(k)w_\iota(k) < 0. \tag{5.27}$$

According to zero-initial conditions, one can obtain $V(\delta(0)) = 0$ and $V(\delta(\infty)) \geq 0$, which implies that

$$\sum_{k=0}^{\infty} e_c^T(k)e_c(k) - \gamma^2 \sum_{k=0}^{\infty} w_\iota^T(k)w_\iota(k) < 0. \tag{5.28}$$

Thus, one can conclude that the closed-loop system (5.17) has a specified $\mathcal{H}_\infty$ tracking performance $\gamma > 0$ if the condition (5.18) is satisfied. The proof is completed.

Then, according to the analysis results proposed in Theorem 5.1, the design problem of the event-triggered output feedback tracking controller will be addressed. More specifically, by using two-step uncertainty elimination method combined with the extended matrix inequality decoupling strategy, the sufficient design conditions for the event-triggered quantized output feedback tracking controller defined in (5.14) will be proposed in terms of LMIs.

Theorem 5.2 *Consider the fuzzy system (5.4), the reference model (5.5), and the output feedback tracking controller (5.14). Suppose that the scalars* $0 \leq b_y < 1$, $0 \leq b_{\hat{y}} < 1$, *and* λ *are known. The closed-loop system (5.17) is asymptotically stable with a specified* $\mathcal{H}_\infty$ *tracking performance* $\gamma > 0$, *if there exist matrices* P_1, P_2, P_3, $\mathbf{J}_y > 0$, $\mathbf{J}_{\hat{y}} > 0$, S_1, S_2, N_1, N_2, R_1, *and* R_2, *scalars* $\vartheta_u > 0$, $\vartheta_y > 0$, *and* $\vartheta_{\hat{y}} > 0$ *satisfying*

$$\begin{bmatrix} P_1 & * \\ P_2 & P_3 \end{bmatrix} > 0, \tag{5.29}$$

$$\begin{bmatrix} \Omega_{11i} & * & * & * & * & * & * & * \\ 0 & -\gamma^2 I & * & * & * & * & * & * \\ \Omega_{31i} & \Omega_{32i} & \Omega_{33} & * & * & * & * & * \\ \Omega_{41i} & 0 & 0 & -I & * & * & * & * \\ \Omega_{51i} & 0 & \Omega_{53i} & 0 & -\vartheta_u I & * & * & * \\ \Omega_{61i} & 0 & \Omega_{63} & 0 & \Omega_{65} & -\vartheta_y I & * & * \\ \Omega_{71} & 0 & \Omega_{73} & 0 & \Omega_{75} & 0 & -\vartheta_{\hat{y}} I & * \\ \Omega_{81i} & 0 & \Omega_{83i} & 0 & \Omega_{85} & \Omega_{86} & \Omega_{87} & \Omega_{88} \end{bmatrix} < 0, \ i = 1, 2, \ldots, r, \tag{5.30}$$

where

$$\Omega_{11i} = \begin{bmatrix} -P_1 + b_y C_i^T \mathbf{J}_y C_i & * & * & * \\ -P_2 & -P_3 + b_{\hat{y}} C_\iota^T \mathbf{J}_{\hat{y}} C_\iota & * & * \\ -b_y \mathbf{J}_y C_i & 0 & (b_y - 1)\mathbf{J}_y & * \\ 0 & -b_{\hat{y}} \mathbf{J}_{\hat{y}} C_\iota & 0 & (b_{\hat{y}} - 1)\mathbf{J}_{\hat{y}} \end{bmatrix},$$

$$\Omega_{31i} = \begin{bmatrix} S_1 A_i + \Theta_1 N_1 C_i & \Theta_2 N_2 C_\iota & -\Theta_1 N_1 & -\Theta_2 N_2 \\ 0 & S_2 A_\iota & 0 & 0 \end{bmatrix}, \ \Omega_{32i} = \begin{bmatrix} S_1 D_i & 0 \\ 0 & S_2 B_\iota \end{bmatrix},$$

$$\Omega_{33} = \begin{bmatrix} P_1 - S_1 - S_1^T & * \\ P_2 & P_3 - S_2 - S_2^T \end{bmatrix}, \ \Omega_{41i} = \begin{bmatrix} C_i & -C_\iota & 0 & 0 \end{bmatrix},$$

$$\Omega_{51i} = \begin{bmatrix} 0 & 0 & 0 & 0 \\ N_1 C_i & N_2 C_\iota & -N_1 & -N_2 \end{bmatrix}, \ \Omega_{53i} = \begin{bmatrix} (S_1 B_i)^T & 0 \\ 0 & 0 \end{bmatrix},$$

$$\Omega_{61i} = \begin{bmatrix} 0 & 0 & 0 & 0 \\ \vartheta_y \beta_y C_i & 0 & -\vartheta_y \beta_y & 0 \end{bmatrix}, \ \Omega_{63} = \begin{bmatrix} (\Theta_1 N_1)^T & 0 \\ 0 & 0 \end{bmatrix}, \ \Omega_{65} = \begin{bmatrix} 0 & N_1^T \\ 0 & 0 \end{bmatrix},$$

$$\Omega_{71} = \begin{bmatrix} 0 & 0 & 0 & 0 \\ 0 & \vartheta_{\widehat{y}} \beta_{\widehat{y}} C_\iota & 0 & -\vartheta_{\widehat{y}} \beta_{\widehat{y}} \end{bmatrix}, \ \Omega_{73} = \begin{bmatrix} (\Theta_2 N_2)^T & 0 \\ 0 & 0 \end{bmatrix}, \ \Omega_{75} = \begin{bmatrix} 0 & N_2^T \\ 0 & 0 \end{bmatrix},$$

$$\Omega_{81i} = \begin{bmatrix} N_1 C_i & 0 & -N_1 & 0 \\ 0 & N_2 C_\iota & 0 & -N_2 \end{bmatrix}, \ \Omega_{83i} = \begin{bmatrix} \lambda(S_1 B_i - \Theta_1 R_1)^T & 0 \\ \lambda(S_1 B_i - \Theta_2 R_2)^T & 0 \end{bmatrix},$$

$$\Omega_{85} = \begin{bmatrix} 0 & \lambda(\vartheta_u \beta_u - R_1)^T \\ 0 & \lambda(\vartheta_u \beta_u - R_2)^T \end{bmatrix}, \ \Omega_{86} = \begin{bmatrix} N_1 & 0 \\ 0 & 0 \end{bmatrix},$$

$$\Omega_{87} = \begin{bmatrix} 0 & 0 \\ N_2 & 0 \end{bmatrix}, \ \Omega_{88} = \begin{bmatrix} -\lambda(R_1 + R_1^T) & * \\ 0 & -\lambda(R_2 + R_2^T) \end{bmatrix}.$$

Moreover, the parameters for the tracking controller defined in (5.14) can be obtained as

$$G_1 = R_1^{-1} N_1, \ \ G_2 = R_2^{-1} N_2. \tag{5.31}$$

Proof First of all, Lemma 1.2 will be used to eliminate the quantization error terms $\Xi_u(k)$, $\Xi_y(k)$, and $\Xi_{\widehat{y}}(k)$, and the two-step uncertainty elimination method will be used to deal with the crossing terms between $\Xi_u(k)$ and $\Xi_y(k)\big(\Xi_{\widehat{y}}(k)\big)$. Specifically, the quantization error term $\Xi_u(k)$ will be separated and eliminated in the first step and the quantization error terms $\Xi_y(k)$ and $\Xi_{\widehat{y}}(k)$ will be separated and eliminated in the second step.

The condition in (5.18) can be expressed as

$$\begin{bmatrix} -\widehat{\mathbf{J}}_y - \widehat{\mathbf{J}}_{\widehat{y}} - \widehat{P} + b_y \Sigma_y^T \mathbf{J}_y \Sigma_y + b_{\widehat{y}} \Sigma_{\widehat{y}}^T \mathbf{J}_{\widehat{y}} \Sigma_{\widehat{y}} & * & * & * \\ 0 & -\gamma^2 I & * & * \\ \widehat{A}_c(\rho) & B_c(\rho) & -P^{-1} & * \\ C_c(\rho) & 0 & 0 & -I \end{bmatrix}$$
$$+ He\left\{\begin{bmatrix} 0 & 0 & Q_1^T(\rho) & 0 \end{bmatrix}^T \frac{\Xi_u(k)}{\beta_u} \begin{bmatrix} Q_2(\rho) & 0 & 0 & 0 \end{bmatrix}\right\} < 0, \tag{5.32}$$

where

$$\widehat{A}_c(\rho) = \begin{bmatrix} \widehat{\Delta}_1(\rho) & \widehat{\Delta}_2(\rho) & \widehat{\Delta}_3(\rho) & \widehat{\Delta}_4(\rho) \\ 0 & A_\iota & 0 & 0 \end{bmatrix}, \ Q_1(\rho) = \begin{bmatrix} B(\rho) \\ 0 \end{bmatrix},$$
$$Q_2(\rho) = \begin{bmatrix} M_1(\rho) & M_2 & M_3 & M_4 \end{bmatrix},$$
$$\widehat{\Delta}_1(\rho) = A(\rho) + B(\rho) G_1 (I + \Xi_y(k)) C(\rho),$$
$$\widehat{\Delta}_2(\rho) = B(\rho) G_2 (I + \Xi_{\widehat{y}}(k)) C_\iota,$$

$$
\begin{aligned}
\widehat{\Delta}_3(\rho) &= -B(\rho)G_1(I + \Xi_y(k)),\\
\widehat{\Delta}_4(\rho) &= -B(\rho)G_2(I + \Xi_{\widehat{y}}(k)),\\
M_1(\rho) &= \beta_u G_1(I + \Xi_y(k))C(\rho),\\
M_2 &= \beta_u G_2(I + \Xi_{\widehat{y}}(k))C_\iota,\\
M_3 &= -\beta_u G_1(I + \Xi_y(k)),\\
M_4 &= -\beta_u G_2(I + \Xi_{\widehat{y}}(k)).
\end{aligned}
$$

According to Lemma 1.2, it can be obtained that the inequality in (5.32) is true if and only if there exists a scalar $\vartheta_u > 0$ such that the following inequality

$$
\begin{bmatrix}
-\widehat{\mathbf{J}}_y - \widehat{\mathbf{J}}_{\widehat{y}} - \widehat{P} + b_y \Sigma_y^T \mathbf{J}_y \Sigma_y + b_{\widehat{y}} \Sigma_{\widehat{y}}^T \mathbf{J}_{\widehat{y}} \Sigma_{\widehat{y}} & * & * & * & * \\
0 & -\gamma^2 I & * & * & * \\
\widehat{A}_c(\rho) & B_c(\rho) & -P^{-1} & * & * \\
C_c(\rho) & 0 & 0 & -I & * \\
W_{51} & 0 & W_{53} & 0 & -\vartheta_u I
\end{bmatrix} < 0
\tag{5.33}
$$

is satisfied, where

$$
W_{51} = \begin{bmatrix} 0 \\ \vartheta_u Q_2(\rho) \end{bmatrix}, \quad W_{53} = \begin{bmatrix} Q_1^T(\rho) \\ 0 \end{bmatrix}.
$$

In the first step, the quantization error term $\Xi_u(k)$ has been separated and eliminated. In this step, the quantization error term $\Xi_y(k)$ and $\Xi_{\widehat{y}}(k)$ will be separated and eliminated. The condition in (5.33) can be rewritten as

$$
\begin{aligned}
&\begin{bmatrix}
-\widehat{\mathbf{J}}_y - \widehat{\mathbf{J}}_{\widehat{y}} - \widehat{P} + b_y \Sigma_y^T \mathbf{J}_y \Sigma_y + b_{\widehat{y}} \Sigma_{\widehat{y}}^T \mathbf{J}_{\widehat{y}} \Sigma_{\widehat{y}} & * & * & * & * \\
0 & -\gamma^2 I & * & * & * \\
\overline{A}_c(\rho) & B_c(\rho) & -P^{-1} & * & * \\
C_c(\rho) & 0 & 0 & -I & * \\
\overline{W}_{51} & 0 & W_{53} & 0 & -\vartheta_u I
\end{bmatrix}\\
&+ He\left\{[0 \;\; 0 \;\; Q_3^T(\rho) \;\; 0 \;\; Q_4^T]^T \left(\frac{\Xi_y(k)}{\beta_y}\right) [Q_5(\rho) \;\; 0 \;\; 0 \;\; 0 \;\; 0]\right\}\\
&+ He\left\{[0 \;\; 0 \;\; Q_6^T(\rho) \;\; 0 \;\; Q_7^T]^T \left(\frac{\Xi_{\widehat{y}(k)}}{\beta_{\widehat{y}}}\right) [Q_8 \;\; 0 \;\; 0 \;\; 0 \;\; 0]\right\} < 0,
\end{aligned}
\tag{5.34}
$$

where

$$\overline{A}_c(\rho) = \begin{bmatrix} A(\rho) + B(\rho)G_1C(\rho) & B(\rho)G_2C_\iota & -B(\rho)G_1 & -B(\rho)G_2 \\ 0 & A_\iota & 0 & 0 \end{bmatrix},$$

$$\overline{W}_{51} = \begin{bmatrix} 0 & 0 & 0 & 0 \\ \vartheta_u\beta_uG_1C(\rho) & \vartheta_u\beta_uG_2C_\iota & -\vartheta_u\beta_uG_1 & -\vartheta_u\beta_uG_2 \end{bmatrix},$$

$$Q_3(\rho) = \begin{bmatrix} B(\rho)G_1 \\ 0 \end{bmatrix}, \; Q_4 = \begin{bmatrix} 0 \\ \vartheta_u\beta_uG_1 \end{bmatrix},$$

$$Q_6(\rho) = \begin{bmatrix} B(\rho)G_2 \\ 0 \end{bmatrix}, \; Q_7 = \begin{bmatrix} 0 \\ \vartheta_u\beta_uG_2 \end{bmatrix},$$

$$Q_5(\rho) = \begin{bmatrix} \beta_yC(\rho) & 0 & -\beta_y & 0 \end{bmatrix}, \; Q_8 = \begin{bmatrix} 0 & \beta_{\widehat{y}}C_\iota & 0 & -\beta_{\widehat{y}} \end{bmatrix}.$$

According to Lemma 1.2, it can be obtained that the inequality in (5.34) is true if and only if there exist two scalars $\vartheta_y > 0$ and $\vartheta_{\widehat{y}} > 0$ such that the following inequality

$$\begin{bmatrix} \Sigma_p & * & * & * & * & * & * \\ 0 & -\gamma^2I & * & * & * & * & * \\ \overline{A}_c(\rho) & B_c(\rho) & -P^{-1} & * & * & * & * \\ C_c(\rho) & 0 & 0 & -I & * & * & * \\ \overline{W}_{51} & 0 & W_{53} & 0 & -\vartheta_uI & * & * \\ W_{61} & 0 & W_{63} & 0 & W_{65} & -\vartheta_yI & * \\ W_{71} & 0 & W_{73} & 0 & W_{75} & 0 & -\vartheta_{\widehat{y}}I \end{bmatrix} < 0 \tag{5.35}$$

is satisfied, where

$$\Sigma_p = -\widehat{\mathbf{J}}_y - \widehat{\mathbf{J}}_{\widehat{y}} - \widehat{P} + b_y\Sigma_y^T\mathbf{J}_y\Sigma_y + b_{\widehat{y}}\Sigma_{\widehat{y}}^T\mathbf{J}_{\widehat{y}}\Sigma_{\widehat{y}},$$

$$W_{61} = \begin{bmatrix} 0 \\ \vartheta_yQ_5(\rho) \end{bmatrix}, \; W_{63} = \begin{bmatrix} Q_3^T(\rho) \\ 0 \end{bmatrix}, \; W_{65} = \begin{bmatrix} Q_4^T \\ 0 \end{bmatrix},$$

$$W_{71} = \begin{bmatrix} 0 \\ \vartheta_{\widehat{y}}Q_8 \end{bmatrix}, \; W_{73} = \begin{bmatrix} Q_6^T(\rho) \\ 0 \end{bmatrix}, \; W_{75} = \begin{bmatrix} Q_7^T \\ 0 \end{bmatrix}.$$

Performing the congruence transformation to the inequality of (5.35) with $diag\{I, I, S^T, I, I, I, I\}$. Then, one can obtain that

$$\begin{bmatrix} \Sigma_p & * & * & * & * & * & * \\ 0 & -\gamma^2I & * & * & * & * & * \\ S^T\overline{A}_c(\rho) & S^TB_c(\rho) & -S^TP^{-1}S & * & * & * & * \\ C_c(\rho) & 0 & 0 & -I & * & * & * \\ \overline{W}_{51} & 0 & W_{53}S & 0 & -\vartheta_uI & * & * \\ W_{61} & 0 & W_{63}S & 0 & W_{65} & -\vartheta_yI & * \\ W_{71} & 0 & W_{73}S & 0 & W_{75} & 0 & -\vartheta_{\widehat{y}}I \end{bmatrix} < 0. \tag{5.36}$$

In terms of $-(P-S)^T P^{-1}(P-S) \leq 0$ and $P > 0$, it can be obtained that $-S^T P^{-1} S \leq -S^T - S + P$. Then, the inequality (5.36) can be guaranteed by

$$\begin{bmatrix} \Sigma_p & * & * & * & * & * & * \\ 0 & -\gamma^2 I & * & * & * & * & * \\ S^T \overline{A}_c(\rho) & S^T B_c(\rho) & -S^T - S + P & * & * & * & * \\ C_c(\rho) & 0 & 0 & -I & * & * & * \\ \overline{W}_{51} & 0 & W_{53}S & 0 & -\vartheta_u I & * & * \\ W_{61} & 0 & W_{63}S & 0 & W_{65} & -\vartheta_y I & * \\ W_{71} & 0 & W_{73}S & 0 & W_{75} & 0 & -\vartheta_{\widehat{y}} I \end{bmatrix} < 0. \tag{5.37}$$

By choosing $P = \begin{bmatrix} P_1 & * \\ P_2 & P_3 \end{bmatrix}$ and $S^T = \begin{bmatrix} S_1 & 0 \\ 0 & S_2 \end{bmatrix}$, the condition in (5.37) can be expressed as

$$\begin{bmatrix} F_{11} & * & * & * & * & * & * \\ 0 & -\gamma^2 I & * & * & * & * & * \\ F_{31} & F_{32} & \Omega_{33} & * & * & * & * \\ F_{41} & 0 & 0 & -I & * & * & * \\ \overline{W}_{51} & 0 & F_{53} & 0 & -\vartheta_u I & * & * \\ F_{61} & 0 & F_{63} & 0 & F_{65} & -\vartheta_y I & * \\ \Omega_{71} & 0 & F_{73} & 0 & F_{75} & 0 & -\vartheta_{\widehat{y}} I \end{bmatrix} < 0, \tag{5.38}$$

where Ω_{33} and Ω_{71} have been defined in (5.30) and $F_{41} = \begin{bmatrix} C(\rho) & -C_\iota & 0 & 0 \end{bmatrix}$,

$$F_{11} = \begin{bmatrix} -P_1 + b_y C^T(\rho)\mathbf{J}_y C(\rho) & * & * & * \\ -P_2 & -P_3 + b_{\widehat{y}} C_\iota^T \mathbf{J}_{\widehat{y}} C_\iota & * & * \\ -b_y \mathbf{J}_y C(\rho) & 0 & (b_y - 1)\mathbf{J}_y & * \\ 0 & -b_{\widehat{y}} \mathbf{J}_{\widehat{y}} C_\iota & 0 & (b_{\widehat{y}} - 1)\mathbf{J}_{\widehat{y}} \end{bmatrix},$$

$$F_{31} = \begin{bmatrix} S_1 A(\rho) + S_1 B(\rho) G_1 C(\rho) & S_1 B(\rho) G_2 C_\iota & -S_1 B(\rho) G_1 & -S_1 B(\rho) G_2 \\ 0 & S_2 A_\iota & 0 & 0 \end{bmatrix},$$

$$F_{32} = \begin{bmatrix} S_1 D(\rho) & 0 \\ 0 & S_2 B_\iota \end{bmatrix},\ F_{41} = \begin{bmatrix} C(\rho) & -C_\iota & 0 & 0 \end{bmatrix},\ F_{53} = \begin{bmatrix} (S_1 B(\rho))^T & 0 \\ 0 & 0 \end{bmatrix},$$

$$F_{61} = \begin{bmatrix} 0 & 0 & 0 & 0 \\ \vartheta_y \beta_y C(\rho) & 0 & -\vartheta_y \beta_y & 0 \end{bmatrix},\ F_{63} = \begin{bmatrix} (S_1 B(\rho) G_1)^T & 0 \\ 0 & 0 \end{bmatrix},$$

$$F_{65} = \begin{bmatrix} 0 & \vartheta_u (\beta_u G_1)^T \\ 0 & 0 \end{bmatrix},\ F_{73} = \begin{bmatrix} (S_1 B(\rho) G_2)^T & 0 \\ 0 & 0 \end{bmatrix},\ F_{75} = \begin{bmatrix} 0 & \vartheta_u (\beta_u G_2)^T \\ 0 & 0 \end{bmatrix}.$$

However, it is worth noting that the condition in (5.38) is still non-convex with respect to the parameters of the event-triggered output feedback tracking controller defined in (5.14). Therefore, the matrix inequality decoupling strategy developed in [9, 10] will be extended to transform the condition in (5.38) into convex linear matrix inequality form.

To this end, suppose that $G_1 = R_1^{-1}N_1$ and $G_2 = R_2^{-1}N_2$. Then, the condition of (5.38) can be reorganized as

$$\begin{bmatrix} F_{11} & * & * & * & * & * & * \\ 0 & -\gamma^2 I & * & * & * & * & * \\ \widehat{F}_{31} & F_{32} & \Omega_{33} & * & * & * & * \\ F_{41} & 0 & 0 & -I & * & * & * \\ \widehat{F}_{51} & 0 & F_{53} & 0 & -\vartheta_u I & * & * \\ F_{61} & 0 & \Omega_{63} & 0 & \Omega_{65} & -\vartheta_y I & * \\ \Omega_{71} & 0 & \Omega_{73} & 0 & \Omega_{75} & 0 & -\vartheta_{\tilde{y}} I \end{bmatrix}$$
$$+He\left\{\begin{bmatrix} 0 & 0 & \Phi_1^T & 0 & \Phi_3^T & 0 & 0 \\ 0 & 0 & \Phi_2^T & 0 & \Phi_4^T & 0 & 0 \end{bmatrix}^T \begin{bmatrix} R_1^{-1} & * \\ 0 & R_2^{-1} \end{bmatrix} \begin{bmatrix} \Phi_5 & 0 & 0 & 0 & 0 & \Phi_6 & 0 \\ \Phi_7 & 0 & 0 & 0 & 0 & 0 & \Phi_8 \end{bmatrix}\right\} < 0, \tag{5.39}$$

where Ω_{33}, Ω_{63}, Ω_{65}, Ω_{71}, Ω_{73}, and Ω_{75} have been defined in (5.30) and

$$\widehat{F}_{31} = \begin{bmatrix} S_1 A(\rho) + \Theta_1 N_1 C(\rho) & \Theta_2 N_2 C_\iota & -\Theta_1 N_1 & -\Theta_2 N_2 \\ 0 & S_2 A_\iota & 0 & 0 \end{bmatrix},$$
$$\widehat{F}_{51} = \begin{bmatrix} 0 & 0 & 0 & 0 \\ N_1 C(\rho) & N_2 C_\iota & -N_1 & -N_2 \end{bmatrix}, \quad \Phi_1 = \begin{bmatrix} S_1 B(\rho) - \Theta_1 R_1 \\ 0 \end{bmatrix},$$
$$\Phi_2 = \begin{bmatrix} S_1 B(\rho) - \Theta_2 R_2 \\ 0 \end{bmatrix}, \quad \Phi_3 = \begin{bmatrix} 0 \\ \vartheta_u \beta_u - R_1 \end{bmatrix}, \quad \Phi_4 = \begin{bmatrix} 0 \\ \vartheta_u \beta_u - R_2 \end{bmatrix},$$
$$\Phi_5 = \begin{bmatrix} N_1 C(\rho) & 0 & -N_1 & 0 \end{bmatrix}, \quad \Phi_6 = \begin{bmatrix} N_1 & 0 \end{bmatrix},$$
$$\Phi_7 = \begin{bmatrix} 0 & N_2 C_\iota & 0 & -N_2 \end{bmatrix}, \quad \Phi_8 = \begin{bmatrix} N_2 & 0 \end{bmatrix}.$$

By applying Lemma 1.4, it can be concluded that the condition of (5.39) is true if and only if there exists a scalar λ such that the following inequality

$$\begin{bmatrix} F_{11} & * & * & * & * & * & * & * \\ 0 & -\gamma^2 I & * & * & * & * & * & * \\ \widehat{F}_{31} & F_{32} & \Omega_{33} & * & * & * & * & * \\ F_{41} & 0 & 0 & -I & * & * & * & * \\ \widehat{F}_{51} & 0 & F_{53} & 0 & -\vartheta_u I & * & * & * \\ F_{61} & 0 & \Omega_{63} & 0 & \Omega_{65} & -\vartheta_y I & * & * \\ \Omega_{71} & 0 & \Omega_{73} & 0 & \Omega_{75} & 0 & -\vartheta_{\tilde{y}} I & * \\ F_{81} & 0 & F_{83} & 0 & \Omega_{85} & \Omega_{86} & \Omega_{87} & \Omega_{88} \end{bmatrix} < 0 \tag{5.40}$$

is satisfied, where Ω_{33}, Ω_{63}, Ω_{65}, Ω_{71}, Ω_{73}, Ω_{75}, Ω_{85}, Ω_{86}, Ω_{87} and Ω_{88} have been defined in (5.30) and

$$F_{81} = \begin{bmatrix} \Phi_5 \\ \Phi_7 \end{bmatrix}, \quad F_{83} = \begin{bmatrix} \lambda \Phi_1^T \\ \lambda \Phi_2^T \end{bmatrix}.$$

Moreover, the condition in (5.40) can be reformulated as

$$\sum_{i=1}^{r} \rho_i(\epsilon(k)) \begin{bmatrix} \Omega_{11i} & * & * & * & * & * & * & * \\ 0 & -\gamma^2 I & * & * & * & * & * & * \\ \Omega_{31i} & \Omega_{32i} & \Omega_{33} & * & * & * & * & * \\ \Omega_{41i} & 0 & 0 & -I & * & * & * & * \\ \Omega_{51i} & 0 & \Omega_{53i} & 0 & -\vartheta_u I & * & * & * \\ \Omega_{61i} & 0 & \Omega_{63} & 0 & \Omega_{65} & -\vartheta_y I & * & * \\ \Omega_{71} & 0 & \Omega_{73} & 0 & \Omega_{75} & 0 & -\vartheta_{\widehat{y}} I & * \\ \Omega_{81i} & 0 & \Omega_{83i} & 0 & \Omega_{85} & \Omega_{86} & \Omega_{87} & \Omega_{88} \end{bmatrix} < 0. \tag{5.41}$$

Then, according to the characteristic of the basis functions presented in (5.3), it can be obtained that the inequality of (5.41) is true if inequality (5.30) is satisfied. The proof is completed.

For the event-triggered quantized output feedback tracking controller design strategy proposed in this section, the remark is given as follows:

Remark 5.3 If the scalars b_y, $b_{\widehat{y}}$, λ and the dimension adjustment matrices Θ_1, Θ_2 are given, then, the design conditions for the static output feedback tracking controller proposed in Theorem 5.2 are strictly LMIs which is able to be solved easily by the LMI toolbox in Matlab. In general, the scalars b_y and $b_{\widehat{y}}$ are given for the event-triggered conditions, otherwise, the co-design approach given in [1, 11] can be used to deal with the case that the scalars b_y and $b_{\widehat{y}}$ are unknown. The scalar λ introduced by the extended matrix inequality decoupling strategy is able to be obtained by using the *fminsearch* function in optimization toolbox of Matlab. Besides, we can choose the dimension adjustment matrices Θ_1 and Θ_2 as $\Theta_1 = \Theta_2 = B_i$ or $\Theta_1 = \Theta_2 = \begin{bmatrix} I \in \mathbb{R}^{n_u \times n_u} & 0 \in \mathbb{R}^{n_u \times (n_x - n_u)} \end{bmatrix}^T$ with $n_x > n_u$.

5.1.3 Simulation Example

Let us consider a discrete-time T–S fuzzy model in the form of (5.1), where

$$A_1 = \begin{bmatrix} 0.6985 & -0.2942 \\ 0.2547 & 0.9533 \end{bmatrix},\ B_1 = \begin{bmatrix} 0.3665 \\ 0.0582 \end{bmatrix},\ D_1 = \begin{bmatrix} 0.1576 \\ -0.0832 \end{bmatrix},$$
$$A_2 = \begin{bmatrix} 0.6985 & -0.2942 \\ 0.2547 & 0.9533 \end{bmatrix},\ B_2 = \begin{bmatrix} 0.1430 \\ 0.0227 \end{bmatrix},\ D_2 = \begin{bmatrix} -0.0128 \\ 0.1324 \end{bmatrix},$$
$$C_1 = \begin{bmatrix} 1 & 0 \end{bmatrix},\ C_2 = \begin{bmatrix} 1 & 0 \end{bmatrix}.$$

Then, the reference model is assumed to be

$$\widehat{x}(k+1) = 0.2\widehat{x}(k) - 0.2\iota(k),$$
$$\widehat{y}(k) = 0.2\widehat{x}(k).$$

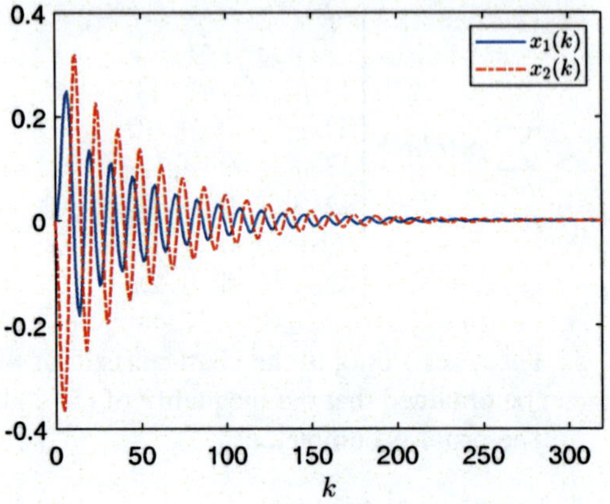

Fig. 5.1 Response of the system state $x(k)$

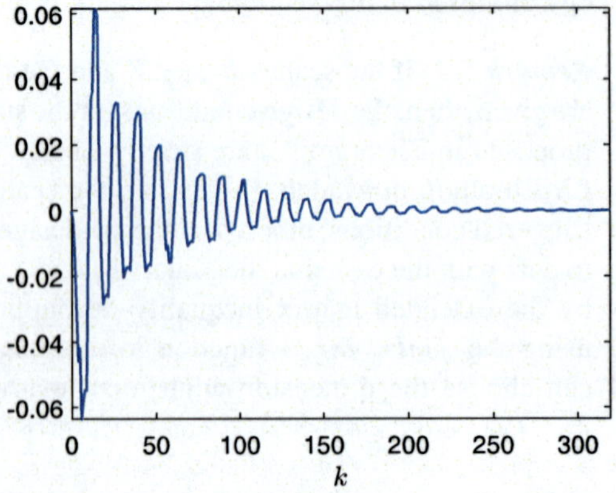

Fig. 5.2 Response of the control input $u(k)$

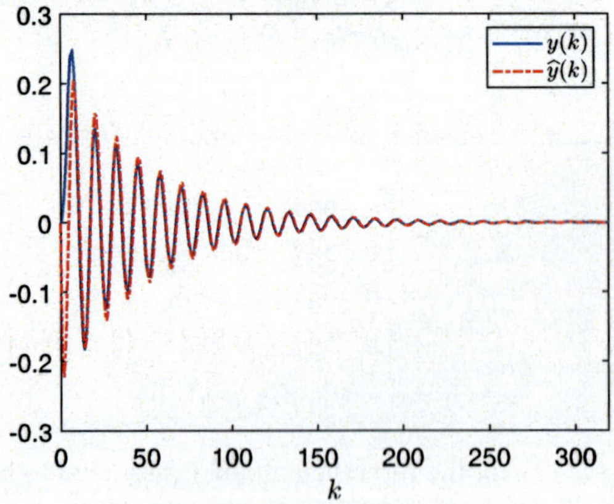

Fig. 5.3 Responses of $y(k)$ and $\widehat{y}(k)$

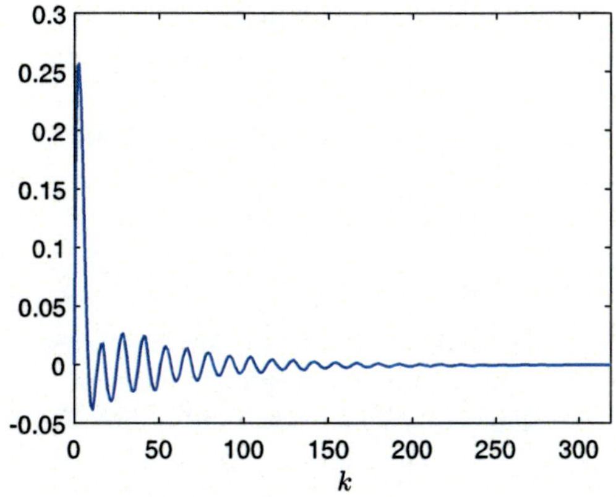

Fig. 5.4 Response of $e_c(k)$

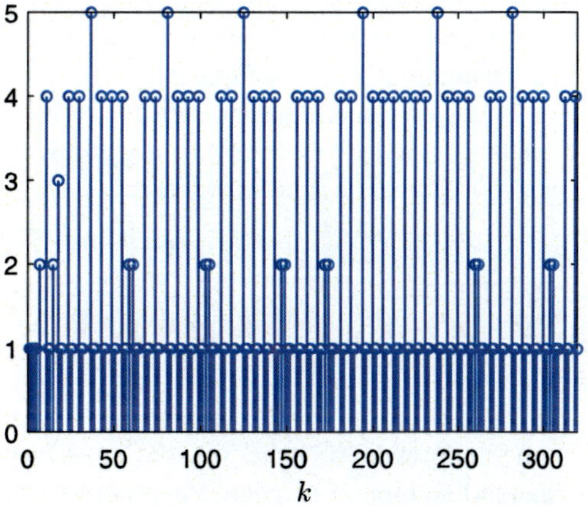

Fig. 5.5 Release instants and release intervals of the communication channel from the plant to the controller

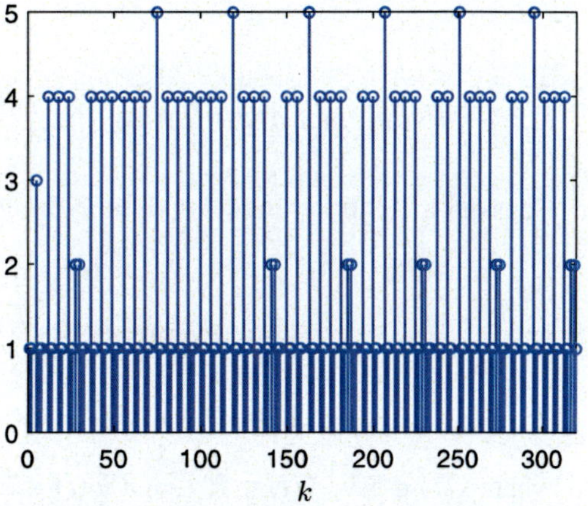

Fig. 5.6 Release instants and release intervals of the communication channel from the reference model to the controller

By applying Theorem 5.2 with $\gamma = 1.7$, $b_y = 0.25$, $b_{\widehat{y}} = 0.25$, $\lambda = 0.5$, $\beta_y = 0.25$, $\beta_{\widehat{y}} = 0.25$, $\beta_u = 0.25$, $\Theta_1 = \begin{bmatrix} 1 & 0 \end{bmatrix}^T$, and $\Theta_2 = \begin{bmatrix} 1 & 0 \end{bmatrix}^T$, it can be obtained that

$$G_1 = -0.2858,\ G_2 = 0.0042,\ \mathbf{J}_y = 7.2937,\ \mathbf{J}_{\widehat{y}} = 52.1520.$$

Moreover, assume that $x(0) = \begin{bmatrix} 0 & 0 \end{bmatrix}^T$, $\widehat{x}(0) = 0$, $w(k) = -\sin(0.5k)e^{-0.02k}$, $\Xi_u(k) = 0.2\sin(0.2k)$, $\Xi_y(k) = 0.2\sin(0.5k)$, $\Xi_{\widehat{y}}(k) = 0.2\sin(0.1k)$, $\iota(k) = 5\cos(0.5k - 0.7)e^{-0.02k}$, $\rho_1(\epsilon(k)) = \sin^2(x_1(k))$, $\rho_2(\epsilon(k)) = 1 - \rho_1(\epsilon(k))$.

The simulation results of the closed-loop system (5.17) are given in Figs. 5.1, 5.2, 5.3, 5.4, 5.5 and 5.6. The trajectories of $x_1(k)$ and $x_2(k)$ are shown in Fig. 5.1. The trajectory of $u(k)$ is shown in Fig. 5.2. The trajectories of $y(k)$ and $\widehat{y}(k)$ are shown in Fig. 5.3. The trajectory of $e_c(k)$ is shown in Fig. 5.4. The simulation result on release instants and release intervals of the communication channel from the plant to the controller is depicted in Fig. 5.5. The simulation result on release instants and release intervals of the communication channel from the reference model to the controller is depicted in Fig. 5.6.

Based on the simulation results given in Figs. 5.1, 5.2, 5.3, 5.4, 5.5, and 5.6, it can be concluded that the event-triggered output feedback tracking controller design approach developed in this section is effective.

5.2 $\mathscr{L}_2$–$\mathscr{L}_\infty$ Filtering with Time Delays and DoS Attacks

In this section, the event-triggered $\mathscr{L}_2$–$\mathscr{L}_\infty$ filtering issue for T–S fuzzy systems with time-delay and DoS attacks is studied. A piecewise Lyapunov–Krasovskii functional method is applied to judge the stability of the T–S fuzzy closed-loop system and ensure the T–S fuzzy closed-loop system having a prescribed $\mathscr{L}_2$–$\mathscr{L}_\infty$ performance.

5.2.1 *Problem Formulation*

Consider a T–S fuzzy model with time-delay as follows:

Plant Rule i_{th}: **IF** $q_1(t)$ is R_{1i}, $q_2(t)$ is R_{2i}, and, $\cdots$, and $q_p(t)$ is R_{pi}, **THEN**

$$\begin{aligned} \dot{x}(t) &= A_i x(t) + B_i x(t - \bar{\rho}) + E_i w(t), \\ \tilde{y}(t) &= C_i x(t), \end{aligned} \tag{5.42}$$

where $x(t) \in \mathbb{R}^{n_x}$, $\tilde{y}(t) \in \mathbb{R}^{n_y}$, and $w(t) \in \mathbb{R}^{n_w}$ are the state vector, the output vector, and the arbitrary disturbance belonging to $\mathscr{L}_2[0, \infty)$, respectively. $q(t) = [\,q_1(t),\ q_2(t),\ \ldots,\ q_p(t)\,]$, where $q_d(t)$, $d = 1, 2, \ldots, p$ are measurable premise variables; R_{di}, $i = 1, 2, \ldots, r$, $d = 1, 2, \ldots, p$ are the fuzzy sets; r is the number of fuzzy rules. $A_i \in \mathbb{R}^{n_x \times n_x}$, $B_i \in \mathbb{R}^{n_x \times n_x}$, $E_i \in \mathbb{R}^{n_x \times n_w}$ and $C_i \in \mathbb{R}^{n_y \times n_x}$ are known constant matrices for $i = 1, 2, \ldots, r$; $\bar{\rho}$ is time delay.

The fuzzy basic function is provided as follows:

$$\theta_i(q(t)) = \frac{\prod_{d=1}^{p} R_{di}(q_d(t))}{\sum_{i=1}^{r} \prod_{d=1}^{p} R_{di}(q_d(t))},$$

where $R_{di}(q_d(t))$ is the grade of the membership of $q_d(t)$ in R_{di}. Thus, it is easily get that $\theta_i(q(t)) \geq 0$ and $\sum_{i=1}^{r} \theta_i(q(t)) = 1$.

Then system (5.42) can be denoted as:

$$\begin{aligned} \dot{x}(t) &= A(\theta)x(t) + B(\theta)x(t - \bar{\rho}) + E(\theta)w(t), \\ \tilde{y}(t) &= C(\theta)x(t), \end{aligned} \tag{5.43}$$

where

$$\begin{aligned} A(\theta) &= \sum_{i=1}^{r} \theta_i(q(t))A_i, \ B(\theta) = \sum_{i=1}^{r} \theta_i(q(t))B_i, \\ E(\theta) &= \sum_{i=1}^{r} \theta_i(q(t))E_i, \ C(\theta) = \sum_{i=1}^{r} \theta_i(q(t))C_i. \end{aligned}$$

According to system (5.43), the following filter is proposed

$$\begin{aligned} \dot{x}_f(t) &= A(\theta)x_f(t) + B(\theta)x_f(t - \bar{\rho}) + L(\tilde{y}(t) - y_f(t)), \\ y_f(t) &= C(\theta)x_f(t), \end{aligned} \tag{5.44}$$

where $x_f(t) \in \mathbb{R}^{n_x}$, $y_f(t) \in \mathbb{R}^{n_y}$, and $L \in \mathbb{R}^{n_x \times n_y}$ are the state vector of the filter, the output vector of the filter, and the gain matrix of the filter. Define the filtering error $z(t) = x(t) - x_f(t)$ and the output error $y(t) = \tilde{y}(t) - y_f(t)$, then the filtering error system is represented as follows:

$$\begin{aligned} \dot{z}(t) &= A(\theta)z(t) + B(\theta)z(t - \bar{\rho}) + E(\theta)w(t) - Ly(t), \\ y(t) &= C(\theta)z(t). \end{aligned} \tag{5.45}$$

Consider system (5.42) with prescribed DoS attacks in this section. The DoS model is given by [12]:

$$\mathfrak{F}_{DoS}(t) = \begin{cases} 1, \ t \in [hT, \ hT + T_{off}), \\ 0, \ t \in [hT + T_{off}, \ (h+1)T), \end{cases} \tag{5.46}$$

where $h \in \mathbb{N}$ is the periodic count; $\{hT\}_{h \in \mathbb{N}}$ is the time instants which means that there is no DoS attack. The sets $\cup_{h \in \mathbb{N}}[hT, hT + T_{off})$ are DoS-free interval, in which DoS attacks are stopping and signals are sent, while the sets $\cup_{h\in\mathbb{N}}[hT+T_{off},\,(h+1)T)$ denote the set of DoS interval, where DoS attacks are active and no signal is sent. $T_{off}\in\mathbb{R}_{>0}(T_{off}<T)$ signifies the period where the DoS attacks are asleeping. Assuming that T_{off} is time-varying and exists real scalars $T_{off}^{\min}\in\mathbb{R}_{>0}$ and $T_{off}^{\max}\in\mathbb{R}_{>0}$ such that $T_{off}^{\min}\leq T_{off}<T_{off}^{\max}<T<\infty$ holds.

Consider $y(t)$ is sampled at time $n\tau$ $(n=1,2,\ldots)$ with $\tau>0$. Then we predefined an event-triggering condition [13], only signals that meet the specified conditions are sent. Firstly, when there are no DoS attacks, the delivery of data is decided by the following triggering condition:

$$[y(t_{\iota_n}\tau)-y(t_\iota\tau)]^T H[y(t_{\iota_n}\tau)-y(t_\iota\tau)]>\sigma y^T(t_{\iota_n}\tau)Hy(t_{\iota_n}\tau),\, n\in\mathbb{N}, \tag{5.47}$$

where $\sigma\in(0,1)$ is a threshold and $H>0$ is a weighting matrix. $t_\iota\tau$ is the last event-triggering instant and $t_{\iota_n}\tau=t_\iota\tau+n\tau$ $(\iota,n\in\mathbb{N})$ is the sampling instant. However, considering the impact of DoS attacks, the event-triggered mechanism in (5.47) is not able to be used directly. Thus, to offset the DoS attacks in (5.46), the event-triggered mechanism in (5.47) needs to be revised. Similar to [14], for $\forall\, h\in\mathbb{N}$, set $T_{off}\equiv T_{off}^{\min}$, then the event triggering instant

$$t_{\iota,h+1}\tau=\left\{t_{\iota_n}\tau \text{ satisfying (5.47)}\,\middle|\,t_{\iota_n}\tau\in[hT,\ hT+T_{off}^{\min})\right\}\cup\{(h+1)T\}, \tag{5.48}$$

in which h, t_{ι_n}, ι_n, and $n\in\mathbb{N}$. ι represents the amount of triggering times arising in $(h+1)$-th period.

In the effect of DoS attacks, the $y(t)$ is successfully sent to the filter, i.e., $\bar{y}(t)=y(t_{\iota,h+1}\tau)$, else the $y(t)$ is zero, i.e., $\bar{y}(t)=0$. Thereby, the following results are obtained

$$\bar{y}(t)=\begin{cases}y(t_{\iota,h+1}\tau), & t\in[t_{\iota,h+1}\tau,\ t_{\iota+1,h+1}\tau)\cap[hT,\ hT+T_{off}^{\min}),\\ 0, & t\in[hT+T_{off}^{\min},\ hT+T),\end{cases} \tag{5.49}$$

where $\{t_{\iota,h}\tau\}$ represents the sequence of successful transmission instants $((t_{0,h+1}\tau)=hT)$, which is generated by the event-triggered mechanism in (5.48); $\iota\in l(h)\triangleq\{0,1,2,\ldots,\iota(h)\}$ with $h\in\mathbb{N}$ and $\iota(h)=\sup\{\iota\in\mathbb{N}\,\big|\,hT+T_{off}^{\min}\geq t_{\iota,h+1}\tau\}$.

In what follows, for the sake of brevity, define

$$\begin{aligned}
\mathcal{I}_{\iota,h}&=[t_{\iota,h+1}\tau,\ t_{\iota+1,h+1}\tau),\,\iota\in l(h)(h\in\mathbb{N}),\\
\Omega_{1,h}&=[hT,\ hT+T_{off}^{\min}),\\
\Omega_{2,h}&=[hT+T_{off}^{\min},\ (h+1)T),\\
\kappa_{\iota,h}&\triangleq\sup\{k\in\mathbb{N}\,|\,t_{\iota,h+1}\tau+k\tau<t_{\iota+1,\,h+1}\tau\},
\end{aligned}$$

$$\begin{cases} \mathscr{K}_{\iota,h}^{k} = [t_{\iota,h+1}\tau + (k-1)\tau,\ t_{\iota,h+1}\tau + k\tau], k \in \{1, \ldots, \kappa_{\iota,h}\}, \\ \mathscr{K}_{\iota,h}^{\kappa_{\iota,h}+1} = [t_{\iota,h+1}\tau + \kappa_{\iota,h}\tau,\ t_{\iota+1,h+1}\tau]. \end{cases} \tag{5.50}$$

Then $\mathscr{J}_{\iota,h}$ is able to express as

$$\mathscr{J}_{\iota,h} = \cup_{k=1}^{\kappa_{\iota,h}} \mathscr{K}_{\iota,h}^{k} \cup \mathscr{K}_{\iota,h}^{\kappa_{\iota,h}+1}. \tag{5.51}$$

Noting that

$$\Omega_{1,h} = \cup_{\iota=0}^{\iota(h)} \{\mathscr{J}_{\iota,h} \cap \Omega_{1,h}\} \subseteq \cup_{\iota=0}^{\iota(h)} \mathscr{J}_{\iota,h},$$

and combining (5.50) and (5.51), $\Omega_{1,h}$ is reorganized as

$$\Omega_{1,h} = \cup_{\iota=0}^{\iota(h)} \cup_{k=1}^{\kappa_{\iota,h}+1} \{\mathscr{K}_{\iota,h}^{k} \cap \Omega_{1,h}\}.$$

Set

$$\bar{\Omega}_{\iota,h}^{k} = \mathscr{K}_{\iota,h}^{k} \cap \Omega_{1,h}.$$

Then

$$\Omega_{1,h} = \cup_{\iota=0}^{\iota(h)} \cup_{k=1}^{\kappa_{\iota,h}+1} \bar{\Omega}_{\iota,h}^{k}.$$

For $\iota \in l(h)(h \in \mathbb{N})$, giving the following two piecewise functions:

$$\phi_{\iota,h}(t) = \begin{cases} t - t_{\iota,h+1}\tau, & t \in \bar{\Omega}_{\iota,h}^{1}, \\ t - t_{\iota,h+1}\tau - \tau, & t \in \bar{\Omega}_{\iota,h}^{2}, \\ \vdots & \\ t - t_{\iota,h+1}\tau - \kappa_{\iota,h}\tau, & t \in \bar{\Omega}_{\iota,h}^{\kappa_{\iota,h}+1}, \end{cases} \tag{5.52}$$

and

$$\bar{z}_{\iota,h}(t) = \begin{cases} 0, & t \in \bar{\Omega}_{\iota,h}^{1}, \\ z(t_{\iota,h+1}\tau) - z(t_{\iota,h+1}\tau + \tau), & t \in \bar{\Omega}_{\iota,h}^{2}, \\ \vdots & \\ z(t_{\iota,h+1}\tau) - z(t_{\iota,h+1}\tau + \kappa_{\iota,h}\tau), & t \in \bar{\Omega}_{\iota,h}^{\kappa_{\iota,h}+1}. \end{cases} \tag{5.53}$$

Then, according to the above two functions, it can be observed that $\phi_{\iota,h}(t) \in [0, \tau), t \in \mathscr{J}_{\iota,h} \cap \Omega_{1,h}$. The event-triggered sampling signal can be represented as

$$z(t_{\iota,h+1}\tau) = z(t - \phi_{\iota,h}(t)) + \bar{z}_{\iota,h}(t), t \in \mathscr{J}_{\iota,h} \cap \Omega_{1,h},$$

and $\bar{z}_{\iota,h}(t)$ meets

$$\bar{z}_{\iota,h}^{T}(t) H \bar{z}_{\iota,h}(t) \le \sigma z^{T}(t - \phi_{\iota,h}(t)) H z(t - \phi_{\iota,h}(t)). \tag{5.54}$$

Define $L_c(\theta) = LC(\theta)$. Then filtering error system (5.45) can be rewritten as

$$\dot{z}(t) = \begin{cases} A(\theta)z(t) + B(\theta)z(t-\bar{\rho}) + E(\theta)w(t) - L_c(\theta)z(t-\phi_{\iota,h}(t)) \\ -L_c(\theta)\bar{z}_{\iota,h}(t), & t \in \mathcal{I}_{\iota,h} \cap \Omega_{1,h}, \\ A(\theta)z(t) + B(\theta)z(t-\bar{\rho}) + E(\theta)w(t), & t \in \Omega_{2,h}, \end{cases}$$
$$y(t) = C(\theta)z(t),$$
$$z(t) = \varkappa(t), \quad t \in [-\tau, 0], \tag{5.55}$$

where $\bar{z}_{\iota,h}(t)$ satisfies (5.54), $\varkappa(t)$ is an initial value.

Based on the above analysis, the piecewise Lyapunov–Krasovskii functional for filtering error system (5.55) is chosen as following :

$$\begin{aligned} V_{\alpha(t)}(t) = {} & z^T(t)P_{\alpha(t)}z(t) + \int_{t-\bar{\rho}}^{t} z^T(s)q(\cdot)S_{\alpha(t)}z(s)ds \\ & + \int_{t-\tau}^{t} z^T(s)q(\cdot)Q_{\alpha(t)}z(s)ds \\ & + \int_{-\tau}^{0}\int_{t+\epsilon}^{t} \dot{z}^T(s)q(\cdot)U_{\alpha(t)}\dot{z}(s)dsd\epsilon \\ & + \int_{-\tau}^{0}\int_{t+\epsilon}^{t} \dot{z}^T(s)q(\cdot)Z_{\alpha(t)}\dot{z}(s)dsd\epsilon, \end{aligned}$$

where $P_{\alpha(t)}$, $S_{\alpha(t)}$, $Q_{\alpha(t)}$, $U_{\alpha(t)}$, and $Z_{\alpha(t)} \in \mathbb{R}_+^n$; $q(\cdot) \triangleq e^{2(-1)^{\alpha(t)}g_{\alpha(t)}(t-s)}$, in which $\alpha(t) = 1$ for $t \in [-\tau, 0] \cup (\cup_{h\in\mathbb{N}}\Omega_{1,h})$, and $\alpha(t) = 2$ for $t \in \cup_{h\in\mathbb{N}}\Omega_{2,h}$. It implies that $\alpha(t_{m,h}) = m$ and $\alpha(t_{m,h}^-) = 3 - m$.

The event-triggered $\mathcal{L}_2$–$\mathcal{L}_\infty$ filtering problem considered in this section is to design the $\mathcal{L}_2$–$\mathcal{L}_\infty$ filter in the form of (5.44) and a triggering matrix $H > 0$ such that

(1) The filtering error system in (5.55) is exponentially stable when $w(t) = 0$, i.e. there exist scalars b and μ, such that $\|z(t)\| \le be^{-\mu t}\|\varkappa(0)\|$ for $\forall t \ge 0$.
(2) The filtering error system in (5.55) has a prescribed $\mathcal{L}_2$–$\mathcal{L}_\infty$ performance γ, i.e., under the zero-initial condition,

$$\sup_{t\ge 0}\{y^T(t)y(t)\} \le \gamma^2 \int_0^\infty w^T(t)w(t)dt \tag{5.56}$$

is satisfied for any nonzero $w(t) \in \mathcal{L}_2[0, \infty)$.

5.2.2 Main Results

In this subsection, a criterion of the exponential stability and an $\mathscr{L}_2$–$\mathscr{L}_\infty$ performance analysis result will be given for filtering error system (5.55) subject to DoS attacks (5.46). Then, a co-design method for the needed $\mathscr{L}_2$–$\mathscr{L}_\infty$ filtering gain and event-triggering parameter will be provided.

Firstly, the following Lemma is given to derive the main results.

Lemma 5.1 *For given DoS signals (5.46), the filter gain L, and several scalars $\sigma \in (0, 1)$, $g_m \in (0, \infty)$, and $\tau > 0$. If there exist matrices P_m, S_m, Q_m, U_m, Z_m, and $H \in \mathbb{R}^n_+$, and matrices $\Upsilon_m, \bar{\Upsilon}_m, \tilde{\Upsilon}_m$, $m \in \{1, 2\}$ such that the following matrix inequalities hold:*

$$\Gamma_m(\theta) < 0, \tag{5.57}$$

where

$$\Gamma_m(\theta) = \begin{bmatrix} \bar{\Gamma}_m(\theta) + He\{\bar{\Gamma}_m\} & * & * & * & * & * \\ \sqrt{\tau}\bar{\Upsilon}_m^T & -e^{2(-1)^m g_m \beta_m \tau} U_m & * & * & * & * \\ \sqrt{\tau}\tilde{\Upsilon}_m^T & 0 & -e^{2(-1)^m g_m \beta_m \tau} U_m & * & * & * \\ \sqrt{\tau}\Upsilon_m^T & 0 & 0 & -e^{2(-1)^m g_m \beta_m \tau} Z_m & * & * \\ \sqrt{\tau} U_m A_m(\theta) & 0 & 0 & 0 & -U_m & * \\ \sqrt{\tau} Z_m A_m(\theta) & 0 & 0 & 0 & 0 & -Z_m \end{bmatrix},$$

$$\beta_1 \triangleq 1, \;\; \beta_2 \triangleq 0, \;\; A_1(\theta) = \begin{bmatrix} A(\theta) & -L_c(\theta) & 0 & -L_c(\theta) & B(\theta) \end{bmatrix},$$

$$A_2(\theta) = \begin{bmatrix} A(\theta) & 0 & 0 & B(\theta) \end{bmatrix},$$

$$\bar{\Gamma}_1(\theta) = \begin{bmatrix} \tilde{\Gamma}_1(\theta) & * & * & * & * \\ -L_c^T(\theta)P_1^T & \sigma H & * & * & * \\ 0 & 0 & -e^{-2g_1\tau}Q_1 & * & * \\ -L_c^T(\theta)P_1^T & 0 & 0 & -H & * \\ B^T(\theta)P_1^T & 0 & 0 & 0 & -e^{-2g_1\bar{\rho}}S_1 \end{bmatrix},$$

$$\tilde{\Gamma}_1(\theta) = He\{P_1A(\theta)\} + S_1 + Q_1 + 2g_1P_1,$$

$$\bar{\Gamma}_2(\theta) = \begin{bmatrix} He\{P_2A(\theta)\} + S_2 + Q_2 - 2g_2P_2 & * & * & * \\ 0 & 0 & * & * \\ 0 & 0 & -e^{2g_2\tau}Q_2 & * \\ B^T(\theta)P_2^T & 0 & 0 & -e^{2g_2\bar{\rho}}S_2 \end{bmatrix},$$

$$\bar{\Gamma}_1 = [\Upsilon_1 + \bar{\Upsilon}_1 \;\; -\bar{\Upsilon}_1 + \tilde{\Upsilon}_1 \;\; -\Upsilon_1 - \tilde{\Upsilon}_1 \;\; 0 \;\; 0],$$

$$\bar{\Gamma}_2 = [\Upsilon_2 + \bar{\Upsilon}_2 \;\; -\bar{\Upsilon}_2 + \tilde{\Upsilon}_2 \;\; -\Upsilon_2 - \tilde{\Upsilon}_2 \;\; 0],$$

$$\Upsilon_1 = \begin{bmatrix} \Upsilon_{11}^T & \Upsilon_{12}^T & \Upsilon_{13}^T & \Upsilon_{14}^T & \Upsilon_{15}^T \end{bmatrix}^T, \;\; \Upsilon_2 = [\Upsilon_{21}^T \;\; \Upsilon_{22}^T \;\; \Upsilon_{23}^T \;\; \Upsilon_{24}^T]^T,$$

$$\bar{\Upsilon}_1 = \begin{bmatrix} \bar{\Upsilon}_{11}^T & \bar{\Upsilon}_{12}^T & \bar{\Upsilon}_{13}^T & \bar{\Upsilon}_{14}^T & \bar{\Upsilon}_{15}^T \end{bmatrix}^T, \;\; \bar{\Upsilon}_2 = [\bar{\Upsilon}_{21}^T \;\; \bar{\Upsilon}_{22}^T \;\; \bar{\Upsilon}_{23}^T \;\; \bar{\Upsilon}_{24}^T]^T,$$

$$\tilde{\Upsilon}_1 = \begin{bmatrix} \tilde{\Upsilon}_{11}^T & \tilde{\Upsilon}_{12}^T & \tilde{\Upsilon}_{13}^T & \tilde{\Upsilon}_{14}^T & \tilde{\Upsilon}_{15}^T \end{bmatrix}^T, \;\; \tilde{\Upsilon}_2 = [\tilde{\Upsilon}_{21}^T \;\; \tilde{\Upsilon}_{22}^T \;\; \tilde{\Upsilon}_{23}^T \;\; \tilde{\Upsilon}_{24}^T]^T,$$

then along the trajectories of filtering error system (5.55) with $w(t) = 0$, for $h \in \mathbb{N}$, one can obtain that

$$V_m(t) \leq e^{2(-1)^m g_m(t-t_{m,h})} V_m(t_{m,h}), \tag{5.58}$$

in which

$$t \in [t_{m,h}, t_{3-m,h+m-1}],\ m = \{1, 2\},$$

$$t_{m,h} = \begin{cases} hT, & m = 1, \\ hT + T_{off}^{\min}, & m = 2. \end{cases}$$

Proof Define

$$\zeta(t) = \left[z^T(t)\ \ z^T(t - \phi_{\iota,h}(t))\ \ z_\tau^T(t)\ \ \bar{z}_{\iota,h}^T(t)\ \ z_{\bar{\rho}}^T(t)\right]^T,$$
$$z_\tau(t) = z(t - \tau),\ z_\rho(t) = z(t - \bar{\rho}).$$

Then, when $t \in [t_{1,h}, t_{2,h}]$, $V_1(t)$ along the trajectories of filtering error system (5.55), one can get

$$\begin{aligned}
\dot{V}_1(t) &= 2z^T(t)P_1\dot{z}(t) + z^T(t)(S_1 + Q_1)z(t) - z_{\bar{\rho}}^T(t)e^{-2g_1\bar{\rho}}S_1 z_{\bar{\rho}}(t) \\
&\quad - z_\tau^T(t)e^{-2g_1\tau}Q_1 z_\tau(t) + \tau\dot{z}^T(t)(U_1 + Z_1)\dot{z}(t) \\
&\quad - 2g_1\int_{t-\bar{\rho}}^{t} z(s)e^{-2g_1(t-s)}S_1 z(s)ds - 2g_1\int_{t-\tau}^{t} z(s)e^{-2g_1(t-s)}Q_1 z(s)ds \\
&\quad - \int_{-\tau}^{0} \dot{z}^T(t+\epsilon)e^{2g_1\epsilon}U_1\dot{z}(t+\epsilon)d\epsilon - 2g_1\int_{-\tau}^{0}\int_{t+\epsilon}^{t} \dot{z}^T(s)e^{-2g_1(t-s)}U_1\dot{z}(s)dsd\epsilon \\
&\quad - \int_{-\tau}^{0} \dot{z}^T(t+\epsilon)e^{2g_1\epsilon}Z_1\dot{z}(t+\epsilon)d\epsilon - 2g_1\int_{-\tau}^{0}\int_{t+\epsilon}^{t} \dot{z}^T(s)e^{-2g_1(t-s)}Z_1\dot{z}(s)dsd\epsilon,
\end{aligned}$$

by $-\tau < \epsilon < 0$, one can be obtained

$$-\int_{-\tau}^{0} \dot{z}^T(t+\epsilon)e^{2g_1\epsilon}U_1\dot{z}(t+\epsilon)d\epsilon \leq -\int_{t-\tau}^{t} \dot{z}^T(\tau)e^{-2g_1\tau}U_1\dot{z}(\tau)d\tau,$$

$$-\int_{-\tau}^{0} \dot{z}^T(t+\epsilon)e^{2g_1\epsilon}Z_1\dot{z}(t+\epsilon)d\epsilon \leq -\int_{t-\tau}^{t} \dot{z}^T(\tau)e^{-2g_1\tau}Z_1\dot{z}(\tau)d\tau.$$

Then, $\dot{V}_1(t)$ can be reorganized as follows

$$
\begin{aligned}
\dot{V}_1(t) \le\, & 2z^T(t)P_1\dot{z}(t) + z^T(t)(S_1+Q_1)z(t) - z_{\bar{\rho}}^T(t)e^{-2g_1\bar{\rho}}S_1 z_{\bar{\rho}}(t) \\
& - z_\tau^T(t)e^{-2g_1\tau}Q_1 z_\tau(t) + \tau\dot{z}^T(t)(U_1+Z_1)\dot{z}(t) \\
& - 2g_1\int_{t-\bar{\rho}}^{t} z(s)e^{-2g_1(t-s)}S_1 z(s)ds - 2g_1\int_{t-\tau}^{t} z(s)e^{-2g_1(t-s)}Q_1 z(s)ds \\
& - \int_{t-\tau}^{t}\dot{z}^T(\tau)e^{-2g_1\tau}U_1\dot{z}(\tau)d\tau - \int_{t-\tau}^{t}\dot{z}^T(\tau)e^{-2g_1\tau}Z_1\dot{z}(\tau)d\tau \\
& - 2g_1\int_{-\tau}^{0}\int_{t+\epsilon}^{t}\dot{z}^T(s)e^{-2g_1(t-s)}U_1\dot{z}(s)dsd\epsilon \\
& - 2g_1\int_{-\tau}^{0}\int_{t+\epsilon}^{t}\dot{z}^T(s)e^{-2g_1(t-s)}Z_1\dot{z}(s)dsd\epsilon \\
\le\, & -2g_1V_1(t) + z^T(t)2g_1P_1z(t) + \dot{z}^T(t)2P_1z(t) + z^T(t)(S_1+Q_1)z(t) \\
& - z_{\bar{\rho}}^T(t)e^{-2g_1\bar{\rho}}S_1z_{\bar{\rho}}(t) + \tau\dot{z}^T(t)(U_1+Z_1)\dot{z}(t) - z_\tau^T(t)e^{-2g_1\tau}Q_1z_\tau(t) \\
& - \varpi_1 - \varpi_2 - \varpi_3 + 2\zeta^T(t)(\Upsilon_1N_1 + \bar{\Upsilon}_1N_2 + \tilde{\Upsilon}_1N_3),
\end{aligned}
$$

where

$$
\varpi_1 = \int_{t-\tau}^{t}\dot{z}^T(s)e^{-2g_1\tau}Z_1\dot{z}(s)ds,\ \varpi_2 = \int_{t-\phi_{\iota,h}(t)}^{t}\dot{z}^T(s)e^{-2g_1\tau}U_1\dot{z}(s)ds,
$$

$$
\varpi_3 = \int_{t-\tau}^{t-\phi_{\iota,h}(t)}\dot{z}^T(s)e^{-2g_1\tau}U_1\dot{z}(s)ds,\ N_1 = z(t) - z_\tau(t) - \int_{t-\tau}^{t}\dot{z}(s)ds \triangleq 0,
$$

$$
N_2 = z(t) - z(t-\phi_{\iota,h}(t)) - \int_{t-\phi_{\iota,h}(t)}^{t}\dot{z}(s)ds \triangleq 0,
$$

$$
N_3 = z(t-\phi_{\iota,h}(t)) - z_\tau(t) - \int_{t-\tau}^{t-\phi_{\iota,h}(t)}\dot{z}(s)ds \triangleq 0.
$$

According to [15], one can be obtained

$$\begin{aligned} &-2\zeta^T(t)\Upsilon_1\int_{t-\tau}^{t}\dot{z}(s)ds \\ \leq&\tau\zeta^T(t)\Upsilon_1 e^{2g_1\tau}Z_1^{-1}\Upsilon_1^T\zeta(t)+\int_{t-\tau}^{t}\dot{z}^T(s)e^{-2g_1\tau}Z_1\dot{z}(s)ds, \end{aligned} \tag{5.59}$$

$$\begin{aligned} &-2\zeta^T(t)\bar{\Upsilon}_1\int_{t-\phi_{\iota,h}(t)}^{t}\dot{z}(s)ds \\ \leq&\tau\zeta^T(t)\bar{\Upsilon}_1 e^{2g_1\tau}U_1^{-1}\bar{\Upsilon}_1^T\zeta(t)+\int_{t-\phi_{\iota,h}(t)}^{t}\dot{z}^T(s)e^{-2g_1\tau}U_1\dot{z}(s)ds, \end{aligned} \tag{5.60}$$

$$\begin{aligned} &-2\zeta^T(t)\tilde{\Upsilon}_1\int_{t-\tau}^{t-\phi_{\iota,h}(t)}\dot{z}(s)ds \\ \leq&\tau\zeta^T(t)\tilde{\Upsilon}_1 e^{2g_1\tau}U_1^{-1}\tilde{\Upsilon}_1^T\zeta(t)+\int_{t-\tau}^{t-\phi_{\iota,h}(t)}\dot{z}^T(s)e^{-2g_1\tau}U_1\dot{z}(s)ds, \end{aligned} \tag{5.61}$$

by (5.54), (5.59), (5.60), and (5.61), one has

$$\begin{aligned} \dot{V}_1(t)\leq&-2g_1V_1(t)+\zeta^T(t)[\Gamma_{11}^1(\theta)+\tau\Upsilon_1 e^{2g_1\tau}Z_1^{-1}\Upsilon_1^T+\tau\bar{\Upsilon}_1 e^{2g_1\tau}U_1^{-1}\bar{\Upsilon}_1^T \\ &+\tau\tilde{\Upsilon}_1 e^{2g_1\tau}U_1^{-1}\tilde{\Upsilon}_1^T+\tau A_1^T(\theta)(U_1+Z_1)A_1(\theta)]\zeta(t). \end{aligned}$$

Using Lemma 1.1, $\Gamma_1 < 0$ in (5.57) is equivalent to

$$\begin{aligned} &\Gamma_{11}^1(\theta)+\tau\Upsilon_1 e^{2g_1\tau}Z_1^{-1}\Upsilon_1^T+\tau\bar{\Upsilon}_1 e^{2g_1\tau}U_1^{-1}\bar{\Upsilon}_1^T \\ &+\tau\tilde{\Upsilon}_1 e^{2g_1\tau}U_1^{-1}\tilde{\Upsilon}_1^T+\tau A_1^T(\theta)(U_1+Z_1)A_1(\theta)<0, \end{aligned}$$

which implies

$$\dot{V}_1(t)+2g_1V_1(t)\leq 0.$$

When $t\in[t_{2,h},t_{1,h+1})$, $h\in\mathbb{N}$. Along the same line as the proof in $V_1(t)$, one obtains

$$\begin{aligned} \dot{V}_2(t)\leq&2g_2V_2(t)+\zeta^T(t)[\Gamma_{11}^2(\theta)+\tau\Upsilon_2 Z_2^{-1}\Upsilon_2^T+\tau\bar{\Upsilon}_2 U_2^{-1}\bar{\Upsilon}_2^T+\tau\tilde{\Upsilon}_2 U_2^{-1}\tilde{\Upsilon}_2^T \\ &+\tau A_2^T(U_2+Z_2)A_2]\zeta(t). \end{aligned}$$

Using Lemma 1.1, $\Gamma_2 < 0$ in (5.57) is equivalent to

$$\Gamma_{11}^2(\theta) + \tau \Upsilon_2 Z_2^{-1} \Upsilon_2^T + \tau \bar{\Upsilon}_2 U_2^{-1} \bar{\Upsilon}_2^T + \tau \tilde{\Upsilon}_2^T U_2^{-1} \tilde{\Upsilon}_2^T$$
$$+ \tau A_2^T(\theta)(U_2 + Z_2) A_2(\theta) < 0,$$

which implies

$$\dot{V}_2(t) - 2g_2 V_2(t) \leq 0.$$

Therefore, in the light of the above analysis, (5.57) can guarantee (5.58). This completes the proof.

Theorem 5.3 *For given DoS signals (5.46), the filtering gain L, and several scalars $\sigma \in (0,\ 1)$, $g_m \in (0,\ \infty)$, $\pi_m > 1$, and $\tau > 0$. If there exist matrices P_m, S_m, Q_m, U_m, Z_m, and $H \in \mathbb{R}_+^n$, and matrices $\Upsilon_m, \bar{\Upsilon}_m, \tilde{\Upsilon}_m, m \in \{1, 2\}$ such that (5.57) and the following conditions hold:*

$$\begin{cases} P_1 \leq \pi_2 P_2, \\ P_2 \leq e^{2(g_1+g_2)\tau} \pi_1 P_1, \\ Q_m \leq \pi_{3-m} Q_{3-m}, \\ S_m \leq \pi_{3-m} S_{3-m}, \\ U_m \leq \pi_{3-m} U_{3-m}, \\ Z_m \leq \pi_{3-m} Z_{3-m}, \end{cases} \tag{5.62}$$

$$0 < \psi = 2g_1 T_{off}^{\min} - 2g_2(T - T_{off}^{\min}) - 2(g_1 + g_2)\tau - \ln(\pi_1 \pi_2), \tag{5.63}$$

then filtering error system (5.55) is exponentially stable with $w(t) = 0$ and the decay rate $\nu = \frac{\psi}{2T}$ under DoS attacks.

Proof According to (5.58) in Lemma 5.1, for any $t \geq 0$ one gets

$$V(t) \leq \begin{cases} e^{-2g_1(t-t_{1,h})} V_1(t_{1,h}), & t \in [t_{1,h}, t_{2,h}), \\ e^{2g_2(t-t_{2,h})} V_2(t_{2,h}), & t \in [t_{2,h}, t_{1,h+1}). \end{cases} \tag{5.64}$$

From (5.62), one has

$$\begin{cases} V_1(t_{1,h}) \leq \pi_2 V_2(t_{1,h}^-), \\ V_2(t_{2,h}) \leq \pi_1 e^{2(g_1+g_2)\tau} V_1(t_{2,h}^-). \end{cases} \tag{5.65}$$

It can find a $h \in \mathbb{N}$ to ensure $t \in [t_{1,h}, t_{2,h})$ or $t \in [t_{2,h}, t_{1,h+1})$ for $\forall t \geq 0$. Thereupon, the following two cases will be considered:

Case 1: For $t \in [t_{1,h}, t_{2,h})$, it can be obtained from (5.64) and (5.65)

$$\begin{aligned}V(t) \leq & e^{-2g_1(t-t_{1,h})}V_1(t_{1,h}) \\ \leq & \pi_2 e^{-2g_1(t-t_{1,h})}V_2(t_{1,h}^-) \\ \leq & \pi_2 e^{-2g_1(t-t_{1,h})}e^{2g_2(t_{1,h}-t_{2,h-1})}V_2(t_{2,h-1}) \\ \leq & \pi_1\pi_2 e^{2(g_1+g_2)\tau}e^{-2g_1(t-t_{1,h})}e^{2g_2(t_{1,h}-t_{2,h-1})}e^{-2g_1(t_{2,h-1}-t_{1,h-1})}V_1(t_{1,h-1}) \\ & \vdots \\ \leq & (\pi_1\pi_2)^h e^{2h(g_1+g_2)\tau}e^{2hg_2(T-T_{off})}e^{-2hg_1T_{off}^{\min}}e^{-2g_1(t-t_{1,h})}V_1(0) \\ \leq & e^{h[\ln(\pi_1\pi_2)+2(g_1+g_2)\tau+2g_2(T-T_{off}^{\min})-2g_1T_{off}^{\min}]}e^{-2g_1(t-t_{1,h})}V_1(0) \\ \leq & e^{-\psi h}V_1(0).\end{aligned}$$

Notice that $t < t_{2,h} = hT + T_{off}^{\min}$, i.e., $h > [(t - T_{off}^{\min})/T]$, it follows that

$$V(t) \leq V_1(0)e^{\frac{\psi T_{off}^{\min}}{T}}e^{-\frac{\psi}{T}t}. \tag{5.66}$$

Case 2: For $t \in [t_{2,h}, t_{1,h+1})$, one has

$$\begin{aligned}V(t) \leq & e^{2g_2(t-t_{2,h})}V_2(t_{2,h}) \\ \leq & \pi_1 e^{2(g_1+g_2)\tau}e^{2g_2(t-t_{2,h})}V_1(t_{2,h}^-) \\ \leq & \pi_1 e^{2(g_1+g_2)\tau}e^{2g_2(t-t_{2,h})}e^{-2g_1(t_{2,h}-t_{1,h})}V_1(t_{1,h}) \\ & \vdots \\ \leq & \frac{(\pi_1\pi_2)^{h+1}}{\pi_2}e^{2(h+1)(g_1+g_2)\tau}e^{2hg_2(T-T_{off}^{\min})}e^{-2(h+1)g_1T_{off}^{\min}}e^{2g_2(t-t_{2,h})}V_1(0) \\ \leq & \frac{1}{\pi_2}e^{(h+1)\left[\ln(\pi_1\pi_2)+2(g_1+g_2)\tau+2g_2(T-T_{off}^{\min})-2g_1T_{off}^{\min}\right]}V_1(0) \\ \leq & \frac{1}{\pi_2}e^{-\psi(h+1)}V_1(0).\end{aligned}$$

Same as Case 1, one can get that

$$V(t) \leq \frac{V_1(0)}{\pi_2}e^{-\frac{\psi}{T}t}. \tag{5.67}$$

Define $\lambda_0 = \max\left\{e^{\frac{\psi T_{off}^{\min}}{T}}, \frac{1}{\pi_2}\right\}$, $\lambda_1 = \min\{\lambda_{\min}(P_m)\}$, $\lambda_2 = \max\{\lambda_{\max}(P_m)\}$, $\lambda_3 = \lambda_2 + \bar{\rho}\lambda_{\max}(S_1) + \tau\lambda_{\max}(Q_1) + \frac{\tau^2}{2}\lambda_{\max}(U_1 + Z_1)$. Then it can be obtained from (5.66) and (5.67) that

$$V(t) \leq \lambda_0 e^{-\frac{\psi}{T}t}V_1(0). \tag{5.68}$$

According to definition of $V(t)$ in (5.58), it is easy to get

$$V(t) \geq \lambda_1 \|z(t)\|^2, \; V_1(0) \leq \lambda_3 \|\varkappa(0)\|_\tau^2. \tag{5.69}$$

Combining (5.63), (5.68), and (5.69), one can be deduced that

$$\|z(t)\| \leq \sqrt{\frac{\lambda_0\lambda_3}{\lambda_1}} e^{-\nu t} \|\varkappa(0)\|_\tau, \; \forall t \geq 0.$$

Thus, the filtering error system is exponentially stable with $w(t) = 0$ and the decay rate ν under DoS attacks. This completes the proof.

Based on Theorem 5.3, this subsection provides sufficient conditions to ensure the $\mathscr{L}_2$–$\mathscr{L}_\infty$ performance $\bar{\gamma}$ for filtering error system (5.55) under DoS attacks.

Theorem 5.4 *For given DoS signals (5.46), the filter gain L, and several scalars $\sigma \in (0, 1)$, $g_m \in (0, \infty)$, $\pi_m > 1$, $\gamma > 0$, and $\tau > 0$. If there exist matrices P_m, S_m, Q_m, U_m, Z_m, and $H \in \mathbb{R}_+^n$, and matrices Υ_m, $\bar{\Upsilon}_m$, $\tilde{\Upsilon}_m$, $m \in \{1, 2\}$ such that (5.62), (5.63), and the following matrix inequalities hold:*

$$\Sigma_m(\theta) < 0, \tag{5.70}$$

$$C^T(\theta)C(\theta) - P_1 < 0, \tag{5.71}$$

where

$$\Sigma_m(\theta) = \begin{bmatrix} \bar{\Sigma}_m(\theta) + He\{\bar{\Sigma}_m\} & * & * & * & * & * \\ \sqrt{\tau}\bar{\Upsilon}_m^T & -e^{2(-1)^m g_m \beta_m \tau} U_m & * & * & * & * \\ \sqrt{\tau}\tilde{\Upsilon}_m^T & 0 & -e^{2(-1)^m g_m \beta_m \tau} U_m & * & * & * \\ \sqrt{\tau}\Upsilon_m^T & 0 & 0 & -e^{2(-1)^m g_m \beta_m \tau} Z_m & * & * \\ \sqrt{\tau}U_m A_m(\theta) & 0 & 0 & 0 & -U_m & * \\ \sqrt{\tau}Z_m A_m(\theta) & 0 & 0 & 0 & 0 & -Z_m \end{bmatrix},$$

$\beta_1 \triangleq 1$, $\beta_2 \triangleq 0$, $A_1(\theta) = \begin{bmatrix} A(\theta) & -L_c(\theta) & 0 & -L_c(\theta) & E(\theta) & B(\theta) \end{bmatrix}$,

$A_2(\theta) = \begin{bmatrix} A(\theta) & 0 & 0 & E(\theta) & B(\theta) \end{bmatrix}$,

$$\bar{\Sigma}_1(\theta) = \begin{bmatrix} \tilde{\Sigma}_1(\theta) & * & * & * & * & * \\ -L_c^T(\theta)P_1^T & \sigma H & * & * & * & * \\ 0 & 0 & -e^{-2g_1\tau}Q_1 & * & * & * \\ -L_c^T(\theta)P_1^T & 0 & 0 & -H & * & * \\ E^T(\theta)P_1^T & 0 & 0 & 0 & -\gamma^2 I & * \\ B^T(\theta)P_1^T & 0 & 0 & 0 & 0 & -e^{-2g_1\bar{\rho}}S_1 \end{bmatrix},$$

$\tilde{\Sigma}_1(\theta) = He\{P_1 A(\theta)\} + S_1 + Q_1 + 2g_1 P_1,$

$$\bar{\Sigma}_2(\theta) = \begin{bmatrix} He\{P_2A(\theta)\} + S_2 + Q_2 - 2g_2P_2 & * & * & * & * \\ 0 & 0 & * & * & * \\ 0 & 0 & -e^{2g_2\tau}Q_2 & * & * \\ E^T(\theta)P_2^T & 0 & 0 & -\gamma^2 I & * \\ B^T(\theta)P_2^T & 0 & 0 & 0 & -e^{2g_2\bar{\rho}}S_2 \end{bmatrix},$$

$$\bar{\Xi}_1 = [\Upsilon_1 + \tilde{\Upsilon}_1 \quad -\bar{\Upsilon}_1 + \tilde{\Upsilon}_1 \quad -\Upsilon_1 - \tilde{\Upsilon}_1 \quad 0 \; 0 \; 0],$$

$$\bar{\Xi}_2 = [\Upsilon_2 + \tilde{\Upsilon}_2 \quad -\bar{\Upsilon}_2 + \tilde{\Upsilon}_2 \quad -\Upsilon_2 - \tilde{\Upsilon}_2 \quad 0 \; 0],$$

$$\Upsilon_1 = [\Upsilon_{11}^T \; \Upsilon_{12}^T \; \Upsilon_{13}^T \; \Upsilon_{14}^T \; \Upsilon_{15}^T \; \Upsilon_{16}^T]^T, \; \Upsilon_2 = [\Upsilon_{21}^T \; \Upsilon_{22}^T \; \Upsilon_{23}^T \; \Upsilon_{24}^T \; \Upsilon_{25}^T]^T,$$

$$\bar{\Upsilon}_1 = [\bar{\Upsilon}_{11}^T \; \bar{\Upsilon}_{12}^T \; \bar{\Upsilon}_{13}^T \; \bar{\Upsilon}_{14}^T \; \bar{\Upsilon}_{15}^T \; \bar{\Upsilon}_{16}^T]^T, \; \bar{\Upsilon}_2 = [\bar{\Upsilon}_{21}^T \; \bar{\Upsilon}_{22}^T \; \bar{\Upsilon}_{23}^T \; \bar{\Upsilon}_{24}^T \; \bar{\Upsilon}_{25}^T]^T,$$

$$\tilde{\Upsilon}_1 = [\tilde{\Upsilon}_{11}^T \; \tilde{\Upsilon}_{12}^T \; \tilde{\Upsilon}_{13}^T \; \tilde{\Upsilon}_{14}^T \; \tilde{\Upsilon}_{15}^T \; \tilde{\Upsilon}_{16}^T]^T, \; \tilde{\Upsilon}_2 = [\tilde{\Upsilon}_{21}^T \; \tilde{\Upsilon}_{22}^T \; \tilde{\Upsilon}_{23}^T \; \tilde{\Upsilon}_{24}^T \; \tilde{\Upsilon}_{25}^T]^T,$$

then filtering error system (5.55) is exponentially stable with the prescribed $\mathscr{L}_2$–$\mathscr{L}_\infty$ *performance* $\bar{\gamma} = \sqrt{\frac{\eta_{\max}}{\eta_{\min}}}\gamma$ *with* $\eta_{\min} = \{\frac{1}{\pi_2}, 1\}$, $\eta_{\max} = \{\frac{e^{2g_1 T_{off}^{\max}}}{\pi_2}, e^{2g_2(T - T_{off}^{\min})}\}$.

Proof For any $w(t) \in \mathscr{L}_2[0, \infty)$, using a same proof way as Lemma 5.1, it follows from (5.70) that

$$\begin{cases} \dot{V}_1(t) + 2g_1V_1(t) - \gamma^2 w^T(t)w(t) \le 0, & t \in [t_{1,h}, t_{2,h}), \\ \dot{V}_2(t) - 2g_2V_2(t) - \gamma^2 w^T(t)w(t) \le 0, & t \in [t_{2,h}, t_{1,h+1}). \end{cases} \tag{5.72}$$

For any $t \in [0, (h+1)T)$, $h \in \mathbb{N}$, one can be get from (5.72)

$$\begin{aligned} &\sum_{j=0}^{h}\Big(\int_{t_{1,j}}^{t_{2,j}} \frac{e^{-2g_1(t_{1,j}-t)}}{\pi_2}\gamma^2 w^T(t)w(t)dt + \int_{t_{2,j}}^{t_{1,j+1}} e^{2g_2(t_{1,j+1}-t)}\gamma^2 w^T(t)w(t)dt\Big) \\ \ge &\sum_{j=0}^{h}\Big(\int_{t_{1,j}}^{t_{2,j}} \frac{e^{-2g_1(t_{1,j}-t)}}{\pi_2}(\dot{V}_1(t) + 2g_1V_1(t))dt + \int_{t_{2,j}}^{t_{1,j+1}} e^{2g_2(t_{1,j+1}-t)}(\dot{V}_2(t) - 2g_2V_2(t))dt\Big) \\ \ge &\frac{V_1(t_{1,h+1}) - V_1(0)}{\pi_2} + \sum_{j=0}^{h} V_1(t_{2,j})\left[\frac{e^{2g_1 T_{off}^{\min}}}{\pi_2} - \pi_1 e^{2(g_1+g_2)\tau + 2g_2(T - T_{off}^{\min})}\right]. \end{aligned}$$

From (5.63), one can be obtained that $\frac{e^{2g_1 T_{off}^{\min}}}{\pi_2} - \pi_1 e^{2(g_1+g_2)\tau + 2g_2(T - T_{off}^{\min})} > 0$. Note that $V_1(t_{1,h+1}) > 0$ and $V_1(t_{2,j}) > 0$, $j \in \{0, 1, \ldots, h\}$, $h \in \mathbb{N}$, then, under zero initial condition, i.e., $V_1(0) = 0$, it can be rewritten from the above inequality that

$$\sum_{j=0}^{h}\Big(\int_{t_{1,j}}^{t_{2,j}} \frac{e^{-2g_1(t_{1,j}-t)}}{\pi_2}\gamma^2 w^T(t)w(t)dt + \int_{t_{2,j}}^{t_{1,j+1}} e^{2g_2(t_{1,j+1}-t)}\gamma^2 w^T(t)w(t)dt\Big)$$
$$\geq \frac{V_1(t_{1,h+1})}{\pi_2}.$$

For $\forall t \in [t_{1,j},\ t_{2,j}),\ j \in \{0, 1, \ldots, h\},\ h \in \mathbb{N}$, there is

$$1 \leq e^{-2g_1(t_{1,j}-t)} \leq e^{2g_1 T_{off}^{\max}}.$$

For $\forall t \in [t_{2,j},\ t_{1,j+1}),\ j \in \{0, 1, \ldots, h\},\ h \in \mathbb{N}$, there is

$$1 \leq e^{2g_2(t_{1,j+1}-t)} \leq e^{2g_2(T-T_{off}^{\min})}.$$

Set $\eta_{\min} = \{\frac{1}{\pi_2}, 1\}$ and $\eta_{\max} = \{\frac{e^{2g_1 T_{off}^{\max}}}{\pi_2}, e^{2g_2(T-T_{off}^{\min})}\}$. Then

$$\sum_{j=0}^{h}\int_{t_{1,j}}^{t_{1,j+1}} \eta_{\max}\gamma^2 w^T(t)w(t)dt \geq \eta_{\min} V_1(t_{1,h+1}).$$

By (5.71), it is easy to get

$$\begin{aligned}\int_{0}^{t_{1,j+1}} \gamma^2\frac{\eta_{\max}}{\eta_{\min}} w^T(t)w(t)dt \geq & V_1(t_{1,h+1}) \\ \geq & z^T(t)P_1 z(t) \\ \geq & z^T(t)C^T(\theta)C(\theta)z(t) \\ \geq & y^T(t)y(t).\end{aligned}$$

When $h \to \infty$, i.e., $t_{1,j+1} \to \infty$, it can be derived that

$$\sup_{t\geq 0}\{y^T(t)y(t)\} \leq \int_{0}^{\infty} \bar{\gamma}^2 w^T(t)w(t)dt,$$

which means system (5.55) possesses an $\mathcal{L}_2$–$\mathcal{L}_\infty$ performance index $\bar{\gamma}$ in the light of (5.56). The proof is completed.

Based on the $\mathscr{L}_2$–$\mathscr{L}_\infty$ performance analysis result in Theorem 5.4, a joint method of the $\mathscr{L}_2$–$\mathscr{L}_\infty$ filtering gain and weighting matrix H in event-triggered mechanism (5.48) is presented.

Theorem 5.5 *For given DoS signals (5.46), several scalars $\sigma \in (0,\ 1)$, $g_m \in (0,\ \infty)$, $\pi_m > 1$, $\delta_m > 0$, $\gamma > 0$, and $\tau > 0$. If there exist matrices P_m, S_m, Q_m and $H \in \mathbb{R}^n_+$, and matrices M, Υ_m, $\bar{\Upsilon}_m$, $\tilde{\Upsilon}_m$, $m \in \{1, 2\}$ such that (5.63) and the following matrix inequalities hold:*

$$\begin{bmatrix} -P_1 & * \\ C_r & -I \end{bmatrix} < 0, \tag{5.73}$$

$$\begin{cases} P_1 \le \pi_2 P_2, \\ P_2 \le e^{2(g_1+g_2)\tau} \pi_1 P_1, \\ Q_m \le \pi_{3-m} Q_{3-m}, \\ S_m \le \pi_{3-m} S_{3-m}, \\ \delta_m P_m \le \pi_{3-m} \delta_{3-m} P_{3-m}, \end{cases} \tag{5.74}$$

$$\Psi_i^m < 0, \tag{5.75}$$

for $i = 1, \cdots, r$, where

$$\Psi_i^m = \begin{bmatrix} \bar{\Psi}_{mi} + He\{\bar{\Sigma}_m\} & * & * & * & * & * \\ \sqrt{\tau}\tilde{\Upsilon}_m^T & -e^{2(-1)^m g_m \beta_m \tau}\delta_m P_m & * & * & * & * \\ \sqrt{\tau}\bar{\Upsilon}_m^T & 0 & -e^{2(-1)^m g_m \beta_m \tau}\delta_m P_m & * & * & * \\ \sqrt{\tau}\Upsilon_m^T & 0 & 0 & -e^{2(-1)^m g_m \beta_m \tau}\delta_m P_m & * & * \\ \delta_m\sqrt{\tau}\bar{A}_{mi} & 0 & 0 & 0 & -\delta_m P_m & * \\ \delta_m\sqrt{\tau}\bar{A}_{mi} & 0 & 0 & 0 & 0 & -\delta_m P_m \end{bmatrix},$$

$$\bar{A}_{1i} = \begin{bmatrix} P_1A_i & -MC_i & 0 & -MC_i & P_1E_i & P_1B_i \end{bmatrix},$$
$$\bar{A}_{2i} = \begin{bmatrix} P_2A_i & 0 & 0 & P_2E_i & P_2B_i \end{bmatrix},\ \varsigma_1 = 1,\ \varsigma_2 = 0,$$

$$\bar{\Psi}_{1i} = \begin{bmatrix} He\{P_1A_i\} + S_1 + Q_1 + 2g_1P_1 & * & * & * & * & * \\ -C^T(\theta)M^T & \sigma H & * & * & * & * \\ 0 & 0 & -e^{-2g_1\tau}Q_1 & * & * & * \\ -C_i^TM^T & \tilde{C}_i^T & 0 & -H & * & * \\ E_i^TP_1^T & 0 & 0 & 0 & -\gamma^2I & * \\ B_i^TP_1^T & 0 & 0 & 0 & 0 & -e^{-2g_1\bar{\rho}}S_1 \end{bmatrix},$$

$$\bar{\Psi}_{2i} = \begin{bmatrix} He\{P_2A_i\} + S_2 + Q_2 - 2g_2P_2 & * & * & * & * \\ 0 & 0 & * & * & * \\ 0 & 0 & -e^{2g_2\tau}Q_2 & * & * \\ E_i^TP_2^T & 0 & 0 & -\gamma^2I & * \\ B_i^TP_2^T & 0 & 0 & 0 & -e^{2g_1\bar{\rho}}S_2 \end{bmatrix},$$

and the other notations are the same as those in Theorem 5.4. Then filtering error system (5.55) is exponentially stable with the prescribed $\mathcal{L}_2$–$\mathcal{L}_\infty$ performance $\bar{\gamma}$ and the needed gain matrix can be chosen as $L = P_1^{-1}M$.

Proof Define $U_m = Z_m = \delta_m P_m$. Then by $M = P_1L$, Σ_1 in (5.70) with $m = 1$ can be equivalently expressed as Ψ_1 in (5.74), which ensures that $\Sigma_1 < 0$. The proof is completed.

5.2.3' *Simulation Example*

In this subsection, a numerical example is given to demonstrate the effectiveness of the proposed event-triggered $\mathcal{L}_2$–$\mathcal{L}_\infty$ filter design method for the error system in (5.55) under DoS attacks.

The parameters of nonlinear system (5.55) are chosen as follows:

$$A_1 = \begin{bmatrix} -2.5 & 0.1 \\ 0.2 & -1.5 \end{bmatrix},\ B_1 = \begin{bmatrix} -0.1 & 0 \\ 0 & -0.1 \end{bmatrix},$$
$$A_2 = \begin{bmatrix} -2.1 & 0.2 \\ 0.5 & -2 \end{bmatrix},\ B_2 = \begin{bmatrix} -0.2 & 0 \\ 0 & -0.2 \end{bmatrix},$$
$$E_1 = \begin{bmatrix} 0.3 & 0.2 \\ 0 & 0.4 \end{bmatrix},\ C_1 = \begin{bmatrix} 1 & 0 \\ 0 & 1 \end{bmatrix},$$
$$E_2 = \begin{bmatrix} 0.1 & 0 \\ 0.2 & 0.4 \end{bmatrix},\ C_2 = \begin{bmatrix} 0.5 & 0 \\ 0 & 0.5 \end{bmatrix},$$

and the membership functions are given as $\theta_1(q(t))= \sin^2(t)$, $\theta_2(q(t)) = 1 - \theta_1(q(t))$.

Set

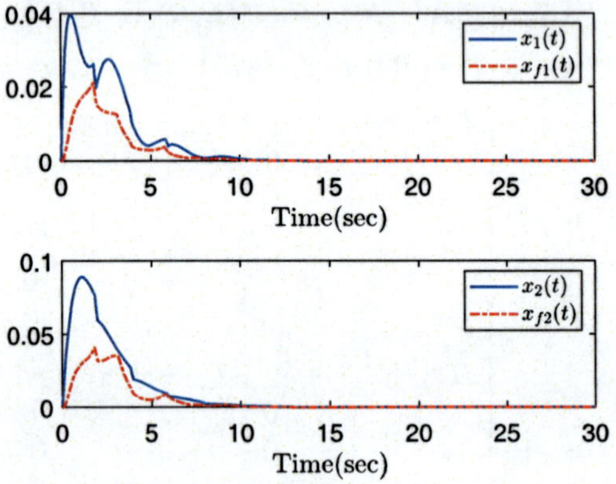

Fig. 5.7 Responses of $x(t)$ and $x_f(t)$ under the DoS attacks

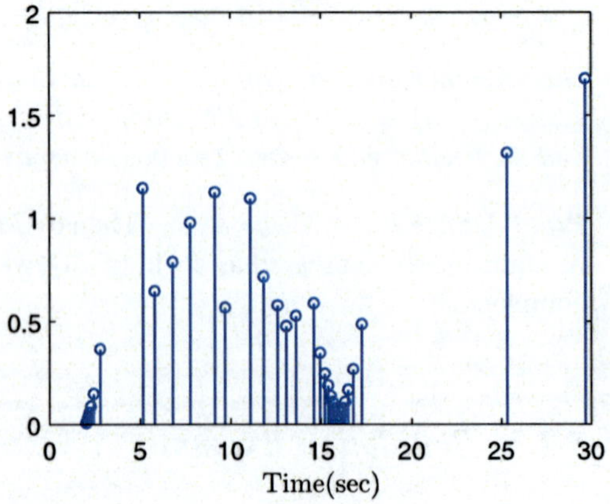

Fig. 5.8 Release time intervals under the DoS attacks

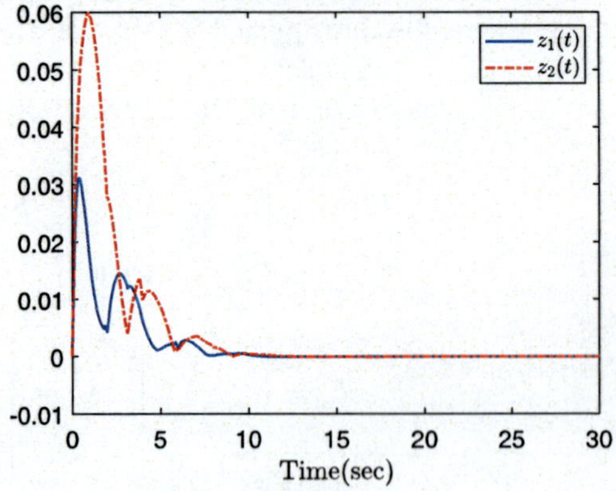

Fig. 5.9 Response of $z(t)$ under the DoS attacks

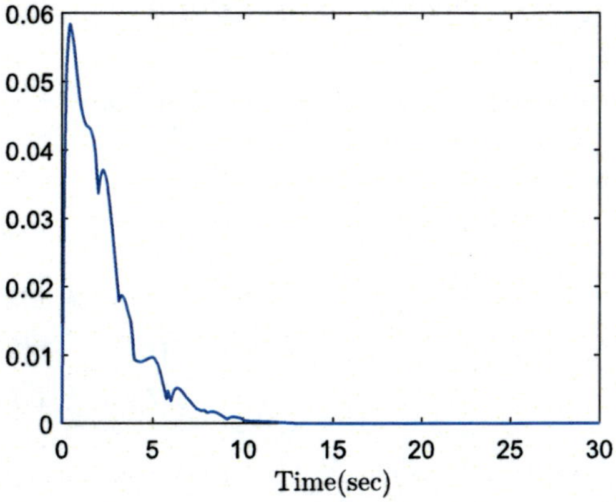

Fig. 5.10 History of $\sqrt{y^T(t)y(t)/\int_0^\infty w^T(s)w(s)ds}$

$$g_1 = 0.05, \; g_2 = 0.5, \; \tau = 0.01, \; \bar{\rho} = 2, \; T = 2, \; T_{off}^{\min} = 1.85,$$
$$T_{off}^{\max} = 1.90, \; \gamma = 0.28, \; \pi_1 = \pi_2 = 1.01, \; \sigma = 0.1.$$

Then, by solving LMIs in Theorem 5.5, the corresponding gain and trigger matrices can be obtained as follows:

$$L = \begin{bmatrix} 0.6344 & 0.9557 \\ 0.6299 & 1.9466 \end{bmatrix}, \quad H = \begin{bmatrix} 2.4941 & 0.3786 \\ 0.3786 & 3.2784 \end{bmatrix}.$$

The initial conditions are chosen as $x(0) = \begin{bmatrix} 0 & 0 \end{bmatrix}^T$, $x_f(0) = \begin{bmatrix} 0 & 0 \end{bmatrix}^T$, and $w(t) = e^{-0.4t}\cos(0.1t)$. Then the trajectories of state $x(t)$ and its estimate $x_f(t)$ are displayed in Fig. 5.7. The event-triggered instant and event time of error system (5.55) is depicted in Fig. 5.8. The trajectory of state $z(t)$ is given in Fig. 5.9. The plot of $\sqrt{y^T(t)y(t)/\int_0^\infty w^T(s)w(s)ds}$ versus time with the zero-initial condition is given in Fig. 5.10, from which one can be observed that the maximum of $\sqrt{y^T(t)y(t)/\int_0^\infty w^T(s)w(s)ds}$ is 0.0584, which is less than the prescribed $\mathcal{L}_2$–$\mathcal{L}_\infty$ performance index $\bar{\gamma} = 0.3079$.

5.3 $\mathcal{H}_\infty$ Filtering with Dynamic Quantization and Deception Attacks

In this section, the problem of event-triggered filtering for discrete-time T–S fuzzy systems is investigated. In the presence of dynamic quantization and stochastic deception attacks, the full- and reduced-order event-triggered filters are designed such that the filtering error system is stochastically stable with the prescribed $\mathcal{H}_\infty$ filtering performance.

5.3.1 Problem Formulation

Consider a nonlinear system represented by T–S fuzzy model, in which the ith rule is described as follows:

Plant Rule i_{th} : **IF** $\epsilon_1(k)$ *is* M_{1i}, $\epsilon_2(k)$ *is* M_{2i}, *and*, $\cdots$, *and* $\epsilon_d(k)$ *is* M_{di}, **THEN**

$$\begin{aligned} x(k+1) &= A_i x(k) + B_i w(k), \\ y(k) &= C_i x(k) + D_i w(k), \\ z(k) &= E_i x(k) + F_i w(k), \end{aligned} \tag{5.76}$$

where $x(k) \in \mathbb{R}^{n_x}$ stands for the state variable, $w(k) \in \mathbb{R}^{n_w}$ is the noise input that is assumed to be the arbitrary signal in $\ell_2[0, \infty)$, $y(k) \in \mathbb{R}^{n_y}$ means the measurement output, and $z(k) \in \mathbb{R}^{n_z}$ is the performance output. $M_{pi}, i = 1, 2, ..., r, p = 1, 2, ..., d$ is the fuzzy sets, where r is the number of fuzzy rules. The premise variables are denoted by $\epsilon_1(k)$, $\epsilon_2(k)$, ..., $\epsilon_d(k)$. The matrices $A_i \in \mathbb{R}^{n_x \times n_x}$, $B_i \in \mathbb{R}^{n_x \times n_w}$, $C_i \in \mathbb{R}^{n_y \times n_x}$, $D_i \in \mathbb{R}^{n_y \times n_w}$, $E_i \in \mathbb{R}^{n_z \times n_x}$, and $F_i \in \mathbb{R}^{n_z \times n_w}$ with $i = 1, 2, ..., r$ indicate given system matrices.

As in Sect. 5.1, the T–S fuzzy model (5.76) can be deduced as follows:

$$\begin{aligned} x(k+1) &= A(\rho)x(k) + B(\rho)w(k), \\ y(k) &= C(\rho)x(k) + D(\rho)w(k), \\ z(k) &= E(\rho)x(k) + F(\rho)w(k), \end{aligned} \tag{5.77}$$

with

$$\begin{aligned} A(\rho) &= \sum_{i=1}^{r} \rho_i(\epsilon(k))A_i, \; B(\rho) = \sum_{i=1}^{r} \rho_i(\epsilon(k))B_i, \\ C(\rho) &= \sum_{i=1}^{r} \rho_i(\epsilon(k))C_i, \; D(\rho) = \sum_{i=1}^{r} \rho_i(\epsilon(k))D_i, \\ E(\rho) &= \sum_{i=1}^{r} \rho_i(\epsilon(k))E_i, \; F(\rho) = \sum_{i=1}^{r} \rho_i(\epsilon(k))F_i. \end{aligned}$$

The output of the fuzzy model (5.77), i.e., $y(k)$ will be transmitted to the dynamic quantizer $q(\star)$ only if

$$\xi_y^T(k)\mathbf{J}\xi_y(k) - b(y(k) - \xi_y(k))^T\mathbf{J}(y(k) - \xi_y(k)) \geq 0 \tag{5.78}$$

is satisfied. Here, $\xi_y(k) = y(k) - y(k_s)$, $y(k)$ is the present data, and $y(k_s)$ represents the data at the last triggering instant. $b \in [0, 1)$ is a predetermined scalar, $\mathbf{J} > 0$ stands for a weighting matrix to be determined.

In this section, to further reduce the amount of data in the communication of network and realize the effective use of the limited network communication resources, $y(k_s)$ is quantized by using the same form of dynamic quantizer in (2.7). According to the definition in Sect. 2.1, the quantized $y(k_s)$ can be described as:

$$\mathbb{Q}(y(k_s)) = \mu_{y_s}(k)q\left(\frac{y(k_s)}{\mu_{y_s}(k)}\right). \tag{5.79}$$

Assume that the quantized output $\mathbb{Q}(y(k_s))$ is transmitted to the filter via imperfect communication network where the deception attacks will be encountered unavoidably and randomly.

According to [23, 24], in this section, the quantized output with stochastic deception attacks can be modeled as

$$\overline{y}(k) = h(k)C(\rho)\alpha(x(k)) + (1 - h(k))\mathbb{Q}(y(k_s)). \tag{5.80}$$

In Eq. (5.80), $\alpha(x(k))$ is used to stand for a nonlinear function about $x(k)$, which satisfies

$$x^T(k)Q^T Qx(k) - \alpha^T(x(k))\alpha(x(k)) \geq 0, \tag{5.81}$$

here, Q stands for a known matrix parameter for $\alpha(x(k))$. In addition, $h(k)$ is a Bernoulli process, which satisfies

$$\begin{aligned} Prob\{h(k) = 1\} &= \mathbf{E}\{h(k)\} = \overline{h}, \\ Prob\{h(k) = 0\} &= 1 - \overline{h}, \end{aligned} \tag{5.82}$$

here, $0 \leq \overline{h} \leq 1$ refers to a given scalar for $h(k)$.

More specifically, $h(k) = 1$ implies that the signal received by the filter is the deception signal, i.e., $\overline{y}(k) = C(\rho)\alpha(x(k))$. $h(k) = 0$ implies that the signal received by the filter is the quantized output, i.e., $\overline{y}(k) = \mathbb{Q}(y(k_s))$.

Moreover, the signal received by the filter can be further rewritten as

$$\begin{aligned} \overline{y}(k) &= h(k)C(\rho)\alpha(x(k)) + (1 - h(k))\mathbb{Q}(y(k_s)) \\ &= h(k)C(\rho)\alpha(x(k)) + (1 - h(k))\left(\mu_{y_s}(k)\left(q\left(\frac{y(k_s)}{\mu_{y_s}(k)}\right) - \frac{y(k_s)}{\mu_{y_s}(k)}\right) + y(k_s)\right) \\ &= h(k)C(\rho)\alpha(x(k)) + (1 - h(k))(\beta(k) + y(k_s)) \\ &= h(k)C(\rho)\alpha(x(k)) + (1 - h(k))(\beta(k) + y(k) - \xi_y(k)), \end{aligned} \tag{5.83}$$

where $\beta(k) = \mu_{y_s}(k)\left(q\left(\frac{y(k_s)}{\mu_{y_s}(k)}\right) - \frac{y(k_s)}{\mu_{y_s}(k)}\right)$.

The filter discussed in this section is presented as

$$\begin{aligned} x_f(k+1) &= A_f x_f(k) + B_f \overline{y}(k), \\ z_f(k) &= E_f x_f(k), \end{aligned} \tag{5.84}$$

where $x_f(k) \in \mathbb{R}^{n_{x_f}}$ stands for the state of the filter, $z_f(k) \in \mathbb{R}^{n_z}$ denotes the output of the filter. The matrices A_f, B_f, and E_f represent the parameters for the filter which need to be designed. Moreover, for $n_{x_f} = n_x$, the filter model in (5.84) stands for a full-order filter; for $1 \leq n_{x_f} < n_x$, the filter model in (5.84) refers to a reduced-order filter.

By defining $\widehat{x}^T(k) = [\ x^T(k) \quad x_f^T(k)\ ]$, $e(k) = z(k) - z_f(k)$, the filtering error system can be obtained as

$$\begin{aligned} \widehat{x}(k+1) &= (\widehat{A}_1 + \widehat{h}(k)\widehat{A}_2)\widehat{x}(k) + (\widehat{B}_1 + \widehat{h}(k)\widehat{B}_2)w(k) \\ &\quad + (S_1 + \widehat{h}(k)S_2)\alpha(x(k)) + (S_3 + \widehat{h}(k)S_4) \\ &\quad \times \beta(k) - (S_3 + \widehat{h}(k)S_4)\xi_y(k), \\ e(k) &= \widehat{E}\widehat{x}(k) + \widehat{F}w(k), \end{aligned} \tag{5.85}$$

where

$$\begin{aligned} \widehat{A}_1 &= \begin{bmatrix} A(\rho) & 0 \\ (1-\bar{h})B_f C(\rho) & A_f \end{bmatrix}, \quad \widehat{A}_2 = \begin{bmatrix} 0 & 0 \\ -B_f C(\rho) & 0 \end{bmatrix}, \\ \widehat{B}_1 &= \begin{bmatrix} B(\rho) \\ (1-\bar{h})B_f D(\rho) \end{bmatrix}, \quad \widehat{B}_2 = \begin{bmatrix} 0 \\ -B_f D(\rho) \end{bmatrix}, \\ S_1 &= \begin{bmatrix} 0 \\ \bar{h}B_f C(\rho) \end{bmatrix}, \quad S_2 = \begin{bmatrix} 0 \\ B_f C(\rho) \end{bmatrix}, \\ S_3 &= \begin{bmatrix} 0 \\ (1-\bar{h})B_f \end{bmatrix}, \quad S_4 = \begin{bmatrix} 0 \\ -B_f \end{bmatrix}, \\ \widehat{E} &= \begin{bmatrix} E(\rho) & -E_f \end{bmatrix}, \quad \widehat{F} = F(\rho), \quad \widehat{h}(k) = h(k) - \bar{h}. \end{aligned}$$

The following important definition on stochastic stability will be used in the analysis of event-triggered filtering.

Definition 1 [18, 24] For any initial condition $\widehat{x}(0)$, when $w(k) = 0$, it can be concluded that the filtering error system (5.85) is stochastically stable, if there exists a matrix $\mathbf{W} > 0$ such that the following inequality

$$\mathbf{E}\left\{ \sum_{k=0}^{\infty} \|\widehat{x}(k)\|^2 \middle| \widehat{x}(0) \right\} < \widehat{x}^T(0)\mathbf{W}\widehat{x}(0) \tag{5.86}$$

is satisfied.

Then the event-triggered filtering problem considered in this section is to design the event-triggered filter in (5.84) such that

(1) The filtering error system (5.85) is stochastically stable when $w(k) = 0$ in the sense of Definition 1.
(2) The filtering error system (5.85) has a prescribed $\mathscr{H}_\infty$ filtering performance $\gamma > 0$, i.e., under the zero-initial condition,

$$\mathbf{E}\left\{\sum_{k=0}^{\infty} e^T(k)e(k)\right\} < \gamma^2 \sum_{k=0}^{\infty} w^T(k)w(k)$$

is satisfied for any nonzero $w(k) \in \ell_2[0, \infty)$.

5.3.2 Main Results

In this subsection, we shall to investigate under what conditions, such that, in the presence of event-triggered communication scheme, dynamic quantization, and stochastic cyber attacks, the filtering error system (5.85) is stochastically stable with the prescribed $\mathscr{H}_\infty$ filtering performance $\gamma > 0$ to give filter gain matrices. Based on the Lyapunov theory, a useful $\mathscr{H}_\infty$ filtering performance analysis criterion will be formulated in the following theorem, which can ensure the stochastic stability and the $\mathscr{H}_\infty$ filtering performance of the filtering error system (5.85).

Theorem 5.6 *Consider the fuzzy system (5.77) and the filter (5.84). Suppose that the scalars $0 \le b < 1$, $\bar{h}$, quantization range bound M_{y_s}, and quantization error Δ are known. The filtering error system (5.85) is stochastically stable with the prescribed $\mathscr{H}_\infty$ filtering performance $\gamma > 0$, if there exist matrices $P > 0$, $\mathbf{J} > 0$, and M, scalars $\xi > 0$, $f_1 > 0$, and $f_2 > 0$ satisfying*

$$M_{y_s} - \frac{1}{\xi} > 0, \tag{5.87}$$

$$\begin{bmatrix} \Omega_{11} & * & * & * & * \\ \Omega_{21} & P - M - M^T & * & * & * \\ \Omega_{31} & 0 & P - M - M^T & * & * \\ \Omega_{41} & 0 & 0 & -I & * \\ \Omega_{51} & 0 & 0 & 0 & -f_1 I \end{bmatrix} < 0, \tag{5.88}$$

where

$$\Omega_{11} = \begin{bmatrix} -P + b\widehat{C}^T\mathbf{J}\widehat{C} + f_2\widehat{Q}^T\widehat{Q} & * & * & * & * \\ bD^T(\rho)\mathbf{J}\widehat{C} & -\gamma^2 I + bD^T(\rho)\mathbf{J}D(\rho) & * & * & * \\ 0 & 0 & -f_2 I & * & * \\ 0 & 0 & 0 & -f_1 I & * \\ -b\mathbf{J}\widehat{C} & -b\mathbf{J}D(\rho) & 0 & 0 & (b-1)\mathbf{J} \end{bmatrix},$$

$$\Omega_{21} = \begin{bmatrix} M^T\widehat{A}_1 & M^T\widehat{B}_1 & M^T S_1 & M^T S_3 & -M^T S_3 \end{bmatrix},$$

$$\Omega_{31} = \begin{bmatrix} \widehat{h}M^T\widehat{A}_2 & \widehat{h}M^T\widehat{B}_2 & \widehat{h}M^T S_2 & \widehat{h}M^T S_4 & -\widehat{h}M^T S_4 \end{bmatrix},$$

$$\Omega_{41} = \begin{bmatrix} \widehat{E} & \widehat{F} & 0 & 0 & 0 \end{bmatrix}, \ \Omega_{51} = \begin{bmatrix} 2f_1\xi\Delta\widehat{C} & 2f_1\xi\Delta D(\rho) & 0 & 0 & -2f_1\xi\Delta I \end{bmatrix},$$

$$\widehat{C} = \begin{bmatrix} C(\rho) & 0 \end{bmatrix}, \ \widehat{Q} = \begin{bmatrix} Q & 0 \end{bmatrix}, \ \widehat{h} = \left(\bar{h}(1-\bar{h})\right)^{1/2},$$

and the parameter of the dynamic quantizer defined in (5.79), i.e., $\mu_{y_s}(k)$ obeys the following online adjusting strategy:

$$\xi \|y(k_s)\| \leq \mu_{y_s}(k) \leq 2\xi \|y(k_s)\|. \tag{5.89}$$

Proof Firstly, we will prove that when the conditions (5.87) and (5.88) are satisfied, for $w(k) = 0$, the filtering error system (5.85) is stochastically stable with the online adjusting strategy for the dynamic quantizer's parameter $\mu_{y_s}(k)$ given in (5.89).

When $w(k) = 0$, the inequality in (5.88) can be rewritten as

$$\begin{bmatrix} \widehat{\Omega}_{11} & * & * & * \\ \widehat{\Omega}_{21} & P - M - M^T & * & * \\ \widehat{\Omega}_{31} & 0 & P - M - M^T & * \\ \widehat{\Omega}_{51} & 0 & 0 & -f_1 I \end{bmatrix} < 0, \tag{5.90}$$

where

$$\widehat{\Omega}_{11} = \begin{bmatrix} -P + b\widehat{C}^T \mathbf{J} \widehat{C} + f_2 \widehat{Q}^T \widehat{Q} & * & * & * \\ 0 & -f_2 I & * & * \\ 0 & 0 & -f_1 I & * \\ -b\mathbf{J}\widehat{C} & 0 & 0 & (b-1)\mathbf{J} \end{bmatrix},$$

$$\widehat{\Omega}_{21} = \begin{bmatrix} M^T \widehat{A}_1 & M^T S_1 & M^T S_3 & -M^T S_3 \end{bmatrix},$$

$$\widehat{\Omega}_{31} = \begin{bmatrix} \widehat{h} M^T \widehat{A}_2 & \widehat{h} M^T S_2 & \widehat{h} M^T S_4 & -\widehat{h} M^T S_4 \end{bmatrix},$$

$$\widehat{\Omega}_{51} = \begin{bmatrix} 2 f_1 \xi \Delta \widehat{C} & 0 & 0 & -2 f_1 \xi \Delta I \end{bmatrix}.$$

Based on the inequality $(P - M)^T P^{-1}(P - M) \geq 0$ and $P > 0$, it can be obtained that $P - M - M^T \geq -M^T P^{-1} M$. Then, the inequality of (5.90) implies that

$$\begin{bmatrix} \widehat{\Omega}_{11} & * & * & * \\ \widehat{\Omega}_{21} & -M^T P^{-1} M & * & * \\ \widehat{\Omega}_{31} & 0 & -M^T P^{-1} M & * \\ \widehat{\Omega}_{51} & 0 & 0 & -f_1 I \end{bmatrix} < 0. \tag{5.91}$$

Performing congruence transformation to (5.91) by $diag\{I,\ M^{-T},\ M^{-T},\ f_1^{-1} I\}$, the following conclusions can be drawn

$$\begin{bmatrix} \widehat{\Omega}_{11} & * & * & * \\ \overline{\Omega}_{21} & -P^{-1} & * & * \\ \overline{\Omega}_{31} & 0 & -P^{-1} & * \\ \overline{\Omega}_{51} & 0 & 0 & -f_1^{-1} I \end{bmatrix} < 0, \tag{5.92}$$

where

$$\overline{\Omega}_{21} = \left[\widehat{A}_1 \;\; S_1 \;\; S_3 \;\; -S_3\right],$$
$$\overline{\Omega}_{31} = \left[\widehat{h}\widehat{A}_2 \;\; \widehat{h}S_2 \;\; \widehat{h}S_4 \;\; -\widehat{h}S_4\right],$$
$$\overline{\Omega}_{51} = \left[2\xi\Delta\widehat{C} \;\; 0 \;\; 0 \;\; -2\xi\Delta I\right].$$

By using Lemma 1.1 to (5.92), it can be obtained that

$$\Upsilon_0 + \Upsilon_1 + f_1\Upsilon_2 + f_2\Upsilon_3 < 0, \tag{5.93}$$

where

$$\begin{aligned}
\Upsilon_0 &= \left[\widehat{A}_1 \;\; S_1 \;\; S_3 \;\; -S_3\right]^T P\left[\widehat{A}_1 \;\; S_1 \;\; S_3 \;\; -S_3\right] + \widehat{h}^2\left[\widehat{A}_2 \;\; S_2 \;\; S_4 \;\; -S_4\right]^T \\
&\quad \times P\left[\widehat{A}_2 \;\; S_2 \;\; S_4 \;\; -S_4\right] - diag\{P,\ 0,\ 0,\ 0\}, \\
\Upsilon_1 &= b\left[\widehat{C} \;\; 0 \;\; 0 \;\; -I\right]^T \mathbf{J}\left[\widehat{C} \;\; 0 \;\; 0 \;\; -I\right] - diag\{0,\ 0,\ 0,\ \mathbf{J}\}, \\
\Upsilon_2 &= \left[2\xi\Delta\widehat{C} \;\; 0 \;\; 0 \;\; -2\xi\Delta I\right]^T\left[2\xi\Delta\widehat{C} \;\; 0 \;\; 0 \;\; -2\xi\Delta I\right] - diag\{0,\ 0,\ I,\ 0\}, \\
\Upsilon_3 &= \left[\widehat{Q} \;\; 0 \;\; 0 \;\; 0\right]^T\left[\widehat{Q} \;\; 0 \;\; 0 \;\; 0\right] - diag\{0,\ I,\ 0,\ 0\}.
\end{aligned}$$

Now, the Lyapunov function is constructed as

$$V\left(\widehat{x}(k)\right) = \widehat{x}^T(k)P\widehat{x}(k), \;\; P > 0. \tag{5.94}$$

Then, by considering the fact $\mathbf{E}\left\{\widehat{h}^2(k)\right\} = \overline{h}(1-\overline{h}) = \widehat{h}^2$, the following result can be haven

$$\begin{aligned}
&\mathbf{E}\{V\left(\widehat{x}(k+1)\right)\} - V\left(\widehat{x}(k)\right) \\
&= \mathbf{E}\left\{\widehat{x}^T(k+1)P\widehat{x}(k+1)\right\} - \widehat{x}^T(k)P\widehat{x}(k) \\
&= \mathbf{E}\Big\{\Big(\Big(\widehat{A}_1 + \widehat{h}(k)\widehat{A}_2\Big)\widehat{x}(k) + \Big(S_1 + \widehat{h}(k)S_2\Big)\alpha(x(k)) \\
&\quad + \Big(S_3 + \widehat{h}(k)S_4\Big)\beta(k) - \Big(S_3 + \widehat{h}(k)S_4\Big)\xi_y(k)\Big)^T P \\
&\quad \times \Big(\Big(\widehat{A}_1 + \widehat{h}(k)\widehat{A}_2\Big)\widehat{x}(k) + \Big(S_1 + \widehat{h}(k)S_2\Big)\alpha(x(k)) \\
&\quad + \Big(S_3 + \widehat{h}(k)S_4\Big)\beta(k) - \Big(S_3 + \widehat{h}(k)S_4\Big)\xi_y(k)\Big)\Big\} - \widehat{x}^T(k)P\widehat{x}(k) \\
&= \widetilde{x}^T(k)\mathbf{E}\Big\{\left(\left[\widehat{A}_1 \;\; S_1 \;\; S_3 \;\; -S_3\right] + \widehat{h}(k)\left[\widehat{A}_2 \;\; S_2 \;\; S_4 \;\; -S_4\right]\right)^T \\
&\quad \times P\left(\left[\widehat{A}_1 \;\; S_1 \;\; S_3 \;\; -S_3\right] + \widehat{h}(k)\left[\widehat{A}_2 \;\; S_2 \;\; S_4 \;\; -S_4\right]\right)\Big\}\widetilde{x}(k) - \widetilde{x}^T(k)\mathbf{P}_1\widetilde{x}(k) \\
&= \widetilde{x}^T(k)\Upsilon_0\widetilde{x}(k),
\end{aligned} \tag{5.95}$$

where $\widetilde{x}^T(k) = \left[\widehat{x}^T(k) \;\; \alpha^T(x(k)) \;\; \beta^T(k) \;\; \xi_y^T(k)\right]$, $\mathbf{P}_1 = diag\{P,\ 0,\ 0,\ 0\}$.

According to the event-triggered condition given in (5.78), one can be obtained that $b(y(k) - \xi_y(k))^T\mathbf{J}(y(k) - \xi_y(k)) - \xi_y^T(k)\mathbf{J}\xi_y(k) \geq 0$, which is equivalent to

$$\widetilde{x}^T(k)\Upsilon_1\widetilde{x}(k) \geq 0. \tag{5.96}$$

The condition in (5.89) can be expressed as

$$\frac{1}{2\xi}\mu_{y_s}(k) \leq \|y(k_s)\| \leq \frac{1}{\xi}\mu_{y_s}(k). \tag{5.97}$$

By considering the condition in (5.87), the condition in (5.97) can be rewritten as

$$\frac{1}{2\xi}\mu_{y_s}(k) \leq \|y(k_s)\| \leq M_{y_s}\mu_{y_s}(k). \tag{5.98}$$

Based on the condition in (5.98), $\left\|\frac{y(k_s)}{\mu_{y_s}(k)}\right\| \leq M_{y_s}$ can be gained, which implies that

$$\left\|q\left(\frac{y(k_s)}{\mu_{y_s}(k)}\right) - \frac{y(k_s)}{\mu_{y_s}(k)}\right\| \leq \Delta. \tag{5.99}$$

In addition, according to the homogeneity property of Euclidean norm, it can be obtained that

$$\begin{aligned}
\|\beta(k)\| &= \left\|\mu_{y_s}(k)\left(q\left(\frac{y(k_s)}{\mu_{y_s}(k)}\right) - \frac{y(k_s)}{\mu_{y_s}(k)}\right)\right\| \\
&= \mu_{y_s}(k)\left\|q\left(\frac{y(k_s)}{\mu_{y_s}(k)}\right) - \frac{y(k_s)}{\mu_{y_s}(k)}\right\| \\
&\leq \mu_{y_s}(k)\Delta \\
&\leq 2\xi\Delta\|y(k_s)\|,
\end{aligned} \tag{5.100}$$

i.e.,

$$\beta^T(k)\beta(k) \leq 4\xi^2\Delta^2(y(k) - \xi_y(k))^T(y(k) - \xi_y(k)). \tag{5.101}$$

The inequality in (5.101) can be further rewritten as

$$\widetilde{x}^T(k)\Upsilon_2\widetilde{x}(k) \geq 0. \tag{5.102}$$

Moreover, the inequality in (5.81) can be further expressed as

$$\widetilde{x}^T(k)\Upsilon_3\widetilde{x}(k) \geq 0. \tag{5.103}$$

By using Lemma 1.3 with inequalities (5.95), (5.96), (5.102), (5.103), and scalars $f_1 > 0$, $f_2 > 0$, it's easy to conclude that if the inequality in (5.93) is satisfied, then, $\widetilde{x}^T(k)\Upsilon_0\widetilde{x}(k) < 0$, i.e.,

$$\mathbf{E}\{V(\widehat{x}(k+1))\} - V(\widehat{x}(k)) < 0, \tag{5.104}$$

which implies that

$$\mathbf{E}\{V(\widehat{x}(k+1))\} - V(\widehat{x}(k)) \leq -\lambda_{\min}(-\Upsilon_0)\widetilde{x}^T(k)\widetilde{x}(k). \tag{5.105}$$

Then, taking mathematical expectation for both sides of the inequality (5.105), for any $c \geq 1$, summing up the inequality (5.105) on both sides from $k = 0$ to $k = c$, the following conclusion can be given

$$\mathbf{E}\left\{\widehat{x}^T(c+1)P\widehat{x}(c+1)\right\} - \widehat{x}^T(0)P\widehat{x}(0) \leq -\lambda_{\min}(-\Upsilon_0)\mathbf{E}\left\{\sum_{k=0}^{c}\|\widetilde{x}(k)\|^2\right\}, \tag{5.106}$$

where $\widehat{x}(0)$ is the initial condition. The inequality (5.106) can be expressed as

$$\mathbf{E}\left\{\sum_{k=0}^{c}\|\widetilde{x}(k)\|^2\right\} \leq (\lambda_{\min}(-\Upsilon_0))^{-1}\left(\widehat{x}^T(0)P\widehat{x}(0) - \mathbf{E}\left\{\widehat{x}^T(c+1)P\widehat{x}(c+1)\right\}\right). \tag{5.107}$$

When $c=1,\ldots,\infty$, considering $\mathbf{E}\left\{\widehat{x}^T(\infty)P\widehat{x}(\infty)\right\} \geq 0$ and $\mathbf{E}\left\{\sum_{k=0}^{\infty}\|\widetilde{x}(k)\|^2\right\} \geq \mathbf{E}\left\{\sum_{k=0}^{\infty}\|\widehat{x}(k)\|^2\right\}$,

$$\begin{aligned}\mathbf{E}\left\{\sum_{k=0}^{\infty}\|\widehat{x}(k)\|^2\right\} &\leq (\lambda_{\min}(-\Upsilon_0))^{-1}\widehat{x}^T(0)P\widehat{x}(0)\\ &= \widehat{x}^T(0)\left(\lambda_{\min}(-\Upsilon_0)\right)^{-1}P\widehat{x}(0)\\ &= \widehat{x}^T(0)\mathbf{W}\widehat{x}(0),\end{aligned}$$

where $\mathbf{W} = (\lambda_{\min}(-\Upsilon_0))^{-1}P$.

Based on $\widetilde{x}^T(k)\Upsilon_0\widetilde{x}(k) < 0$, $\Upsilon_0 < 0$ can be known, $\lambda_{\min}(-\Upsilon_0) > 0$. Thus, $\mathbf{W} = (\lambda_{\min}(-\Upsilon_0))^{-1}P > 0$. Then, according to Definition 1, it can be concluded that when $w(k) = 0$, the filtering error system (5.85) is stochastically stable with the online adjusting strategy for the dynamic quantizer's parameter $\mu_{y_s}(k)$ given in (5.89) if the conditions given in (5.87) and (5.88) are satisfied.

Next, for $w(k) \neq 0$, the $\mathscr{H}_\infty$ filtering performance constraints will be considered of the filtering error system (5.85) with zero-initial condition.

Similar to the above proof process of stochastic stability, the inequality in (5.88) implies that

$$\begin{bmatrix}\Omega_{11} & * & * & * & *\\ \Lambda_{21} & -P^{-1} & * & * & *\\ \Lambda_{31} & 0 & -P^{-1} & * & *\\ \Omega_{41} & 0 & 0 & -I & *\\ \Lambda_{51} & 0 & 0 & 0 & -f_1^{-1}I\end{bmatrix} < 0, \tag{5.108}$$

where

$$\begin{aligned}\Lambda_{21} &= \left[\widehat{A}_1\ \ \widehat{B}_1\ \ S_1\ \ S_3\ \ -S_3\right],\\ \Lambda_{31} &= \left[\widehat{hA_2}\ \ \widehat{hB_2}\ \ \widehat{h}S_2\ \ \widehat{h}S_4\ \ -\widehat{h}S_4\right],\\ \Lambda_{51} &= \left[2\xi\Delta\widehat{C}\ \ 2\xi\Delta D(\rho)\ \ 0\ \ 0\ \ -2\xi\Delta I\right].\end{aligned}$$

According to Lemma 1.1, the inequality in (5.108) can be indicated as

$$\widehat{\Upsilon}_0 + \widehat{\Upsilon}_1 + f_1\widehat{\Upsilon}_2 + f_2\widehat{\Upsilon}_3 < 0, \tag{5.109}$$

where

$$\begin{aligned}
\widehat{\Upsilon}_0 &= \left[\widehat{A}_1\ \widehat{B}_1\ S_1\ S_3\ -S_3\right]^T P\left[\widehat{A}_1\ \widehat{B}_1\ S_1\ S_3\ -S_3\right] \\
&\quad + \widehat{h}^2\left[\widehat{A}_2\ \widehat{B}_2\ S_2\ S_4\ -S_4\right]^T P\left[\widehat{A}_2\ \widehat{B}_2\ S_2\ S_4\ -S_4\right] \\
&\quad + \left[\widehat{E}\ \widehat{F}\ 0\ 0\ 0\right]^T\left[\widehat{E}\ \widehat{F}\ 0\ 0\ 0\right] - diag\{P,\ \gamma^2 I,\ 0,\ 0,\ 0\}, \\
\widehat{\Upsilon}_1 &= b\left[\widehat{C}\ D(\rho)\ 0\ 0\ -I\right]^T \mathbf{J}\left[\widehat{C}\ D(\rho)\ 0\ 0\ -I\right] - diag\{0,\ 0,\ 0,\ 0,\ \mathbf{J}\}, \\
\widehat{\Upsilon}_2 &= \left[2\xi\Delta\widehat{C}\ 2\xi\Delta D(\rho)\ 0\ 0\ -2\xi\Delta I\right]^T\left[2\xi\Delta\widehat{C}\ 2\xi\Delta D(\rho)\ 0\ 0\ -2\xi\Delta I\right] \\
&\quad - diag\{0,\ 0,\ 0,\ I,\ 0\}, \\
\widehat{\Upsilon}_3 &= \left[\widehat{Q}\ 0\ 0\ 0\ 0\right]^T\left[\widehat{Q}\ 0\ 0\ 0\ 0\right] - diag\{0,\ 0,\ I,\ 0,\ 0\}.
\end{aligned}$$

Based on the filtering error system (5.85) and the Lyapunov function defined in (5.94), it can be obtained that

$$\begin{aligned}
&\mathbf{E}\{V(\widehat{x}(k+1))\} - V(\widehat{x}(k)) + e^T(k)e(k) - \gamma^2 w^T(k)w(k) \\
&= \mathbf{E}\Big\{\Big(\big(\widehat{A}_1 + \widehat{h}(k)\widehat{A}_2\big)\widehat{x}(k) + \big(\widehat{B}_1 + \widehat{h}(k)\widehat{B}_2\big)w(k) \\
&\quad + \big(S_1 + \widehat{h}(k)S_2\big)\alpha(x(k)) + \big(S_3 + \widehat{h}(k)S_4\big)\beta(k) \\
&\quad - \big(S_3 + \widehat{h}(k)S_4\big)\xi_y(k)\Big)^T P\Big(\big(\widehat{A}_1 + \widehat{h}(k)\widehat{A}_2\big)\widehat{x}(k) \\
&\quad + \big(\widehat{B}_1 + \widehat{h}(k)\widehat{B}_2\big)w(k) + \big(S_1 + \widehat{h}(k)S_2\big)\alpha(x(k)) \\
&\quad + \big(S_3 + \widehat{h}(k)S_4\big)\beta(k) - \big(S_3 + \widehat{h}(k)S_4\big)\xi_y(k)\Big)\Big\} \\
&\quad + \big(\widehat{E}\widehat{x}(k) + \widehat{F}w(k)\big)^T\big(\widehat{E}\widehat{x}(k) + \widehat{F}w(k)\big) \\
&\quad - \widehat{x}^T(k)P\widehat{x}(k) - \gamma^2 w^T(k)w(k) \\
&= \overrightarrow{x}^T(k)\mathbf{E}\Big\{\Big(\left[\widehat{A}_1\ \widehat{B}_1\ S_1\ S_3\ -S_3\right] + \widehat{h}(k)\left[\widehat{A}_2\ \widehat{B}_2\ S_2\ S_4\ -S_4\right]\Big)^T \\
&\quad \times P\Big(\left[\widehat{A}_1\ \widehat{B}_1\ S_1\ S_3\ -S_3\right] + \widehat{h}(k)\left[\widehat{A}_2\ \widehat{B}_2\ S_2\ S_4\ -S_4\right]\Big)\Big\}\overrightarrow{x}(k) \\
&\quad + \overrightarrow{x}^T(k)\left[\widehat{E}\ \widehat{F}\ 0\ 0\ 0\right]^T\left[\widehat{E}\ \widehat{F}\ 0\ 0\ 0\right]\overrightarrow{x}(k) \\
&\quad - \overrightarrow{x}^T(k)\mathbf{P}_2\overrightarrow{x}(k) \\
&= \overrightarrow{x}^T(k)\widehat{\Upsilon}_0\overrightarrow{x}(k),
\end{aligned} \tag{5.110}$$

where

$$\begin{aligned}
\overrightarrow{x}^T(k) &= \left[\widehat{x}^T(k)\ w^T(k)\ \alpha^T(x(k))\ \beta^T(k)\ \xi_y^T(k)\right], \\
\mathbf{P}_2 &= diag\{P,\ \gamma^2 I,\ 0,\ 0,\ 0\}.
\end{aligned}$$

For $w(k) \neq 0$, the inequality in (5.96) can be rewritten as

$$\vec{x}^T(k)\widehat{\Upsilon}_1\vec{x}(k) \geq 0, \tag{5.111}$$

the inequality in (5.102) can be modified as

$$\vec{x}^T(k)\widehat{\Upsilon}_2\vec{x}(k) \geq 0, \tag{5.112}$$

and the inequality in (5.103) can be expressed as

$$\vec{x}^T(k)\widehat{\Upsilon}_3\vec{x}(k) \geq 0. \tag{5.113}$$

Then, by using Lemma 1.3 with inequalities (5.110), (5.111), (5.112), and (5.113) and scalars $f_1 > 0$ and $f_2 > 0$, it can be concluded that if the inequality in (5.109) is satisfied, then, $\vec{x}^T(k)\widehat{\Upsilon}_0\vec{x}(k) < 0$, i.e.,

$$\mathbf{E}\{V(\widehat{x}(k+1))\} - V(\widehat{x}(k)) + e^T(k)e(k) - \gamma^2 w^T(k)w(k) < 0. \tag{5.114}$$

Moreover, summing up (5.114) for $k = 0, 1, 2, \ldots, \infty$, then

$$\mathbf{E}\{V(\widehat{x}(\infty))\} - V(\widehat{x}(0)) + \sum_{k=0}^{\infty} e^T(k)e(k) - \gamma^2 \sum_{k=0}^{\infty} w^T(k)w(k) < 0. \tag{5.115}$$

By considering the fact that $V(\widehat{x}(0)) = 0$ and $\mathbf{E}\{V(\widehat{x}(\infty))\} \geq 0$, it can be concluded that

$$\sum_{k=0}^{\infty} e^T(k)e(k) < \gamma^2 \sum_{k=0}^{\infty} w^T(k)w(k), \tag{5.116}$$

which implies that if the conditions given in (5.87) and (5.88) are satisfied, the specified $\mathscr{H}_\infty$ filtering performance for the filtering error system (5.85) can be ensured by using the online adjustment strategy for the dynamic quantizer's parameter $\mu_{y_s}(k)$ given in (5.89). The proof is completed.

According to the analysis results developed in Theorem 5.6, the design problem of event-triggered full-order $\mathscr{H}_\infty$ filter will be addressed in this subsection. More specifically, the sufficient design conditions for the full-order filter defined in (5.84) and the additional parameter of the online adjusting strategy for the quantizer's dynamic parameter defined in (5.89) will be proposed in terms of LMIs.

Theorem 5.7 *Consider the fuzzy system (5.77) and the filter (5.84). Suppose that the scalars $0 \leq b < 1$, $\overline{h}$, quantization range bound M_{y_s}, quantization error Δ and matrix $\Theta = I_{n_x \times n_x}$ are known. The filtering error system (5.85) is stochastically stable with a prescribed $\mathscr{H}_\infty$ filtering performance $\gamma > 0$, if there exist matrices $P_1 > 0$, P_2, $P_3 > 0$, M_1, M_2, M_3, $\mathbf{J} > 0$, $\widehat{A}_f$, $\widehat{B}_f$, and $\widehat{E}_f$, scalars $\varphi > 0$, $f_1 > 0$, and $f_2 > 0$ satisfying*

$$\varphi M_{y_s} - f_1 > 0, \tag{5.117}$$

$$\begin{bmatrix} K_{11i} & * & * & * & * \\ K_{21i} & K_{22} & * & * & * \\ K_{31i} & 0 & K_{22} & * & * \\ K_{41i} & 0 & 0 & -I & * \\ K_{51i} & 0 & 0 & 0 & -f_1 I \end{bmatrix} < 0,\ i = 1, 2, \ldots, r, \tag{5.118}$$

where

$$K_{11i} = \begin{bmatrix} T_{11i} & * & * & * & * \\ T_{21i} & -\gamma^2 I + bD_i^T \mathbf{J} D_i & * & * & * \\ 0 & 0 & -f_2 I & * & * \\ 0 & 0 & 0 & -f_1 I & * \\ T_{51i} & -b\mathbf{J}D_i & 0 & 0 & (b-1)\mathbf{J} \end{bmatrix},$$

$$K_{21i} = \begin{bmatrix} T_{61i} & T_{62i} & T_{63i} & T_{64} & -T_{64} \end{bmatrix},$$

$$K_{22} = \begin{bmatrix} P_1 - M_1 - M_1^T & * \\ P_2 - M_2 - M_3^T \Theta^T & P_3 - M_3 - M_3^T \end{bmatrix},$$

$$K_{31i} = \begin{bmatrix} T_{71i} & T_{72i} & T_{73i} & T_{74} & -T_{74} \end{bmatrix},\ K_{41i} = \begin{bmatrix} T_{81i} & F_i & 0 & 0 & 0 \end{bmatrix},$$

$$K_{51i} = \begin{bmatrix} T_{91i} & 2\varphi\Delta D_i & 0 & 0 & -2\varphi\Delta I \end{bmatrix},$$

$$T_{11i} = \begin{bmatrix} -P_1 + bC_i^T \mathbf{J} C_i + f_2 Q^T Q & * \\ -P_2 & -P_3 \end{bmatrix},$$

$$T_{21i} = \begin{bmatrix} bD_i^T \mathbf{J} C_i & 0 \end{bmatrix},\ T_{51i} = \begin{bmatrix} -b\mathbf{J}C_i & 0 \end{bmatrix},$$

$$T_{61i} = \begin{bmatrix} M_1 A_i + (1-\bar{h})\Theta \widehat{B}_f C_i & \Theta \widehat{A}_f \\ M_2 A_i + (1-\bar{h})\widehat{B}_f C_i & \widehat{A}_f \end{bmatrix},$$

$$T_{62i} = \begin{bmatrix} M_1 B_i + (1-\bar{h})\Theta \widehat{B}_f D_i \\ M_2 B_i + (1-\bar{h})\widehat{B}_f D_i \end{bmatrix},\ T_{63i} = \begin{bmatrix} \bar{h}\Theta \widehat{B}_f C_i \\ \bar{h}\widehat{B}_f C_i \end{bmatrix},$$

$$T_{64} = \begin{bmatrix} (1-\bar{h})\Theta \widehat{B}_f \\ (1-\bar{h})\widehat{B}_f \end{bmatrix},\ T_{71i} = \begin{bmatrix} -\widehat{h}\Theta \widehat{B}_f C_i & 0 \\ -\widehat{h}\widehat{B}_f C_i & 0 \end{bmatrix},$$

$$T_{72i} = \begin{bmatrix} -\widehat{h}\Theta \widehat{B}_f D_i \\ -\widehat{h}\widehat{B}_f D_i \end{bmatrix},\ T_{73i} = \begin{bmatrix} \widehat{h}\Theta \widehat{B}_f C_i \\ \widehat{h}\widehat{B}_f C_i \end{bmatrix},$$

$$T_{74} = \begin{bmatrix} -\widehat{h}\Theta \widehat{B}_f \\ -\widehat{h}\widehat{B}_f \end{bmatrix},\ T_{81i} = \begin{bmatrix} E_i & -\widehat{E}_f \end{bmatrix},\ T_{91i} = \begin{bmatrix} 2\varphi\Delta C_i & 0 \end{bmatrix},$$

and the parameter of the dynamic quantizer defined in (5.79), i.e., $\mu_{y_s}(k)$ *obeys the online adjusting strategy presented in (5.89) with* $\xi = \varphi / f_1$.

Moreover, the parameters for the full-order filter defined in (5.84) can be obtained as

$$A_f = M_3^{-1}\widehat{A}_f,\ B_f = M_3^{-1}\widehat{B}_f,\ E_f = \widehat{E}_f. \tag{5.119}$$

Proof Assume that the related matrices M^T and P have the form as

$$M^T = \begin{bmatrix} M_1 & \Theta M_3 \\ M_2 & M_3 \end{bmatrix}, \ P = \begin{bmatrix} P_1 & * \\ P_2 & P_3 \end{bmatrix}, \tag{5.120}$$

where M_3 is nonsingular. Moreover, by defining $\widehat{A}_f = M_3 A_f$, $\widehat{B}_f = M_3 B_f$, $\widehat{E}_f = E_f$, and $\varphi = f_1 \xi$, then, the inequality in (5.117) can be obtained based on (5.87) and the inequality in (5.88) can be rewritten as

$$\sum_{i=1}^{r} \rho_i(\epsilon(k)) \begin{bmatrix} K_{11i} & * & * & * & * \\ K_{21i} & K_{22} & * & * & * \\ K_{31i} & 0 & K_{22} & * & * \\ K_{41i} & 0 & 0 & -I & * \\ K_{51i} & 0 & 0 & 0 & -f_1 I \end{bmatrix} < 0, \tag{5.121}$$

it is easy to know that if the inequality of (5.118) is satisfied, then the inequality of (5.121) is true. The proof is completed.

According to the analysis results developed in Theorem 5.6, the design problem of event-triggered reduced-order $\mathcal{H}_\infty$ filter will be addressed in this subsection. More specifically, the sufficient design conditions for the reduced-order filter defined in (5.84) and the additional parameter of the online adjusting strategy for the quantizer's dynamic parameter defined in (5.89) will be proposed in terms of LMIs.

Theorem 5.8 *Consider the fuzzy system (5.77) and the filter (5.84). Suppose that the scalars $0 \le b < 1$, $\overline{h}$, quantization range bound M_{y_s}, quantization error Δ and matrix $\Theta = \left[I_{n_{x_f} \times n_{x_f}} \ \ 0_{n_{x_f} \times (n_x - n_{x_f})} \right]^T$ are known. The filtering error system (5.85) is stochastically stable with a prescribed $\mathcal{H}_\infty$ filtering performance $\gamma > 0$, if there exist matrices $P_1 > 0$, P_2, $P_3 > 0$, M_1, M_2, M_3, $\mathbf{J} > 0$, $\widehat{A}_f$, $\widehat{B}_f$, and $\widehat{E}_f$, scalars $\varphi > 0$, $f_1 > 0$, and $f_2 > 0$ satisfying the conditions in (5.117) and (5.118) with the online adjusting strategy for the parameter of the dynamic quantizer given in Theorem 5.7.*

Moreover, the parameters of the reduced-order filter defined in (5.84) can be obtained according to (5.119).

Proof The proof of Theorem 5.8 is similar to Theorem 5.7. Thus, the proof is omitted.

For the event-triggered quantized filter design strategy proposed in this section, some remarks are given as follows:

Remark 5.4 If the scalars b, $\overline{h}$, quantization error Δ, quantization range bound M_{y_s} and the dimension adjustment matrix Θ are given, then, the design conditions for the full- and reduced-order event-triggered $\mathcal{H}_\infty$ filters presented in Theorems 5.7 and 5.8 are strictly LMIs which are able to be solved easily by the LMI toolbox in Matlab. In general, the scalars b, $\overline{h}$, quantization error Δ, and quantization range bound M_{y_s} are given for the event-triggered condition, the occurring probability of stochastic deception attacks, and the dynamic quantizer. The co-design approach given in [25] can be used to deal with the case that the scalar b is unknown. However, how to deal with the filter design problems with the unknown parameters for dynamic quantizer and unknown occurring probability for stochastic deception attacks are still challenging problems, which deserve further study.

Remark 5.5 By introducing a dimension adjustment matrix Θ, both the full- and reduced-order event-triggered $\mathscr{H}_\infty$ filtering problems have been addressed for discrete-time nonlinear networked systems with dynamic quantization and stochastic deception attacks in a unified framework. Moreover, the $\mathscr{H}_\infty$ filtering performance γ can be optimized by

$$\min\ \mu \ \text{ subject to LMIs (5.117) and (5.118) with } \mu = \gamma^2.$$

Then, we have that $\gamma_{\min} = \sqrt{\mu_{\min}}$.

Remark 5.6 In [20–22], the filtering problem has been considered for networked control systems with dynamic quantization. However, it should be noted that the transmission problem of the dynamic quantizer's parameter has not been considered in the above results, i.e., the same dynamic quantizer's parameter can be computed on both sides of the network communication channel according to the received signal, independently (see [20–22] for more details). In fact, according to [19, 26], the parameter of the dynamic quantizer also should be transmitted via the unreliable network communication channel for practical applications. Therefore, as in [26], a feasible adjusting rule for the dynamic quantizer's parameter considered in this section, i.e., $\mu_{y_s}(k)$ is given as follows:

$$\mu_{y_s}(k) = \begin{cases} \text{floor}\left(2\xi|y(k_s)| \times 10^s\right) \times 10^{-s}, & 0 \le \xi|y(k_s)| < \frac{1}{2}, \\ 1, & \frac{1}{2} \le \xi|y(k_s)| < 1, \\ \text{floor}\left(2\xi|y(k_s)|\right), & 1 \le \xi|y(k_s)|, \end{cases}$$

where $s = \min\left\{s \in \mathbf{N}^+ \middle| (2\xi|y(k_s)| \times 10^s) > 1\right\}$ and the scalar $\xi = \varphi/f_1$ can be obtained by solving the linear matrix inequalities in (5.117) and (5.118). Here, the function floor(ϖ) denotes the maximum integer that is not bigger than ϖ.

Remark 5.7 In contrast with the existing results on quantized filtering in [16–18, 20–22] or event-triggered filtering in [25, 27], and [28], the considered filtering problem in this section is more general for two reasons. On the one hand, in this section, both the dynamic quantizer and the event-triggered communication scheme are employed to reduce the number of data transmission in the communication channel from the plant and the filter. On the other hand, in the presence of stochastic deception attacks, both the full- and reduced-order filtering problem have been solved in a unified framework.

5.3.3 Simulation Example

Let us consider a discrete-time T–S fuzzy model in the form of (5.76), where

$$A_1 = \begin{bmatrix} 0.7819 & 1.1899 & 1.1899 \\ -0.0238 & 0.7231 & -0.0177 \\ -0.2380 & -0.1769 & 0.5640 \end{bmatrix},\ B_1 = \begin{bmatrix} 0.0196 \\ 0.0257 \\ -0.0019 \end{bmatrix},$$

$$A_2 = \begin{bmatrix} 0.6749 & 1.1093 & 1.1093 \\ -0.0222 & 0.7239 & -0.0169 \\ -0.2219 & -0.1688 & 0.5720 \end{bmatrix},\ B_2 = \begin{bmatrix} 0.0187 \\ 0.0257 \\ -0.0018 \end{bmatrix},$$

$$C_1 = \begin{bmatrix} 1 & 0 & 0 \end{bmatrix},\ D_1 = 1,\ E_1 = \begin{bmatrix} 1 & 0 & 0 \end{bmatrix},\ F_1 = 0.5,$$

$$C_2 = \begin{bmatrix} 1 & 0 & 0 \end{bmatrix},\ D_2 = 1,\ E_2 = \begin{bmatrix} 1 & 0 & 0 \end{bmatrix},\ F_2 = 0.5.$$

By applying Theorem 5.7 with $M_{y_s} = 200$, $\Delta = 0.5$, $b = 0.25$, $\bar{h} = 0.5$, and $Q = diag\{0.1,\ 0.2,\ 0.1\}$, it can be obtained that $\gamma = 0.5747$, $\mathbf{J} = 0.0490$, $\xi = 0.0050$, and

$$A_f = \begin{bmatrix} 0.6702 & 1.1430 & 1.1433 \\ -0.0611 & 0.7202 & -0.0205 \\ -0.1951 & -0.1644 & 0.5761 \end{bmatrix},\ B_f = \begin{bmatrix} -0.0595 \\ -0.0384 \\ 0.0322 \end{bmatrix},$$

$$E_f = \begin{bmatrix} -1.0831 & -0.1568 & -0.1502 \end{bmatrix}.$$

Moreover, initial conditions are assumed as $x(0) = \begin{bmatrix} 0 & 0 & 0 \end{bmatrix}^T$, $x_f(0) = \begin{bmatrix} 0 & 0 & 0 \end{bmatrix}^T$, $w(k) = 10\cos(0.5k)e^{-0.5k}$, $\rho_1(\epsilon(k)) = 1 - \rho_2(\epsilon(k))$, $\rho_2(\epsilon(k)) = \frac{x_1^2(k)}{9}$, where $-3 \le x_1(k) \le 3$, and

$$\alpha(x(k)) = \begin{bmatrix} -\tanh(0.2x_2(k)) \\ -\tanh(0.1x_1(k)) \\ -\tanh(0.1x_3(k)) \end{bmatrix}.$$

The simulation results of the filtering error system (5.85) are given in Figs. 5.11, 5.12, 5.13, and 5.14. The trajectories of $z(k)$ and $z_f(k)$ are shown in Fig. 5.11. The trajectory of filtering error $e(k)$ is shown in Fig. 5.12. The trajectory the dynamic quantizer's parameter $\mu_{y_s}(k)$ is shown in Fig. 5.13. The simulation result on release instants and release intervals is depicted in Fig. 5.14.

Based on the simulation results given in Figs. 5.11, 5.12, 5.13, and 5.14, it can be concluded that the event-triggered full-order $\mathcal{H}_\infty$ filter design approach developed in this section is effective.

In the following, the event-triggered reduced-order $\mathcal{H}_\infty$ filter design problem will be studied for this example. By applying Theorem 5.8 with $n_{x_f} = 2$, the following variables are obtained

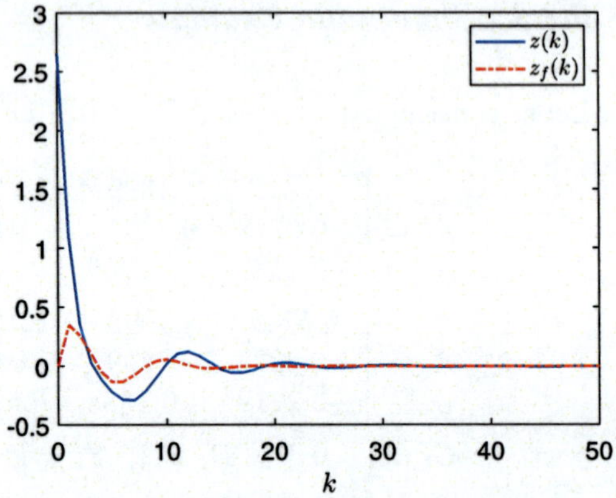

Fig. 5.11 Responses of $z(k)$ and $z_f(k)$

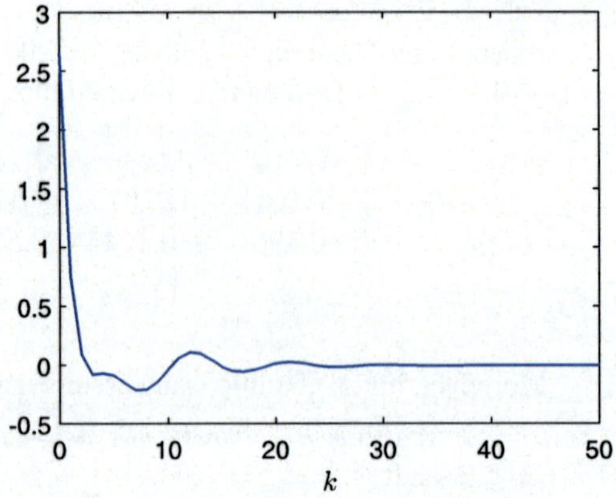

Fig. 5.12 Response of the filtering error $e(k)$

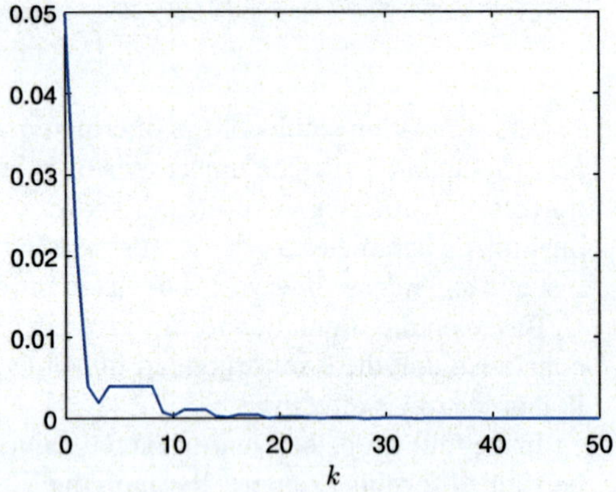

Fig. 5.13 Dynamic quantizer's parameter $\mu_{y_s}(k)$

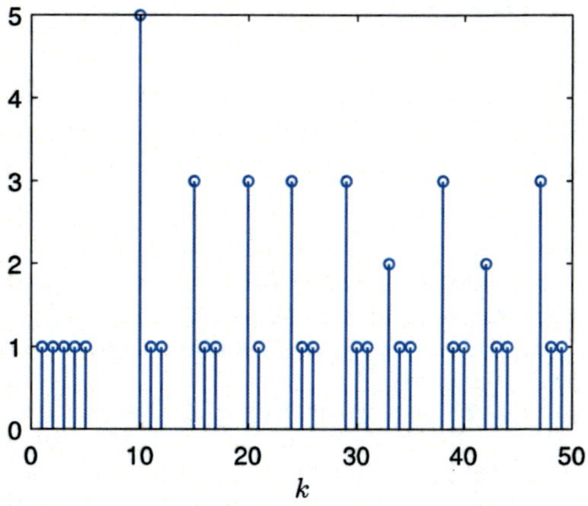

Fig. 5.14 Release instants and release intervals

$$A_f = \begin{bmatrix} 0.3613 & -0.4773 \\ -0.0318 & 0.7510 \end{bmatrix}, \; B_f = \begin{bmatrix} -0.1231 \\ -0.0307 \end{bmatrix},$$
$$E_f = \begin{bmatrix} -0.4246 \; 0.2102 \end{bmatrix}, \; \gamma = 0.6077.$$

Moreover, by applying Theorem 5.8 with $n_{x_f} = 1$, then $\gamma = 0.6098$, $A_f = 0.1171$, $B_f = -0.1288$, and $E_f = -0.4404$ can be obtained.

5.4 Conclusion

In this chapter, the event-triggered output feedback tracking control problem and filtering problem for T–S fuzzy systems have been addressed. For discrete-time fuzzy systems, the event-triggered $\mathscr{H}_\infty$ tracking controller has been designed based on the tracking controller output quantization and the plant output quantization to maintain the prescribed $\mathscr{H}_\infty$ performance. For the continuous-time fuzzy systems, the event-triggered $\mathscr{L}_2$–$\mathscr{L}_\infty$ filtering has been studied with time-delay and DoS attacks. A piecewise Lyapunov–Krasovskii functional method has been proposed to ensure the closed-loop system is exponentially stable and has a prescribed $\mathscr{L}_2$–$\mathscr{L}_\infty$ performance. For discrete-time case, the event-triggered $\mathscr{H}_\infty$ filtering has been investigated with dynamic quantization and stochastic deception attacks. In order to ensure the filtering error system is stochastically stable with the prescribed $\mathscr{H}_\infty$ performance, the full- and reduced-order filters have been designed by LMIs, respectively. Moreover, some simulation results have been given to verify the effectiveness of these control and filtering design strategies.

References

1. Zhang D, Han QL, Jia X (2015) Network-based output tracking control for T–S fuzzy systems using an event-triggered communication scheme. Fuzzy Sets Syst 273:26–48
2. Pan Y, Yang GH (2019) Event-based output tracking control for fuzzy networked control systems with network-induced delays. Appl Math Comput 346:513–530
3. Gu Z, Yue D, Liu J, Ding Z (2017) $\mathscr{H}_\infty$ tracking control of nonlinear networked systems with a novel adaptive event-triggered communication scheme. J Franklin Inst 354:3540–3553
4. Fu M, Xie L (2005) The sector bound approach to quantized feedback control. IEEE Trans Autom Control 50:1698–1711
5. Gao H, Chen T (2008) A new approach to quantized feedback control systems. Automatica 44:534–542
6. Li ZM, Chang XH, Park JH (2021) Quantized static output feedback fuzzy tracking control for discrete-time nonlinear networked systems with asynchronous event-triggered constraints. IEEE Trans Syst, Man, Cybern: Syst 51:3820–3831
7. Liu S, Wei G, Song Y, Liu Y (2016) Error-constrained reliable tracking control for discrete time-varying systems subject to quantization effects. Neurocomputing 174:897–905
8. Liu Z, Wang F, Zhang Y, Chen CLP (2016) Fuzzy adaptive quantized control for a class of stochastic nonlinear uncertain systems. IEEE Trans Cybern 46:524–534
9. Chang XH (2014) Robust output feedback H-infinity control and filtering for uncertain linear systems. Berlin Heidelberg, Springer
10. Chang XH, Yang GH (2014) New results on output feedback $\mathscr{H}_\infty$ control for linear discrete-time systems. IEEE Trans Autom Control 59:1355–1359
11. Yue D, Tian E, Han QL (2013) A delay system method for designing event-triggered controllers of networked control systems. IEEE Trans Autom Control 58:475–481
12. Hu S, Yue D, Xie X, Chen X, Yin X (2018) Resilient event-triggered controller synthesis of networked control systems under periodic DoS jamming attacks. IEEE Trans Cybern 49:4271–4281
13. Yue D, Tian E, Han Q (2012) A delay system method for designing event-triggered controllers of networked control systems. IEEE Trans Autom Control 58:475–481
14. Shisheh FH, Martínez S (2016) On triggering control of single-input linear systems under pulse-width modulated DoS signals. SIAM J Control Optim 54:3084–3105
15. Zhang X, Han Q (2013) Novel delay-derivative-dependent stability criteria using new bounding techniques. Int J Robust Nonlinear Control 23:1419–1432
16. Chang XH, Wang YM (2018) Peak-to-peak filtering for networked nonlinear DC motor systems with quantization. IEEE Trans Industr Inf 14:5378–5388
17. Dong S, Su H, Shi P, Lu R, Wu ZG (2017) Filtering for discrete-time switched fuzzy systems with quantization. IEEE Trans Fuzzy Syst 25:1616–1628
18. Zhang C, Feng G, Gao H, Qiu J (2010) $\mathscr{H}_\infty$ filtering for nonlinear discrete-time systems subject to quantization and packet dropouts. IEEE Trans Fuzzy Syst 19:353–365
19. Niu Y, Ho DWC (2014) Control strategy with adaptive quantizer's parameters under digital communication channels. Automatica 50:2665–2671
20. Che WW, Yang GH (2013) $\mathscr{H}_\infty$ filter design for continuous-time systems with quantised signals. Int J Syst Sci 44:265–274
21. Che WW, Yang G H (2009) Quantised $\mathscr{H}_\infty$ filter design for discrete-time systems. Int J Control 82:195–206
22. Chang XH, Li ZM, Park JH (2018) Fuzzy generalized $\mathscr{H}_2$ filtering for nonlinear discrete-time systems with measurement quantization. IEEE Trans Syst, Man, Cybern: Syst 48:2419–2430
23. Liu J, Wei L, Xie X, Tian E, Fei S (2018) Quantized stabilization for T–S fuzzy systems with hybrid-triggered mechanism and stochastic cyber-attacks. IEEE Trans Fuzzy Syst 26:3820–3834
24. Li ZM, Xiong J (2022) Event-triggered fuzzy filtering for nonlinear networked systems with dynamic quantization and stochastic cyber attacks. ISA Trans 121:53–62

25. Zhang XM, Han QL, Zhang BL (2017) An overview and deep investigation on sampled-data-based event-triggered control and filtering for networked systems. IEEE Trans Industr Inf 13:4–16
26. Chang XH, Yang C, Xiong J (2019) Quantized fuzzy output feedback $\mathscr{H}_\infty$ control for nonlinear systems with adjustment of dynamic parameters. IEEE Trans Syst, Man, Cybern: Syst 49: 2005–2015
27. Su X, Wen Y, Shi P, Lam HK (2019) Event-triggered fuzzy filtering for nonlinear dynamic systems via reduced-order approach. IEEE Trans Fuzzy Syst 27:1215–1225
28. Liu Y, Guo BZ, Park JH, Lee S (2018) Event-based reliable dissipative filtering for T–S fuzzy systems with asynchronous constraints. IEEE Trans Fuzzy Syst 26:2089–2098

Chapter 6
Fuzzy Fault Detection and Fault-Tolerant Control with Quantization

Abstract In this chapter, the quantized fault detection and fault-tolerant control problems are investigated for T–S fuzzy system. The quantized fault detection problem is first investigated for the uncertain T–S fuzzy system. Particularly, we consider two different situations of the filter parameters with and without uncertain terms. And by using Lyapunov function method for the considered system, both robust and resilient fault detection filters are designed respectively such that the residual system is asymptotically stable and guarantees a prescribed $\mathscr{H}_\infty$ noise attenuation level bound. Next, the problem of quantized fault-tolerant control for T–S fuzzy system with Markov jumps and actuator fault is studied. A time-varying coefficient matrix is introduced to describe the relation of the fault signal and the input signal of the actuator. Then the fault-tolerant controller is designed based on the mode-dependent idea and robust stochastic stability approach, and the existence conditions of such controller are formulated in terms of a set of LMIs. In addition, quantized guaranteed cost fault-tolerant control problem is studied for the T–S fuzzy system, where there has the uncertainty term for the controller parameter. Consider the coupling term caused from the uncertainties of controller and the fault model, a two-step design method is introduced to develop a design scheme which can be solved by standard LMI technique, and for the quantization effect, an adjusting rule for the dynamic quantization parameter is also proposed to ensure the stability and the guaranteed cost performance for the resulting closed-loop system. The effectiveness of the proposed design methods are shown through some simulation examples.

Keywords Fault detection · Fault-tolerant control · Resilient control · Guaranteed cost control · Markov jumps

6.1 Fault Detection with Output Quantization

Considering the problem of fault detection in T–S fuzzy system with uncertainty and quantization, whether the fault detection filter contains uncertain terms is fully considered, the main object is to design the filters such that the corresponding residual systems are asymptotically stable and satisfy the certain $\mathscr{H}_\infty$ performance. By

X. Chang et al., *Control and Filtering of Fuzzy Systems Under Communication Channels*,
https://doi.org/10.1007/978-981-99-4346-3_6

introducing the corresponding lemmas, the fault detection filter design conditions for the designed filter are obtained in the form of LMIs.

6.1.1 Problem Formulation

In this section, to investigate the fault detection filter design problem, we consider the following T–S fuzzy model

Plant Rule i_{th}: **IF** $\varepsilon_1(k)$ is M_{1i}, $\varepsilon_2(k)$ is M_{2i}, and, …, and $\varepsilon_d(k)$ is M_{di}, **THEN**

$$\begin{aligned} x(k+1) &= (A_i + \Delta_{Ai}(k))x(k) + (B_i + \Delta_{Bi}(k))u(k) \\ &\quad + (E_i + \Delta_{Ei}(k))v(k) + (E_{fi} + \Delta_{E_{fi}}(k))f(k), \\ y(k) &= (C_i + \Delta_{Ci}(k))x(k) + (D_i + \Delta_{Di}(k))u(k) \\ &\quad + (F_i + \Delta_{Fi}(k))v(k) + (F_{fi} + \Delta_{F_{fi}}(k))f(k), \end{aligned} \tag{6.1}$$

where $x(k) \in \mathbb{R}^{n_x}$ is the state variable, $u(k) \in \mathbb{R}^{n_u}$ is the control input, $y(k) \in \mathbb{R}^{n_y}$ is the measurement output, $v(k) \in \mathbb{R}^{n_v}$ is the unknown input vector (including external disturbance, uninterested fault as well as some norm-bounded unstructured model uncertainty), $f(k) \in \mathbb{R}^{n_f}$ is the fault to be detected and isolated. $\varepsilon_1(k), \ldots, \varepsilon_d(k)$ are measurable premise variables; M_{pi} represent the fuzzy sets with $p = 1, \ldots, d$, $i = 1, \ldots, r$, and the parameter r means the number of fuzzy rules. $A_i \in \mathbb{R}^{n_x \times n_x}$, $B_i \in \mathbb{R}^{n_x \times n_u}$, $E_i \in \mathbb{R}^{n_x \times n_v}$, $E_{fi} \in \mathbb{R}^{n_x \times n_f}$, $C_i \in \mathbb{R}^{n_y \times n_x}$, $D_i \in \mathbb{R}^{n_y \times n_u}$, $F_i \in \mathbb{R}^{n_y \times n_v}$, and $F_{fi} \in \mathbb{R}^{n_y \times n_f}$ are known system matrices. Without loss of generality, assume $v(k)$ and $f(k)$ are ℓ_2-norm bounded. The uncertain parameter matrices are given by

$$\begin{bmatrix} \Delta_{Ai}(k) & \Delta_{Bi}(k) & \Delta_{Ei}(k) & \Delta_{E_{fi}}(k) \\ \Delta_{Ci}(k) & \Delta_{Di}(k) & \Delta_{Fi}(k) & \Delta_{F_{fi}}(k) \end{bmatrix} = \begin{bmatrix} X_{x1i} \\ X_{x2i} \end{bmatrix} \Delta_x(k) \begin{bmatrix} Y_{x1i} & Y_{x2i} & Y_{x3i} & Y_{x4i} \end{bmatrix},$$

where X_{x1i}, X_{x2i}, Y_{x1i}, Y_{x2i}, Y_{x3i}, and Y_{x4i} are known matrices with appropriate dimensions and $\Delta_x(k)$ is an uncertainty satisfying the condition $\Delta_x^T(k)\Delta_x(k) \le I$.

It is assumed that fuzzy basis function is given by

$$\rho_i(\varepsilon(k)) = \frac{\prod_{p=1}^{d} M_{pi}(\varepsilon_p(k))}{\sum_{i=1}^{r} \prod_{p=1}^{d} M_{pi}(\varepsilon_p(k))}, \tag{6.2}$$

where $M_{pi}(\varepsilon_p(k))$ is the grade of membership function of $\varepsilon_p(k)$ in M_{pi}.

Considering the properties of membership function, there has

$$\rho_i(\varepsilon(k)) \ge 0, \quad \sum_{i=1}^{r} \rho_i(\varepsilon(k)) = 1, \quad i = 1, \ldots, r. \tag{6.3}$$

Then a more compact presentation of T–S fuzzy model (6.1) can be derived as follows:

$$\begin{aligned} x(k+1) &= (A(\rho) + \Delta_A(\rho))x(k) + (B(\rho) + \Delta_B(\rho))u(k) \\ &\quad + (E(\rho) + \Delta_E(\rho))v(k) + (E_f(\rho) + \Delta_{E_f}(\rho))f(k), \\ y(k) &= (C(\rho) + \Delta_C(\rho))x(k) + (D(\rho) + \Delta_D(\rho))u(k) \\ &\quad + (F(\rho) + \Delta_F(\rho))v(k) + (F_f(\rho) + \Delta_{F_f}(\rho))f(k), \end{aligned} \tag{6.4}$$

where

$$A(\rho) = \sum_{i=1}^{r} \rho_i(\varepsilon(k))A_i, \ B(\rho) = \sum_{i=1}^{r} \rho_i(\varepsilon(k))B_i,$$

$$E(\rho) = \sum_{i=1}^{r} \rho_i(\varepsilon(k))E_i, \ E_f(\rho) = \sum_{i=1}^{r} \rho_i(\varepsilon(k))E_{fi},$$

$$C(\rho) = \sum_{i=1}^{r} \rho_i(\varepsilon(k))C_i, \ D(\rho) = \sum_{i=1}^{r} \rho_i(\varepsilon(k))D_i,$$

$$F(\rho) = \sum_{i=1}^{r} \rho_i(\varepsilon(k))F_i, \ F_f(\rho) = \sum_{i=1}^{r} \rho_i(\varepsilon(k))F_{fi},$$

$$\Delta_A(\rho) = X_{x1}(\rho)\Delta_x(k)Y_{x1}(\rho) = \sum_{i=1}^{r} \rho_i(\varepsilon(k))X_{x1i}\Delta_x(k)Y_{x1i},$$

$$\Delta_B(\rho) = X_{x1}(\rho)\Delta_x(k)Y_{x2}(\rho) = \sum_{i=1}^{r} \rho_i(\varepsilon(k))X_{x1i}\Delta_x(k)Y_{x2i},$$

$$\Delta_E(\rho) = X_{x1}(\rho)\Delta_x(k)Y_{x3}(\rho) = \sum_{i=1}^{r} \rho_i(\varepsilon(k))X_{x1i}\Delta_x(k)Y_{x3i},$$

$$\Delta_{E_f}(\rho) = X_{x1}(\rho)\Delta_x(k)Y_{x4}(\rho) = \sum_{i=1}^{r} \rho_i(\varepsilon(k))X_{x1i}\Delta_x(k)Y_{x4i},$$

$$\Delta_C(\rho) = X_{x2}(\rho)\Delta_x(k)Y_{x1}(\rho) = \sum_{i=1}^{r} \rho_i(\varepsilon(k))X_{x2i}\Delta_x(k)Y_{x1i},$$

$$\Delta_D(\rho) = X_{x2}(\rho)\Delta_x(k)Y_{x2}(\rho) = \sum_{i=1}^{r} \rho_i(\varepsilon(k))X_{x2i}\Delta_x(k)Y_{x2i},$$

$$\Delta_F(\rho) = X_{x2}(\rho)\Delta_x(k)Y_{x3}(\rho) = \sum_{i=1}^{r} \rho_i(\varepsilon(k))X_{x2i}\Delta_x(k)Y_{x3i},$$

$$\Delta_{F_f}(\rho) = X_{x2}(\rho)\Delta_x(k)Y_{x4}(\rho) = \sum_{i=1}^{r} \rho_i(\varepsilon(k))X_{x2i}\Delta_x(k)Y_{x4i}.$$

The fault detection of the system is generally to select the fault detection filter and fault weight system to construct a residual, so as to determine the fault through the residual evaluation function.

6.1.1.1 Residual Generation

To obtain the residual term, the fault detection filter of the following form is selected as

$$\begin{aligned} x_f(k+1) &= (A_f + \Delta_{Af}(k))x_f(k) + (B_f + \Delta_{Bf}(k))y_q(k), \\ r(k) &= (C_f + \Delta_{Cf}(k))x_f(k) + (D_f + \Delta_{Df}(k))y_q(k), \end{aligned} \tag{6.5}$$

where $x_f(k) \in \mathbb{R}^{n_{xf}}$ and $r(k) \in \mathbb{R}^{n_r}$ represent the state vector of the fault detection filter and the residual signal. $A_f \in \mathbb{R}^{n_{xf} \times n_{xf}}$, $B_f \in \mathbb{R}^{n_{xf} \times n_y}$, $C_f \in \mathbb{R}^{n_r \times n_{xf}}$, and $D_f \in \mathbb{R}^{n_r \times n_y}$ are the filter gain matrices to be determined. $\Delta_{Af}(k)$, $\Delta_{Bf}(k)$, $\Delta_{Cf}(k)$, and $\Delta_{Df}(k)$ represent the uncertain parameter terms, which are assumed to have the similar form as

$$\begin{bmatrix} \Delta_{Af}(k) & \Delta_{Bf}(k) \\ \Delta_{Cf}(k) & \Delta_{Df}(k) \end{bmatrix} = \begin{bmatrix} X_{f1} \\ X_{f2} \end{bmatrix} \Delta_f(k) \begin{bmatrix} Y_{f1} & Y_{f2} \end{bmatrix},$$

where X_{f1}, X_{f2}, Y_{f1}, and Y_{f2} are known matrices with appropriate dimensions and $\Delta_f(k)$ is an uncertainty satisfying $\Delta_f^T(k)\Delta_f(k) \leq I$. Moreover, $y_q(k)$ is the quantized signal of the measurement output $y(k)$. Consider a dynamic quantization scheme [1] which is described as

$$y_q(k) = q_\mu(y(k)) = \mu(k) q\left(\frac{y(k)}{\mu(k)}\right), \tag{6.6}$$

where $\mu(k)$ is the quantization parameter that follows the change of quantization signals and $q(\cdot)$ is a static quantizer.

For the quantization range M and quantization error bound Δ of the above quantizer, the following conditions should be satisfied

$$\begin{aligned} &\left\| q\left(\frac{y(k)}{\mu(k)}\right) - \frac{y(k)}{\mu(k)} \right\| \leq \Delta, \quad \left\| \frac{y(k)}{\mu(k)} \right\| \leq M, \\ &\left\| q\left(\frac{y(k)}{\mu(k)}\right) - \frac{y(k)}{\mu(k)} \right\| > \Delta, \quad \left\| \frac{y(k)}{\mu(k)} \right\| > M. \end{aligned} \tag{6.7}$$

In fault detection, a reference residual model is usually needed to describe the desired behavior of the residual signal $r(k)$. In this section, we consider the following fault weighting system [2],

$$\begin{aligned} x_w(k+1) &= A_w x_w(k) + B_w f(k), \\ r_w(k) &= C_w x_w(k) + D_w f(k), \end{aligned} \tag{6.8}$$

where $x_w(k)$ is the state vector of the fault weighting system and $r_w(k)$ is the so-called reference model with $r_w(z) = W(z)f(z)$, in which the priori $W(z)$ is given and the choice of W_z is to impose frequency weighting on the spectrum of the fault signal for detection.

6.1.1.2 Residual Evaluation

The residual generated by the above fault detection filter and fault weight system can be compared with the residual characteristics and threshold values, so as to obtain the judgment condition of whether the system has a fault, and the judgment condition can distinguish the external disturbance and uncertainty. The logic for detecting a potential fault is described as:

$$\begin{aligned} \|r\|_{RMS} &> J_{th} \Rightarrow \text{alarm, a fault is detected,} \\ \|r\|_{RMS} &\leq J_{th} \Rightarrow \text{no alarm, fault-free,} \end{aligned} \tag{6.9}$$

where

$$\|r\|_{RMS} \triangleq \frac{1}{K}\sqrt{\sum_{k=k_1}^{k_2} r^T(k)r(k)},\ K = k_2 - k_1 + 1 \tag{6.10}$$

is the evaluation function which measures the average energy of the residual signal $r(k)$ over a time interval (k_1, k_2). In order to detect fault as early as possible, the length of time window is set to be limited, so as to obtain the evaluation function and determine the threshold J_{th} of fault detection.

Remark 6.1 In this section, the main attention is paid to the design of fault detection filter, thus to highlight this topic, we will not discuss how to set the threshold and make the decision. Surely, after generating the residual signal by fault detection filter and formulating the evaluation function, this issue can be easy to be handled according to [3].

6.1.1.3 Fault Detection Problem

In order to design fault detection filter with dynamic quantization, a residual system needs to be designed. Therefore, by defining error signal $e_r(k) = r_w(k) - r(k)$ and combining (6.4), (6.5), (6.6), and (6.8), the resulting residual system can be obtained as

$$\begin{aligned} \phi(k+1) &= \mathscr{A}(\rho)\phi(k) + \mathscr{B}(\rho)w(k) + \mathscr{E}\upsilon_y(k), \\ e_r(k) &= \mathscr{C}(\rho)\phi(k) + \mathscr{D}(\rho)w(k) + \mathscr{F}\upsilon_y(k), \\ y(k) &= \mathscr{C}_y(\rho)\phi(k) + \mathscr{D}_y(\rho)w(k), \end{aligned} \tag{6.11}$$

where

$$\phi^T(k) = \left[x_w^T(k)\ x^T(k)\ x_f^T(k)\right],\ w^T(k) = \left[u^T(k)\ v^T(k)\ f^T(k)\right],\ \upsilon_y(k) = y_q(k) - y(k),$$

$$\mathscr{A}(\rho) = \begin{bmatrix} A_w & 0 & 0 \\ 0 & A(\rho) + \Delta_A(\rho) & 0 \\ 0 & (B_f + \Delta_{Bf})(C(\rho) + \Delta_C(\rho)) & A_f + \Delta_{Af} \end{bmatrix},\ \mathscr{E} = \begin{bmatrix} 0 \\ 0 \\ B_f + \Delta_{Bf} \end{bmatrix},$$

$$\mathscr{B}(\rho) = \begin{bmatrix} 0 & 0 & B_w \\ B(\rho) + \Delta_B(\rho) & E(\rho) + \Delta_E(\rho) & E_f(\rho) + \Delta_{E_f}(\rho) \\ \mathscr{B}_1(\rho) & \mathscr{B}_2(\rho) & \mathscr{B}_3(\rho) \end{bmatrix},$$

$$\mathscr{B}_1(\rho) = (B_f + \Delta_{Bf})(D(\rho) + \Delta_D(\rho)),\ \mathscr{B}_2(\rho) = (B_f + \Delta_{Bf})(F(\rho) + \Delta_F(\rho)),$$

$$\mathscr{B}_3(\rho) = (B_f + \Delta_{Bf})(F_f(\rho) + \Delta_{F_f}(\rho)),\ \mathscr{D}(\rho) = \left[\mathscr{D}_1(\rho)\ \mathscr{D}_2(\rho)\ \mathscr{D}_3(\rho)\right],$$

$$\mathscr{C}(\rho) = \left[C_w\ \ -(D_f + \Delta_{Df})(C(\rho) + \Delta_C(\rho))\ \ -(C_f + \Delta_{Cf})\right],$$

$$\mathscr{D}_1(\rho) = -(D_f + \Delta_{Df})(D(\rho) + \Delta_D(\rho)),\ \mathscr{D}_2(\rho) = -(D_f + \Delta_{Df})(F(\rho) + \Delta_F(\rho)),$$

$$\mathscr{D}_3(\rho) = D_w - (D_f + \Delta_{Df})(F_f(\rho) + \Delta_{F_f}(\rho)),\ \mathscr{F} = -(D_f + \Delta_{Df}),$$

$$\mathscr{C}_y(\rho) = \left[0\ \ C(\rho) + \Delta_C(\rho)\ \ 0\right],$$

$$\mathscr{D}_y(\rho) = \left[D(\rho) + \Delta_D(\rho)\ \ F(\rho) + \Delta_F(\rho)\ \ F_f(\rho) + \Delta_{F_f}(\rho)\right].$$

In this section, in order to fully consider the influence of fuzzy system, dynamic quantization and uncertain terms between system and fault detection filter, the design problem of fault detection filter is transformed into an $\mathscr{H}_\infty$ filtering problem by referring to the method in [3]. In other words, we will find the suitable fault detection filter such that the residual system is asymptotically stable and the following performance index J is made small:

$$J = \sup_{w(k)\in\ell_2, w(k)\neq 0} \frac{\|e_r(k)\|_2}{\|w(k)\|_2}.$$

The fault detection problem in this section is to design the fault detection filter in (6.5) and the dynamic quantizer in (6.7) such that

(1) The residual system (6.11) is asymptotically stable;
(2) The residual system (6.11) has a prescribed performance γ, i.e.,

$$\sum_{k=0}^{\infty} e_r^T(k)e_r(k) < \gamma^2 \sum_{k=0}^{\infty} w^T(k)w(k). \tag{6.12}$$

6.1.2 *Analysis and Synthesis of Quantized Residual System*

There are two kinds of uncertainties in the residual system (6.11), which increases the difficulty of fault detection filter design. In this section, a two-step method is used to design the corresponding filter to make the residual system (6.11) asymptotically stable and preserve the desired $\mathscr{H}_\infty$ performance.

6.1.2.1 Performance Analysis of Residual System

In this subsection, the fault detection filter with no uncertainty is considered as

$$\begin{aligned} x_f(k+1) &= A_f x_f(k) + B_f y_q(k), \\ r(k) &= C_f x_f(k) + D_f y_q(k). \end{aligned} \tag{6.13}$$

Then the corresponding residual system can be presented as

$$\begin{aligned} \phi(k+1) &= \check{\mathscr{A}}(\rho)\phi(k) + \check{\mathscr{B}}(\rho)w(k) + \check{\mathscr{E}} \upsilon_y(k), \\ e_r(k) &= \check{\mathscr{C}}(\rho)\phi(k) + \check{\mathscr{D}}(\rho)w(k) + \check{\mathscr{F}} \upsilon_y(k), \\ y(k) &= \mathscr{C}_y(\rho)\phi(k) + \mathscr{D}_y(\rho)w(k), \end{aligned} \tag{6.14}$$

where

$$\check{\mathscr{A}}(\rho) = \begin{bmatrix} A_w & 0 & 0 \\ 0 & A(\rho) + \Delta_A(\rho) & 0 \\ 0 & B_f(C(\rho) + \Delta_C(\rho)) & A_f \end{bmatrix}, \check{\mathscr{E}} = \begin{bmatrix} 0 \\ 0 \\ B_f \end{bmatrix},$$

$$\check{\mathscr{B}}(\rho) = \begin{bmatrix} 0 & 0 & B_w \\ B(\rho) + \Delta_B(\rho) & E(\rho) + \Delta_E(\rho) & E_f(\rho) + \Delta_{E_f}(\rho) \\ B_f(D(\rho) + \Delta_D(\rho)) & B_f(F(\rho) + \Delta_F(\rho)) & B_f(F_f(\rho) + \Delta_{F_f}(\rho)) \end{bmatrix},$$

$$\check{\mathscr{C}}(\rho) = \begin{bmatrix} C_w & -D_f(C(\rho) + \Delta_C(\rho)) & -C_f \end{bmatrix}, \check{\mathscr{F}} = -D_f,$$

$$\check{\mathscr{D}}(\rho) = \begin{bmatrix} -D_f(D(\rho) + \Delta_D(\rho)) & -D_f(F(\rho) + \Delta_F(\rho)) & D_w - D_f(F_f(\rho) + \Delta_{F_f}(\rho)) \end{bmatrix}.$$

In the following theorem, the analysis conditions for the stability of the residual system (6.14) and the $\mathscr{H}_\infty$ performance can be obtained.

Theorem 6.1 *Consider the residual system (6.14). For known quantization range M and quantization error bound Δ, if there exist matrix $P > 0$, scalars $\theta \geq 1$ and γ such that the following matrix inequality holds:*

$$\begin{bmatrix} -P & * & * & * & * & * \\ 0 & -\gamma^2 I & * & * & * & * \\ 0 & 0 & -\frac{M^2}{\Delta^2} I & * & * & * \\ \check{\mathscr{A}}(\rho) & \check{\mathscr{B}}(\rho) & \check{\mathscr{E}} & -P^{-1} & * & * \\ \check{\mathscr{C}}(\rho) & \check{\mathscr{D}}(\rho) & \check{\mathscr{F}} & 0 & -I & * \\ \theta\mathscr{C}_y(\rho) & \theta\mathscr{D}_y(\rho) & 0 & 0 & 0 & -I \end{bmatrix} < 0, \tag{6.15}$$

then the system is asymptotically stable and preserves the $\mathscr{H}_\infty$ performance.

Proof When the signal $y(k)$ to be quantized is in the quantization range, i.e., $\left\| \frac{y(k)}{\mu(k)} \right\| \leq M$, $\left\| q\left(\frac{y(k)}{\mu(k)}\right) - \frac{y(k)}{\mu(k)} \right\| \leq \Delta$ can be obtained according to the quantizer properties.

Considering the homogeneity property of Euclidean norm, one has

$$\|\upsilon_y(k)\| = \left\| \mu(k)\left(q\left(\frac{y(k)}{\mu(k)}\right) - \frac{y(k)}{\mu(k)}\right) \right\| = \mu(k)\left\| q\left(\frac{y(k)}{\mu(k)}\right) - \frac{y(k)}{\mu(k)} \right\| \leq \mu(k)\Delta. \tag{6.16}$$

By defining $\mu(k) = \frac{\theta}{M}\|y(k)\|$, where θ is a scalar satisfying $\theta \geq 1$. Substituting $\mu(k)$ defined into (6.16), we obtain

$$\upsilon_y^T(k)\upsilon_y(k) \leq \frac{\theta^2\Delta^2}{M^2} y^T(k)y(k). \tag{6.17}$$

Defining $\delta^T(k) = \left[\phi^T(k)\ w^T(k)\ \upsilon_y^T(k)\right]$, the above inequality can be rewritten as

$$\delta^T(k)\Sigma_1(\rho)\delta(k) \geq 0, \tag{6.18}$$

where

$$\Sigma_1(\rho) = \left[\mathscr{C}_y(\rho)\ \mathscr{D}_y(\rho)\ 0\right]^T \left[\mathscr{C}_y(\rho)\ \mathscr{D}_y(\rho)\ 0\right] - diag\left\{0,\ 0,\ \frac{M^2}{\theta^2\Delta^2}I\right\}.$$

Construct a Lyapunov function as

$$V(\phi(k)) = \phi^T(k)P\phi(k),\ P > 0, \tag{6.19}$$

then, we have

$$\begin{aligned} & V(\phi(k+1)) - V(\phi(k)) + e_r^T(k)e_r(k) - \gamma^2 w^T(k)w(k) \\ & = \phi^T(k+1)P\phi(k+1) - \phi^T(k)P\phi(k) + e_r^T(k)e_r(k) - \gamma^2 w^T(k)w(k) \\ & = \left(\check{\mathscr{A}}(\rho)\phi(k) + \check{\mathscr{B}}(\rho)w(k) + \check{\mathscr{E}}\upsilon_y(k)\right)^T P\left(\check{\mathscr{A}}(\rho)\phi(k) + \check{\mathscr{B}}(\rho)w(k) + \check{\mathscr{E}}\upsilon_y(k)\right) \\ & \quad - \phi^T(k)P\phi(k) + \left(\check{\mathscr{C}}(\rho)\phi(k) + \check{\mathscr{D}}(\rho)w(k) + \check{\mathscr{F}}\upsilon_y(k)\right)^T (\check{\mathscr{C}}(\rho)\phi(k) + \check{\mathscr{D}}(\rho)w(k) \\ & \quad + \check{\mathscr{F}}\upsilon_y(k)) - \gamma^2 w^T(k)w(k) \\ & = \delta^T(k)\Sigma_0(\rho)\delta(k), \end{aligned} \tag{6.20}$$

where

$$\begin{aligned} \Sigma_0(\rho) = & \left[\check{\mathscr{A}}(\rho)\ \check{\mathscr{B}}(\rho)\ \check{\mathscr{E}}\right]^T P\left[\check{\mathscr{A}}(\rho)\ \check{\mathscr{B}}(\rho)\ \check{\mathscr{E}}\right] \\ & + \left[\check{\mathscr{C}}(\rho)\ \check{\mathscr{D}}(\rho)\ \check{\mathscr{F}}\right]^T \left[\check{\mathscr{C}}(\rho)\ \check{\mathscr{D}}(\rho)\ \check{\mathscr{F}}\right] - diag\left\{P,\ \gamma^2 I,\ 0\right\}. \end{aligned}$$

Based on the Lemma 1.3 and defining a positive scalar $\varepsilon_1 = \theta^2$ such that

$$\Sigma_0(\rho) + \theta^2 \Sigma_1(\rho) < 0, \tag{6.21}$$

which means

$$\begin{aligned}&\left[\check{\mathscr{A}}(\rho)\ \check{\mathscr{B}}(\rho)\ \check{\mathscr{E}}\right]^T P \left[\check{\mathscr{A}}(\rho)\ \check{\mathscr{B}}(\rho)\ \check{\mathscr{E}}\right] + \left[\check{\mathscr{C}}(\rho)\ \check{\mathscr{D}}(\rho)\ \check{\mathscr{F}}\right]^T \left[\check{\mathscr{C}}(\rho)\ \check{\mathscr{D}}(\rho)\ \check{\mathscr{F}}\right]\\ &- diag\left\{P,\ \gamma^2 I,\ \frac{M^2}{\Delta^2} I\right\} + \theta^2 \left[\mathscr{C}_y(\rho)\ \mathscr{D}_y(\rho)\ 0\right]^T \left[\mathscr{C}_y(\rho)\ \mathscr{D}_y(\rho)\ 0\right] < 0.\end{aligned} \tag{6.22}$$

Applying the Lemma 1.1, the condition (6.15) for system stability and the desired performance can be obtained.

On the other hand, under the condition of $\Sigma_0(\delta) < 0$, we have

$$\sum_{k=0}^{\infty} e_r^T(k) e_r(k) < \gamma^2 \sum_{k=0}^{\infty} w^T(k) w(k),$$

with $V(\phi(\infty)) > 0$, $V(\phi(0)) = 0$, which proves that the residual system (6.14) is asymptotically stable and guarantees the performance defined in (6.12) under the condition (6.15).

6.1.2.2 Design of Robust Fault Detection Filter with Quantization

Due to the existence of nonlinear terms and the uncertainty terms in the matrix inequalities of Theorem 6.1, the LMIs cannot be used directly for filter design, the following theorem gives the design conditions of robust fault detection filter with quantization.

Theorem 6.2 *Consider the residual system (6.14). For known quantization range M and quantization error bound Δ, given slack scalars β_1, β_2, and β_3, if there exist matrices P_{11}, P_{21}, P_{22}, P_{33}, $\mathscr{A}_f$, $\mathscr{B}_f$, $\mathscr{C}_f$, $\mathscr{D}_f$, scalars $\theta \geq 1$, γ, and ε_1 such that the following LMIs hold*

$$\check{\Lambda}_i < 0, i = 1, ..., r, \tag{6.23}$$

where

$$\check{\Lambda}_i = \begin{bmatrix} \check{\Lambda}_{11i} & * & * \\ \check{\Lambda}_{21i} & \check{\Lambda}_{22} & * \\ 0 & \check{\Lambda}_{32i} & -\varepsilon_1 I \end{bmatrix}, \ \check{\Lambda}_{11i} = \begin{bmatrix} \check{\Xi}_{11i} & * & * \\ \check{\Xi}_{21i} & \check{\Xi}_{22i} & * \\ 0 & 0 & -\frac{M^2}{\Delta^2} I \end{bmatrix}, \ \check{\Lambda}_{21i} = \begin{bmatrix} \check{\Theta}_{11i} & \check{\Theta}_{12i} & \check{\Theta}_{13} \\ \bar{\mathscr{C}}_i & \bar{\mathscr{D}}_i & -D_f \\ \theta\bar{\mathscr{C}}_{yi} & \theta\bar{\mathscr{D}}_{yi} & 0 \end{bmatrix},$$

$$\check{\Xi}_{11i} = \begin{bmatrix} -P_{11} & * & * \\ -P_{21} & -P_{22} + \varepsilon_1 Y_{x1i}^T Y_{x1i} & * \\ -\beta_1 P_{33} & -\beta_2 P_{33} & -\beta_3 P_{33} \end{bmatrix}, \ \check{\Xi}_{21i} = \begin{bmatrix} 0 & \varepsilon_1 Y_{x2i}^T Y_{x1i} & 0 \\ 0 & \varepsilon_1 Y_{x3i}^T Y_{x1i} & 0 \\ 0 & \varepsilon_1 Y_{x4i}^T Y_{x1i} & 0 \end{bmatrix},$$

$$\check{\Xi}_{22i} = \begin{bmatrix} -\gamma^2 I + \varepsilon_1 Y_{x2i}^T Y_{x2i} & * & * \\ \varepsilon_1 Y_{x3i}^T Y_{x2i} & -\gamma^2 I + \varepsilon_1 Y_{x3i}^T Y_{x3i} & * \\ \varepsilon_1 Y_{x4i}^T Y_{x2i} & \varepsilon_1 Y_{x4i}^T Y_{x3i} & -\gamma^2 I + \varepsilon_1 Y_{x4i}^T Y_{x4i} \end{bmatrix},$$

$$\check{\Theta}_{11i} = \begin{bmatrix} P_{11} A_w & P_{21}^T A_i + \beta_1 \mathscr{B}_f C_i & \beta_1 \mathscr{A}_f \\ P_{21} A_w & P_{22} A_i + \beta_2 \mathscr{B}_f C_i & \beta_2 \mathscr{A}_f \\ \beta_1 P_{33} A_w & \beta_2 P_{33} A_i + \beta_3 \mathscr{B}_f C_i & \beta_3 \mathscr{A}_f \end{bmatrix}, \ \check{\Theta}_{13} = \begin{bmatrix} \beta_1 \mathscr{B}_f \\ \beta_2 \mathscr{B}_f \\ \beta_3 \mathscr{B}_f \end{bmatrix},$$

$$\check{\Theta}_{12i} = \begin{bmatrix} P_{21}^T B_i + \beta_1 \mathscr{B}_f D_i & P_{21}^T E_i + \beta_1 \mathscr{B}_f F_i & P_{11} B_w + P_{21}^T E_{fi} + \beta_1 \mathscr{B}_f F_{fi} \\ P_{22} B_i + \beta_2 \mathscr{B}_f D_i & P_{22} E_i + \beta_2 \mathscr{B}_f F_i & P_{21} B_w + P_{22} E_{fi} + \beta_2 \mathscr{B}_f F_{fi} \\ \beta_2 P_{33} B_i + \beta_3 \mathscr{B}_f D_i & \beta_2 P_{33} E_i + \beta_3 \mathscr{B}_f F_i & \beta_1 P_{33} B_w + \beta_2 P_{33} E_{fi} + \beta_3 \mathscr{B}_f F_{fi} \end{bmatrix},$$

$$\check{\Lambda}_{22} = -diag\{P, I, I\}, \ \check{\Lambda}_{32i} = \begin{bmatrix} \check{\Gamma}_{13i} & -X_{x2i}^T \mathscr{D}_f^T & \theta X_{x2i}^T \end{bmatrix},$$

$$\check{\Gamma}_{13i} = \begin{bmatrix} \left(P_{21}^T X_{x1i} + \beta_1 \mathscr{B}_f X_{x2i}\right)^T & (P_{22} X_{x1i} + \beta_2 \mathscr{B}_f X_{x2i})^T & (\beta_2 P_{33} X_{x1i} + \beta_3 \mathscr{B}_f X_{x2i})^T \end{bmatrix},$$

$$\bar{\mathscr{C}}_i = [C_w \ -D_f C_i \ -C_f], \ \bar{\mathscr{D}}_i = [-D_f D_i \ -D_f F_i \ D_w - D_f F_{fi}],$$

$$\bar{\mathscr{C}}_{yi} = [0 \ C_i \ 0], \ \bar{\mathscr{D}}_{yi} = [D_i \ F_i \ F_{fi}],$$

the residual system (6.14) is asymptotically stable with the $\mathscr{H}_\infty$ performance γ. Then, the robust fault detection filter with quantization gain matrices in (6.13) can be obtained as

$$A_f = P_{33}^{-1} \mathscr{A}_f, \ B_f = P_{33}^{-1} \mathscr{B}_f, \ C_f = \mathscr{C}_f, \ D_f = \mathscr{D}_f. \tag{6.24}$$

Proof Pre- and post-multiplying (6.15) by $diag\{I, I, I, P, I, I\}$ and $diag\{I, I, I, P^T, I, I\}$, respectively, gives

$$\begin{bmatrix} -P & * & * & * & * & * \\ 0 & -\gamma^2 I & * & * & * & * \\ 0 & 0 & -\frac{M^2}{\Delta^2} I & * & * & * \\ P\check{\mathscr{A}}(\rho) & P\check{\mathscr{B}}(\rho) & P\check{\mathscr{E}} & -P & * & * \\ \check{\mathscr{C}}(\rho) & \check{\mathscr{D}}(\rho) & \check{\mathscr{F}} & 0 & -I & * \\ \theta\mathscr{C}_y(\rho) & \theta\mathscr{D}_y(\rho) & 0 & 0 & 0 & -I \end{bmatrix} < 0. \tag{6.25}$$

By separating the uncertainty $\Delta_x(k)$ from above inequality, we have

$$\Xi(\rho) + He\{W_1(\rho)\Delta_x(k)V_1(\rho)\} < 0, \tag{6.26}$$

with

$$
\begin{aligned}
W_1^T(\rho) &= \left[0\ 0\ 0\ (PU(\rho))^T\ (-D_f X_{x2}(\rho))^T\ \theta X_{x2}^T(\rho)\right],\\
U^T(\rho) &= \left[0\ X_{x1}^T(\rho)\ (B_f X_{x2}(\rho))^T\right], V_1(\rho) = \left[V_{11}(\rho)\ V_{12}(\rho)\ 0\ 0\ 0\ 0\right],\\
V_{11}(\rho) &= \left[0\ Y_{x1}(\rho)\ 0\right], V_{12}(\rho) = \left[Y_{x2}(\rho)\ Y_{x3}(\rho)\ Y_{x4}(\rho)\right],\\
\bar{\mathscr{C}}(\rho) &= \left[C_w\ \ -D_f C(\rho)\ \ -C_f\right], \bar{\mathscr{D}}(\rho) = \left[-D_f D(\rho)\ \ -D_f F(\rho)\ \ D_w - D_f F_f(\rho)\right],\\
\bar{\mathscr{C}}_y(\rho) &= \left[0\ C(\rho)\ 0\right], \bar{\mathscr{D}}_y(\rho) = \left[D(\rho)\ F(\rho)\ F_f(\rho)\right],\\
\Xi(\rho) &= \begin{bmatrix} -P & * & * & * & * & * \\ 0 & -\gamma^2 I & * & * & * & * \\ 0 & 0 & -\frac{M^2}{\check{\Delta}^2} I & * & * & * \\ P\bar{\mathscr{A}}(\rho) & P\bar{\mathscr{B}}(\rho) & P\check{\mathscr{E}} & -P & * & * \\ \bar{\mathscr{C}}(\rho) & \bar{\mathscr{D}}(\rho) & \check{\mathscr{F}} & 0 & -I & * \\ \theta\bar{\mathscr{C}}_y(\rho) & \theta\bar{\mathscr{D}}_y(\rho) & 0 & 0 & 0 & -I \end{bmatrix},\\
\bar{\mathscr{A}}(\rho) &= \begin{bmatrix} A_w & 0 & 0 \\ 0 & A(\rho) & 0 \\ 0 & B_f C(\rho) & A_f \end{bmatrix}, \bar{\mathscr{B}}(\rho) = \begin{bmatrix} 0 & 0 & B_w \\ B(\rho) & E(\rho) & E_f(\rho) \\ B_f D(\rho) & B_f F(\rho) & B_f F_f(\rho) \end{bmatrix}.
\end{aligned}
$$

Apply Lemmas 1.1 and 1.2, for a scalar parameter ε_1, the matrix inequality (6.25) holds if and only if the following matrix condition is satisfied

$$
\begin{bmatrix} \Xi(\rho) + \varepsilon_1 V_1^T(\rho) V_1(\rho) & * \\ W_1^T(\rho) & -\varepsilon_1 I \end{bmatrix} < 0. \tag{6.27}
$$

Choosing the Lyapunov matrix as $P = \begin{bmatrix} P_{11} & * & * \\ P_{21} & P_{22} & * \\ \beta_1 P_{33} & \beta_2 P_{33} & \beta_3 P_{33} \end{bmatrix}$, and defining $\mathscr{A}_f = P_{33} A_f$, $\mathscr{B}_f = P_{33} B_f$, $\mathscr{C}_f = C_f$, $\mathscr{D}_f = D_f$, $\check{\Lambda}_\rho = \sum_{i=1}^{r} \rho_i(\varepsilon(k))\check{\Lambda}_i < 0$, the condition (6.23) can be rearranged as the one of Theorem 6.2.

6.1.2.3 Design of Resilient Fault Detection Filter with Quantization

Similar to the design process of robust fault detection filter, the system stability and the performance of the residual system (6.11) need to be analyzed first for the design of corresponding resilient filter.

Theorem 6.3 *Consider the residual system (6.11). For known quantization range M and quantization error bound Δ, if there exists matrix $P > 0$, scalars $\theta \geq 1$ and γ such that the following matrix inequality holds:*

$$\begin{bmatrix} -P & * & * & * & * & * \\ 0 & -\gamma^2 I & * & * & * & * \\ 0 & 0 & -\frac{M^2}{\Delta^2} I & * & * & * \\ \mathscr{A}(\rho) & \mathscr{B}(\rho) & \mathscr{E} & -P^{-1} & * & * \\ \mathscr{C}(\rho) & \mathscr{D}(\rho) & \mathscr{F} & 0 & -I & * \\ \theta\mathscr{C}_y(\rho) & \theta\mathscr{D}_y(\rho) & 0 & 0 & 0 & -I \end{bmatrix} < 0, \tag{6.28}$$

then the system is asymptotically stable and preserves the $\mathscr{H}_\infty$ performance.

Proof This performance analysis condition for residual system (6.11) is an extension of the Theorem 6.1 to a more general case with

$$\begin{aligned} \Sigma_2(\rho) = & \left[\mathscr{A}(\rho)\ \mathscr{B}(\rho)\ \mathscr{E}\right]^T P \left[\mathscr{A}(\rho)\ \mathscr{B}(\rho)\ \mathscr{E}\right] \\ & + \left[\mathscr{C}(\rho)\ \mathscr{D}(\rho)\ \mathscr{F}\right]^T \left[\mathscr{C}(\rho)\ \mathscr{D}(\rho)\ \mathscr{F}\right] - diag\left\{P,\ \gamma^2 I,\ 0\right\}, \end{aligned}$$

and its proof is straightforward and hence omitted.

Due to the existence of nonlinear terms in the matrix inequalities, the quantization error term and coupling terms between two types of uncertainties in Theorem 6.3, some existing results [2] do not work, we choose a two-step approach to give the design conditions of resilient fault detection filter with quantization.

Theorem 6.4 *Consider the residual system (6.11). For known quantization range M and quantization error bound Δ, given scalars β_1, β_2, and β_3, if there exist matrices P_{11}, P_{21}, P_{22}, P_{33}, $\mathscr{A}_f$, $\mathscr{B}_f$, $\mathscr{C}_f$, $\mathscr{D}_f$, scalars $\theta \geq 1$, γ, ε_1, and ε_2 such that the following LMIs hold*

$$\Lambda_i < 0, i = 1, ..., r, \tag{6.29}$$

where

$$\Lambda_i = \begin{bmatrix} \Lambda_{11i} & * & * & * \\ \check{\Lambda}_{21i} & \check{\Lambda}_{22} & * & * \\ \Lambda_{31i} & \check{\Lambda}_{32i} & \Lambda_{33i} & * \\ 0 & \Lambda_{42} & 0 & -\varepsilon_2 I \end{bmatrix},\ \Lambda_{11i} = \begin{bmatrix} \Xi_{11i} & * & * \\ \Xi_{21i} & \Xi_{22i} & * \\ \Xi_{31i} & \Xi_{32i} & \Xi_{33} \end{bmatrix},$$

$$\Lambda_{31i} = \left[\Gamma_{11i}\ \ \Gamma_{12i}\ \ \varepsilon_2 X_{x2i}^T Y_{f2}^T Y_{f2}\right],$$

$$\Lambda_{33i} = -\varepsilon_1 I + \varepsilon_2 X_{x2i}^T Y_{f2}^T Y_{f2} X_{x2i},\ \Lambda_{42} = \left[\Gamma_{14}\ \ -X_{f2}^T\ \ 0\right],$$

$$\Xi_{11i} = \begin{bmatrix} -P_{11} & * & * \\ -P_{21} & -P_{22} + \varepsilon_1 Y_{x1i}^T Y_{x1i} + \varepsilon_2 C_i^T Y_{f2}^T Y_{f2} C_i & * \\ -\beta_1 P_{33} & -\beta_2 P_{33} + \varepsilon_2 Y_{f1}^T Y_{f2} C_i & -\beta_3 P_{33} + \varepsilon_2 Y_{f1}^T Y_{f1} \end{bmatrix},$$

$$\Xi_{21i} = \begin{bmatrix} 0 & \varepsilon_1 Y_{x2i}^T Y_{x1i} + \varepsilon_2 D_i^T Y_{f2}^T Y_{f2} C_i & \varepsilon_2 D_i^T Y_{f2}^T Y_{f1} \\ 0 & \varepsilon_1 Y_{x3i}^T Y_{x1i} + \varepsilon_2 F_i^T Y_{f2}^T Y_{f2} C_i & \varepsilon_2 F_i^T Y_{f2}^T Y_{f1} \\ 0 & \varepsilon_1 Y_{x4i}^T Y_{x1i} + \varepsilon_2 F_{fi}^T Y_{f2}^T Y_{f2} C_i & \varepsilon_2 F_{fi}^T Y_{f2}^T Y_{f1} \end{bmatrix},$$

$$\Xi_{22i} = -diag\left\{\gamma^2 I,\ \gamma^2 I,\ \gamma^2 I\right\} +$$

$$\begin{bmatrix} \varepsilon_1 Y_{x2i}^T Y_{x2i} + \varepsilon_2 D_i^T Y_{f2}^T Y_{f2} D_i & * & * \\ \varepsilon_1 Y_{x3i}^T Y_{x2i} + \varepsilon_2 F_i^T Y_{f2}^T Y_{f2} D_i & \varepsilon_1 Y_{x3i}^T Y_{x3i} + \varepsilon_2 F_i^T Y_{f2}^T Y_{f2} F_i & * \\ \varepsilon_1 Y_{x4i}^T Y_{x2i} + \varepsilon_2 F_{fi}^T Y_{f2}^T Y_{f2} D_i & \varepsilon_1 Y_{x4i}^T Y_{x3i} + \varepsilon_2 F_{fi}^T Y_{f2}^T Y_{f2} F_i & \Xi_{221i} \end{bmatrix},$$

$$\Xi_{221i} = \varepsilon_1 Y_{x4i}^T Y_{x4i} + \varepsilon_2 F_{fi}^T Y_{f2}^T Y_{f2} F_{fi}, \ \Xi_{31i} = \begin{bmatrix} 0 & \varepsilon_2 Y_{f2}^T Y_{f2} C_i & \varepsilon_2 Y_{f2}^T Y_{f1} \end{bmatrix},$$

$$\Xi_{32i} = \begin{bmatrix} \varepsilon_2 Y_{f2}^T Y_{f2} D_i & \varepsilon_2 Y_{f2}^T Y_{f2} F_i & \varepsilon_2 Y_{f2}^T Y_{f2} F_{fi} \end{bmatrix}, \ \Xi_{33} = -\frac{M^2}{\Delta^2} I + \varepsilon_2 Y_{f2}^T Y_{f2},$$

$$\Gamma_{11i} = \begin{bmatrix} 0 & \varepsilon_2 \left(Y_{f2} X_{x2i}\right)^T Y_{f2} C_i & \varepsilon_2 \left(Y_{f2} X_{x2i}\right)^T Y_{f1} \end{bmatrix},$$

$$\Gamma_{12i} = \begin{bmatrix} \varepsilon_2 \left(Y_{f2} X_{x2i}\right)^T Y_{f2} D_i & \varepsilon_2 \left(Y_{f2} X_{x2i}\right)^T Y_{f2} F_i & \varepsilon_2 \left(Y_{f2} X_{x2i}\right)^T Y_{f2} F_{fi} \end{bmatrix},$$

$$\Gamma_{14} = \begin{bmatrix} (\beta_1 P_{33} X_{f1})^T & (\beta_2 P_{33} X_{f1})^T & (\beta_3 P_{33} X_{f1})^T \end{bmatrix},$$

the residual system (6.11) is asymptotically stable with the $\mathcal{H}_\infty$ performance γ. Then, the resilient fault detection filter with quantization gain matrices in (6.5) can be obtained with (6.24).

Proof Pre- and post-multiplying (6.28) by $diag\{I,\ I,\ I,\ P,\ I,\ I\}$ and its transpose, respectively, gives

$$\begin{bmatrix} -P & * & * & * & * & * \\ 0 & -\gamma^2 I & * & * & * & * \\ 0 & 0 & -\frac{M^2}{\Delta^2} I & * & * & * \\ P\mathscr{A}(\rho) & P\mathscr{B}(\rho) & P\mathscr{E} & -P & * & * \\ \mathscr{C}(\rho) & \mathscr{D}(\rho) & \mathscr{F} & 0 & -I & * \\ \theta\mathscr{C}_y(\rho) & \theta\mathscr{D}_y(\rho) & 0 & 0 & 0 & -I \end{bmatrix} < 0. \tag{6.30}$$

To eliminate the effects of uncertainties $\Delta_x(k)$ and $\Delta_f(k)$ on the residual system (6.11), a two-step approach is adopted. In the first step, we first eliminate the uncertainty $\Delta_x(k)$. Above inequality can be arranged as

$$\hat{\Xi}(\rho) + He\{\hat{W}_1(\rho)\Delta_x V_1(\rho)\} < 0. \tag{6.31}$$

with

$$\hat{W}_1^T(\rho) = \begin{bmatrix} 0 & 0 & 0 & (P\hat{U}(\rho))^T & (-(D_f + \Delta_{Df}) X_{x2}(\rho))^T & \theta X_{x2}^T(\rho) \end{bmatrix},$$

$$\hat{U}^T(\rho) = \begin{bmatrix} 0 & X_{x1}^T(\rho) & ((B_f + \Delta_{Bf}) X_{x2}(\rho))^T \end{bmatrix},$$

$$\hat{\mathscr{C}}(\rho) = \begin{bmatrix} C_w & -(D_f + \Delta_{Df}) C(\rho) & -(C_f + \Delta_{Cf}) \end{bmatrix},$$

$$\hat{\mathscr{D}}(\rho) = \begin{bmatrix} -(D_f + \Delta_{Df}) D(\rho) & -(D_f + \Delta_{Df}) F(\rho) & D_w - (D_f + \Delta_{Df}) F_f(\rho) \end{bmatrix},$$

$$\hat{\Xi}(\rho) = \begin{bmatrix} -P & * & * & * & * & * \\ 0 & -\gamma^2 I & * & * & * & * \\ 0 & 0 & -\frac{M^2}{\Delta^2} I & * & * & * \\ P\hat{\mathscr{A}}(\rho) & P\hat{\mathscr{B}}(\rho) & P\mathscr{E} & -P & * & * \\ \hat{\mathscr{C}}(\rho) & \hat{\mathscr{D}}(\rho) & \mathscr{F} & 0 & -I & * \\ \theta\bar{\mathscr{C}}_y(\rho) & \theta\bar{\mathscr{D}}_y(\rho) & 0 & 0 & 0 & -I \end{bmatrix},$$

$$\hat{\mathscr{A}}(\rho) = \begin{bmatrix} A_w & 0 & 0 \\ 0 & A(\rho) & 0 \\ 0 & (B_f + \Delta_{Bf})C(\rho) & A_f + \Delta_{Af} \end{bmatrix},$$

$$\hat{\mathscr{B}}(\rho) = \begin{bmatrix} 0 & 0 & B_w \\ B(\rho) & E(\rho) & E_f(\rho) \\ (B_f + \Delta_{Bf})D(\rho) & (B_f + \Delta_{Bf})F(\rho) & (B_f + \Delta_{Bf})F_f(\rho) \end{bmatrix}.$$

Applying the Lemmas 1.1 and 1.2 yields

$$\begin{bmatrix} \hat{\Xi}(\rho) + \varepsilon_1 V_1^T(\rho)V_1(\rho) & * \\ \hat{W}_1^T(\rho) & -\varepsilon_1 I \end{bmatrix} < 0. \tag{6.32}$$

Now, after separating the $\Delta_x(k)$ terms we're going to do the second step to get rid of uncertainty $\Delta_f(k)$. Separating the uncertainty $\Delta_f(k)$ from (6.32) gives

$$\begin{bmatrix} \Xi(\rho) + \varepsilon_1 V_1^T(\rho)V_1(\rho) & * \\ W_1^T(\rho) & -\varepsilon_1 I \end{bmatrix} + He\{\hat{W}_2(\rho)\Delta_f(k)\hat{V}_2(\rho)\} < 0, \tag{6.33}$$

with

$$\hat{W}_2^T(\rho) = \begin{bmatrix} 0 & 0 & 0 & (PW_{21}(\rho))^T & (-X_{f2}(\rho))^T & 0 & 0 \end{bmatrix}, W_{21}^T(\rho) = \begin{bmatrix} 0 & 0 & X_{f1}^T(\rho) \end{bmatrix},$$
$$\hat{V}_2(\rho) = \begin{bmatrix} \hat{V}_{21}(\rho) & \hat{V}_{22}(\rho) & Y_{f2} & 0 & 0 & 0 & Y_{f2}X_{x2}(\rho) \end{bmatrix},$$
$$\hat{V}_{21}(\rho) = \begin{bmatrix} 0 & Y_{f2}C(\rho) & Y_{f1} \end{bmatrix}, \hat{V}_{22}(\rho) = \begin{bmatrix} Y_{f2}D(\rho) & Y_{f2}F(\rho) & Y_{f2}F_f(\rho) \end{bmatrix}.$$

Once again, by applying Lemmas 1.1 and 1.2, for a scalar ε_2, (6.33) holds if and only if

$$\begin{bmatrix} \begin{bmatrix} \Xi(\rho) + \varepsilon_1 V_1^T(\rho)V_1(\rho) & * \\ W_1^T(\rho) & -\varepsilon_1 I \end{bmatrix} + \varepsilon_2 \hat{V}_2^T(\rho)\hat{V}_2(\rho) & * \\ \hat{W}_2^T(\rho) & -\varepsilon_2 I \end{bmatrix} < 0, \tag{6.34}$$

is met.

Then let us choose the Lyapunov matrix as $P = \begin{bmatrix} P_{11} & * & * \\ P_{21} & P_{22} & * \\ \beta_1 P_{33} & \beta_2 P_{33} & \beta_3 P_{33} \end{bmatrix}$, and define $\mathscr{A}_f = P_{33} A_f$, $\mathscr{B}_f = P_{33} B_f$, $\mathscr{C}_f = C_f$, $\mathscr{D}_f = D_f$, and the matrix $\Lambda_\rho = \sum_{i=1}^{r} \rho_i(\varepsilon(k))\Lambda_i < 0$, the condition (6.34) can be rearranged as the one of Theorem 6.4.

Remark 6.2 By solving the LMIs conditions of Theorems 6.2 and 6.4, we can find a feasible solution for the robust fault detection filter with quantization and resilient fault detection filter with quantization, respectively. To obtain an optimal $\mathscr{H}_\infty$ performance bound for corresponding residual system, we can solve the optimization problem

$$\min \mu, \; subject\ to\ (6.23)\ or\ (6.29)\ with\ \mu = \gamma^2.$$

Then, the optimal $\mathscr{H}_\infty$ performance is $\gamma_{\min} = \sqrt{\mu_{\min}}$.

Remark 6.3 It should be noted that, in Theorems 6.2 and 6.4, the scalars $\beta_1, \beta_2, \beta_3$ are free parameters, which need to be determined in advance. In that case, (6.23) and (6.29) are LMIs in terms of unknown variables. To handle this problem, we can apply a numerical optimization algorithm, such as the program "fminsearch" in the optimization toolbox of Matlab, then a locally convergent solution to the problem is obtained.

6.1.3 Simulation Example

In this section, we provide a numerical example to illustrate the efficiency of the design strategies for robust and resilient fault detection filters with quantization. A two-rule polynomial fuzzy system is adopted, which is described in (6.4). The relevant system parameters are given as follows:

$$A_1 = \begin{bmatrix} 0.5 & -0.2 \\ 0.2 & 0.4 \end{bmatrix},\ A_2 = \begin{bmatrix} 0.4 & -0.2 \\ 0.2 & 0.4 \end{bmatrix},\ B_1 = B_2 = \begin{bmatrix} 0.3 \\ 0.2 \end{bmatrix},\ E_1 = E_2 = \begin{bmatrix} 0.3 \\ 0.1 \end{bmatrix},$$

$$E_{f1} = E_{f2} = \begin{bmatrix} 0.3 & 0.6 \\ 0.4 & 0.2 \end{bmatrix},\ C_1 = \begin{bmatrix} -0.3 & -0.1 \\ 0.2 & -0.6 \end{bmatrix},\ C_2 = \begin{bmatrix} 0.5 & -0.1 \\ 0.2 & -0.6 \end{bmatrix},$$

$$D_1 = D_2 = \begin{bmatrix} -0.3 \\ 0.2 \end{bmatrix},\ F_1 = F_2 = \begin{bmatrix} 0.1 \\ -0.3 \end{bmatrix},\ F_{f1} = F_{f2} = \begin{bmatrix} -0.7 & 0.1 \\ 0.1 & -0.5 \end{bmatrix},$$

$$X_{x11} = \begin{bmatrix} 0.1 \\ 0.3 \end{bmatrix},\ X_{x12} = \begin{bmatrix} -0.3 \\ -0.3 \end{bmatrix},\ X_{x21} = X_{x22} = \begin{bmatrix} -0.4 & 0.1 \end{bmatrix}^T,\ Y_{x11} = \begin{bmatrix} -0.3 & 0.1 \end{bmatrix},$$

$$Y_{x12} = \begin{bmatrix} -1 & 0.2 \end{bmatrix},\ Y_{x21} = Y_{x22} = -0.5,\ Y_{x31} = Y_{x32} = -0.3,\ Y_{x41} = Y_{x42} = \begin{bmatrix} 0.4 & 0.2 \end{bmatrix}.$$

The reference residual model in the form (6.8) with

$$A_w = \begin{bmatrix} -0.4 & -1 \\ 0.4 & -0.2 \end{bmatrix},\ B_w = \begin{bmatrix} -0.2 & 0.6 \\ -1 & 0.2 \end{bmatrix},\ C_w = \begin{bmatrix} 0.4 & 0.2 \\ 1 & -0.7 \end{bmatrix},\ D_w = \begin{bmatrix} -0.1 & 0.4 \\ 0.8 & 0.2 \end{bmatrix}.$$

Case I: Design of Robust Fault Detection Filter with Quantization

In the following, we first design the robust fault detection filter with quantization in the form of (6.13) for the above system. For given $\Delta = 0.01$ and $M = 10$, by solving the LMI in Theorem 6.2 with $\beta_1 = -0.7$, $\beta_2 = 0.2$, $\beta_3 = 0.2$, we have $\theta = 5489.3$, and the gain matrices of robust fault detection filter with quantization in (6.13) as

$$A_f = \begin{bmatrix} -0.2443 & -0.6190 \\ 0.3744 & -0.1994 \end{bmatrix},\ B_f = \begin{bmatrix} 0.0013 & -0.0027 \\ 0.0093 & -0.0004 \end{bmatrix},$$
$$C_f = \begin{bmatrix} 0.0789 & 0.0153 \\ 0.1679 & -0.1184 \end{bmatrix},\ D_f = \begin{bmatrix} 0.0011 & -0.0005 \\ 0.0008 & -0.0008 \end{bmatrix}.$$

To demonstrate the effectiveness of the design method, an unknown input is assume as $v(k) = 3\cos(0.1k)e^{-2k}$, and the input $u(k) = 0$. The fault signal $f(k)$ is simulated as a pulse of unit amplitude occurred from $k_1 = 30$ to $k_2 = 60$ (and is zero otherwise) as shown in Fig. 6.1.

Moreover, some additive assumptions are taken as $\Delta_x(k) = 0.3\sin(k)$, the initial condition $x(0) = x_f(0) = x_r(0) = [0 \quad 0]^T$, and

$$\upsilon_y(k) = \frac{\theta\Delta}{M}\sin(y(k)) = \frac{5489.3 \times 0.01}{10}\sin(y(k)).$$

Fig. 6.1 Fault signal $f(k)$

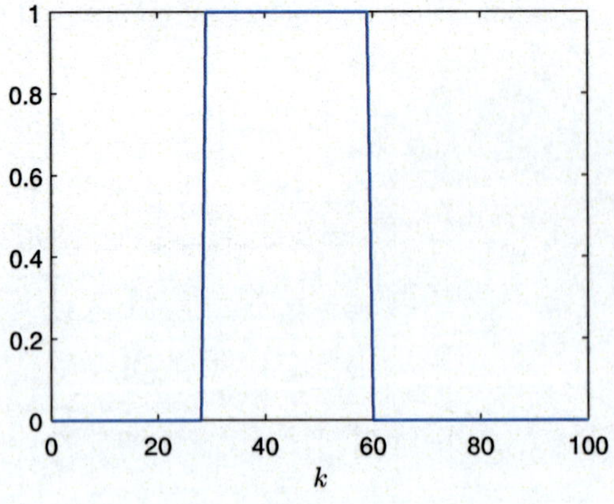

Fig. 6.2 Residual $r(k)$ by robust fault detection filter with quantization

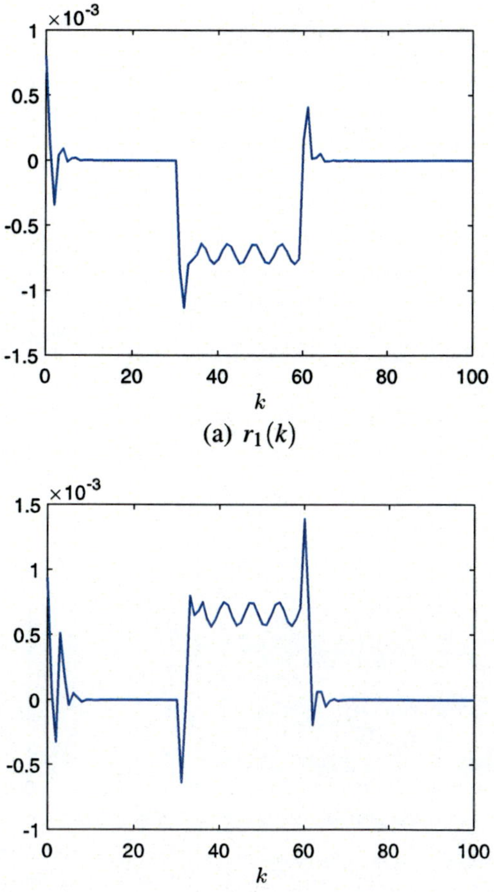

(a) $r_1(k)$

(b) $r_2(k)$

Fig. 6.3 Residual evaluation function $\|r\|_{RMS}$ by robust fault detection filter with quantization

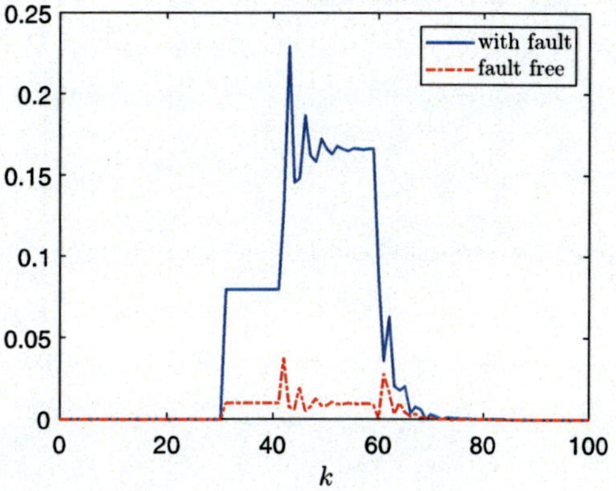

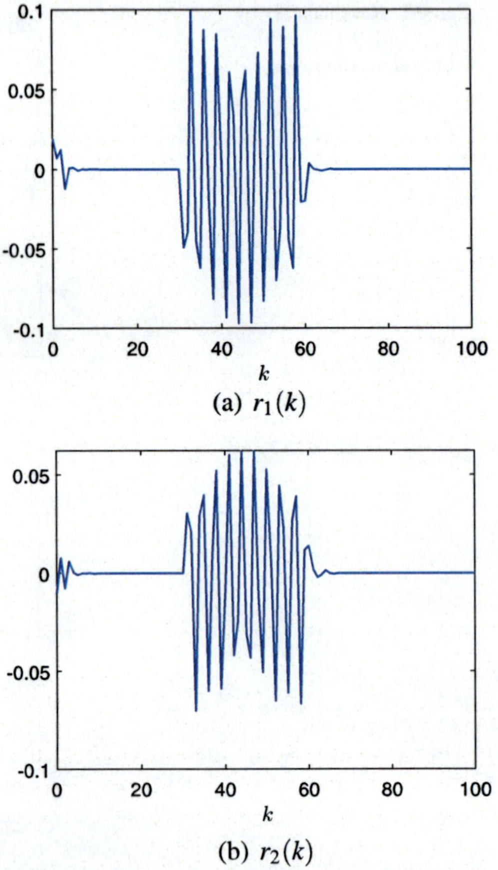

Fig. 6.4 Residual $r(k)$ by resilient fault detection filter with quantization

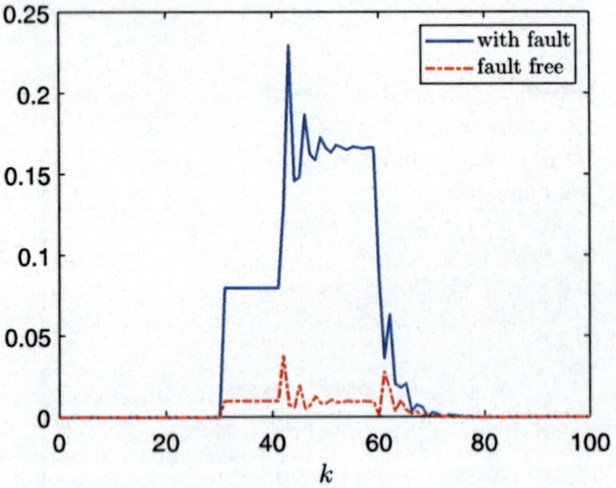

Fig. 6.5 Residual evaluation function $\|r\|_{RMS}$ by resilient fault detection filter with quantization

The response of residual $r(k)$ generated by the robust fault detection filter with quantization is presented in Fig. 6.2. Suppose the time window $K = 10$. Figure 6.3 gives the response of residual evaluation function $\|r\|_{RMS}$. Compared with the case of no fault, the dashed line of the residual evaluation function has an obvious jump, which means the designed robust fault detection filter with quantization can effectively detect a possible fault when its occurrence.

Case II: Design of resilient Fault Detection Filter with Quantization

Now we design the resilient fault detection filter with quantization for the above uncertain discrete-time system. Consider a non-fragile filter in the form of (6.5) with some uncertainties and assume the known parameters are given as

$$X_{f1} = \begin{bmatrix}-0.4 & 0.5\end{bmatrix}^T, \quad X_{f2} = \begin{bmatrix}-0.3 & 0.2\end{bmatrix}^T,$$
$$Y_{f1} = \begin{bmatrix}-0.7 & -0.1\end{bmatrix}, \quad Y_{f2} = \begin{bmatrix}-0.2 & 0.6\end{bmatrix}.$$

Then for given $\Delta = 0.01$ and $M = 10$, solving the LMIs in Theorem 6.4 with $\beta_1 = -0.7$, $\beta_2 = 0.2$, $\beta_3 = 0.2$ gives the gain matrices of resilient fault detection filter with quantization in (6.5) as

$$A_f = \begin{bmatrix}-0.3162 & -0.3517\\ 0.2430 & 0.0343\end{bmatrix}, \quad B_f = \begin{bmatrix}0.0018 & -0.0047\\ 0.0138 & -0.0008\end{bmatrix},$$
$$C_f = \begin{bmatrix}0.0611 & 0.0320\\ 0.0871 & 0.0007\end{bmatrix}, \quad D_f = \begin{bmatrix}0.0018 & -0.0006\\ 0.0018 & -0.0014\end{bmatrix},$$

and $\theta = 5902.9$.

Following the same simulation conditions, i.e., conduct the simulation with the same $d(k)$, $u(k)$, $f(k)$, $\Delta_x(k)$, and initial conditions $x(0)$, $x_f(0)$, $x_r(0)$, we will demonstrate the effectiveness of the designed resilient fault detection filter with quantization by assuming $\Delta_f(k) = 0.5\cos(2k)$ and

$$\upsilon_y(k) = \frac{\theta\Delta}{M}\sin(y(k)) = \frac{5902.9 \times 0.01}{10}\sin(y(k)).$$

Then the simulation results of residual $r(k)$ generated by resilient fault detection filter with quantization are provided in Fig. 6.4 and the response of residual evaluation function $\|r\|_{RMS}$ is illustrated in Fig. 6.5. It can be seen from these figures that the designed resilient fault detection filter with quantization can also effectively detect the fault.

6.2 Fault-Tolerant Control with Input Quantization

In this section, we investigate quantized fault-tolerant controller design problem for T–S fuzzy Markov jump system with actuator fault. The main purpose is to design a mode-dependent controller to ensure the stochastic $\mathscr{H}_\infty$ stability of the closed loop system. And the fault-tolerant controller design scheme is derived based on the mode-dependent idea and robust stochastic stability approach in terms of a set of LMIs.

6.2.1 Problem Formulation

The T–S fuzzy Markov jump system is described as the following form:

Plant Rule i_{th}: **IF** $\varepsilon_1(k)$ is M_{1i}, $\varepsilon_2(k)$ is M_{2i}, and, ..., and $\varepsilon_d(k)$ is M_{di}, **THEN**

$$\begin{aligned} x(k+1) &= A_{i\tau(k)}x(k) + B_{i\tau(k)}\bar{u}(k) + D_{i\tau(k)}w(k), \\ z(k) &= L_{i\tau(k)}x(k), \end{aligned} \tag{6.35}$$

where $i = 1, 2, \ldots, r$ with r being the number of fuzzy rules, M_{pi} are fuzzy sets with $p = 1, \ldots, d$, $i = 1, \ldots, r$, and $\varepsilon_1(k)$, $\varepsilon_2(k)$, ..., $\varepsilon_d(k)$ are the premise variables; $x(k) \in \mathbb{R}^{n_x}$ is the state, $\bar{u}(k) \in \mathbb{R}^{n_u}$ is the control input, $z(k) \in \mathbb{R}^{n_z}$ is the regulated output, and $w(k) \in \mathbb{R}^{n_w}$ is the disturbance vector belonging to $\ell_2[0, \infty)$, respectively; $A_{i\tau(k)}$, $B_{i\tau(k)}$, $D_{i\tau(k)}$ and $L_{i\tau(k)}$ are known constant real matrices with appropriate dimensions; the subscript

$$\tau(k) \in \mathcal{S} \triangleq \{1, 2, \ldots, \ell\}, \tag{6.36}$$

is a Markov chain with the transition probability matrix $\Omega = [\pi_{mn}]$ given by

$$\mathbb{P}\{\tau(k+1) = n | \tau(k) = m\} = \pi_{mn}, \tag{6.37}$$

where $0 \leq \pi_{mn} \leq 1, \forall m, n \in \mathcal{S}$ and $\sum_{n=1}^{\ell} \pi_{mn} = 1, \forall m \in \mathcal{S}$.

System (6.35) can be represented by the following compact form:

$$\begin{aligned} x(k+1) &= A_{\tau(k)}(\rho)x(k) + B_{\tau(k)}(\rho)\bar{u}(k) + D_{\tau(k)}(\rho)w(k), \\ z(k) &= L_{\tau(k)}(\rho)x(k), \end{aligned} \tag{6.38}$$

where

$$A_{\tau(k)}(\rho) = \sum_{i=1}^{r} \rho_i(\varepsilon(k))A_{i\tau(k)},\ B_{\tau(k)}(\rho) = \sum_{i=1}^{r} \rho_i(\varepsilon(k))B_{i\tau(k)},$$
$$D_{\tau(k)}(\rho) = \sum_{i=1}^{r} \rho_i(\varepsilon(k))D_{i\tau(k)},\ L_{\tau(k)}(\rho) = \sum_{i=1}^{r} \rho_i(\varepsilon(k))L_{i\tau(k)}. \tag{6.39}$$

6.2.1.1 Actuator Fault

Now, we will develop a fault-tolerant quantization control strategy. Consider that there exist some types of fault during execution of actuator. Then, we can describe the control input by

$$\bar{u}(k) = \beta(k)\hat{u}(k), \tag{6.40}$$

where $\hat{u}(k)$ represents the transmitted signal over the communication channel, $\beta(k) = diag\{\beta_1(k), \beta_2(k), \ldots, \beta_{n_u}(k)\}$, $0 < \underline{\beta}_p \le \beta_p(k) \le \overline{\beta}_p < 1$, $p = 1, 2, \ldots, n_u$, $\underline{\beta}_p$, and $\overline{\beta}_p$ are known and can be monitored in practice.

Further, define $\beta_1 = diag\left\{\frac{\underline{\beta}_1+\overline{\beta}_1}{2}, \frac{\underline{\beta}_2+\overline{\beta}_2}{2}, \ldots, \frac{\underline{\beta}_{n_u}+\overline{\beta}_{n_u}}{2}\right\}$, $\beta_p = -diag\left\{\frac{\underline{\beta}_1-\overline{\beta}_1}{2}, \frac{\underline{\beta}_2-\overline{\beta}_2}{2}, \ldots, \frac{\underline{\beta}_{n_u}-\overline{\beta}_{n_u}}{2}\right\}$. Then we can describe the faulty input as

$$\bar{u}(k) = (\beta_1 + \beta_p \Delta_f)\hat{u}(k), \tag{6.41}$$

where Δ_f is an uncertainty term satisfying $\Delta_f^T \Delta_f \le I$.

6.2.1.2 Dynamic Quantizer

Next, a dynamic quantizer [1] is introduced to handle the signal to be transmitted, which is described as the following form:

$$\hat{u}(k) = \mu(k) q\left(\frac{u(k)}{\mu(k)}\right), \tag{6.42}$$

where $\mu(k)$ is an adjustable dynamic parameter and $q(\cdot)$ is uniform static quantizer with the quantization range M and quantization error bound Δ satisfying

$$\|q(\iota) - \iota\| \le \Delta,\ if\ \|\iota\| \le M,$$
$$\|q(\iota)\| \ge M - \Delta,\ if\ \|\iota\| \ge M. \tag{6.43}$$

Furthermore, introduce $v(k) \triangleq \hat{u}(k) - u(k)$. Then it follows that

$$\|v(k)\| \le \Delta\mu(k). \tag{6.44}$$

Assuming that there exists a parameter $\theta > 1$ to be designed and the parameter $\mu(k)$ will be adjusted dynamically based on quantizer input to reach the required task. And the adjusting scheme is given as

$$\frac{\sqrt{\theta}}{M}\|u(k)\| \leq \mu(k) \leq \frac{2\sqrt{\theta}}{M}\|u(k)\|. \tag{6.45}$$

According to (6.44) and (6.45), we have

$$\|v(k)\| \leq \frac{2\Delta\sqrt{\theta}}{M}\|u(k)\|. \tag{6.46}$$

Based on the definition of $v(k)$, we can easily obtain that

$$\hat{u}(k) = u(k) + v(k). \tag{6.47}$$

6.2.1.3 Controller

Consider the mode-dependent feedback controller as the following form

$$u(k) = K_{\tau(k)}x(k), \tag{6.48}$$

where $K_{\tau(k)}$, $\tau(k) \in \mathcal{S}$, is the controller gain to be designed later.

6.2.1.4 Closed-Loop System

From (6.38), (6.41), (6.47) and (6.48), we can obtain the following closed-loop system form

$$\begin{aligned} x(k+1) &= \mathcal{A}_{\tau(k)}(\rho)x(k) + \mathcal{B}_{\tau(k)}(\rho)v(k) + D_{\tau(k)}(\rho)w(k), \\ z(k) &= L_{\tau(k)}(\rho)x(k), \end{aligned} \tag{6.49}$$

where $\mathcal{A}_{\tau(k)}(\rho) = A_{\tau(k)}(\rho) + B_{\tau(k)}(\rho)(\beta_1 + \beta_p\Delta_f)K_{\tau(k)}$, $\mathcal{B}_{\tau(k)}(\rho) = B_{\tau(k)}(\rho)(\beta_1 + \beta_p\Delta_f)$.

In the following, $\mathcal{A}_{\tau(k)}(\rho)$, $A_{\tau(k)}(\rho)$, $\mathcal{B}_{\tau(k)}(\rho)$, $B_{\tau(k)}(\rho)$, $D_{\tau(k)}(\rho)$, $L_{\tau(k)}(\rho)$ and $K_{\tau(k)}$ will be represented by $\mathcal{A}_m(\rho)$, $A_m(\rho)$, $\mathcal{B}_m(\rho)$, $B_m(\rho)$, $D_m(\rho)$, $L_m(\rho)$ and K_m, respectively.

Before beginning the main results, we firstly introduce the following definitions.

Definition 6.1 [4] Under the condition that $w(k) = 0$, if the following inequality is satisfied:

$$\sum_{k=0}^{\infty} \mathbb{E}\left\{x^T(k)x(k)\right\} < \infty, \tag{6.50}$$

system (6.49) is stochastically stable.

Definition 6.2 [5] For a given scalar $\gamma > 0$, under zero initial condition, if system (6.49) is stochastically stable, and the following inequality holds

$$\sum_{k=0}^{\infty} \mathbb{E}\left\{z^T(k)z(k)\right\} < \gamma^2 \left\{\sum_{k=0}^{\infty} w^T(k)w(k)\right\}, \tag{6.51}$$

the system in (6.49) is said to be stochastically stable with the $\mathcal{H}_\infty$ performance index γ.

The quantized fault-tolerant controller design problem considered in this section is to design the quantized fault-tolerant controller in (6.48) and the dynamic quantizer in (6.42) such that

(1) The quantized fuzzy closed-loop system in (6.49) is stochastically stable when $w(k) = 0$.
(2) The quantized fuzzy closed-loop system in (6.49) has a prescribed $\mathcal{H}_\infty$ performance γ.

6.2.2 *Main Results*

The stability of system (6.49) with $\mathcal{H}_\infty$ performance is discussed and the analysis conditions are given in the following theorem.

Theorem 6.5 *Consider system (6.49). For known parameters M and Δ, the system is stochastically stable with the required $\mathcal{H}_\infty$ performance γ, if there exist scalar $\theta > 1$, matrices $P_m > 0$ and $U > 0$, $m \in \mathcal{S}$ such that*

$$\begin{bmatrix} -U & * & * & * & * & * \\ 0 & -I & * & * & * & * \\ 0 & 0 & -\gamma^2 I & * & * & * \\ L_m(\rho) & 0 & 0 & -I & * & * \\ K_m & 0 & 0 & 0 & -\frac{M^2}{4\Delta^2\theta} I & * \\ \pi_\ell \mathcal{A}_m(\rho) & \pi_\ell \mathcal{B}_m(\rho) & \pi_\ell D_m(\rho) & 0 & 0 & -\bar{P}_\ell \end{bmatrix} < 0, \tag{6.52}$$

$$U - P_m < 0, \tag{6.53}$$

where $\pi_\ell = [\ \sqrt{\pi_{m1}}\ \sqrt{\pi_{m2}}\ \ldots\ \sqrt{\pi_{m\ell}}\ ]^T$ and $\bar{P}_\ell = diag\left\{P_1^{-1}, P_2^{-1}, \ldots, P_\ell^{-1}\right\}$.

Proof Adopting Lemma 1.1 can obtain that

$$\begin{bmatrix} -U + \frac{4\Delta^2\theta}{M^2} K_m^T K_m & * & * \\ 0 & -I & * \\ 0 & 0 & -\gamma^2 I \end{bmatrix} + [\, \mathscr{A}_m(\rho) \;\; \mathscr{B}_m(\rho) \;\; D_m(\rho) \,]^T \sum_{n\in\mathcal{S}} \pi_{mn} P_n [\, \mathscr{A}_m(\rho) \;\; \mathscr{B}_m(\rho) \;\; D_m(\rho) \,] + [\, L_m(\rho) \;\; 0 \;\; 0 \;\; 0 \,]^T [\, L_m(\rho) \;\; 0 \;\; 0 \;\; 0 \,] < 0. \tag{6.54}$$

Performing congruence transformation with $\left[x^T(k) \; v^T(k) \; w^T(k)\right]^T \neq 0$ to (6.54), we have

$$\begin{aligned} &(\mathscr{A}_m(\rho)x(k) + \mathscr{B}_m(\rho)v(k) + D_m(\rho)w(k))^T \sum_{n\in\mathcal{S}} \pi_{mn} P_n \\ &\times (\mathscr{A}_m(\rho)x(k) + \mathscr{B}_m(\rho)v(k) + D_m(\rho)w(k)) \\ &- x^T(k)Ux(k) + x^T(k)L_m^T(\rho)L_m(\rho)x(k) - \gamma^2 w^T(k)w(k) \\ &+ \frac{4\Delta^2\theta}{M^2} x^T(k)K_m^T K_m x(k) - v^T(k)v(k) < 0. \end{aligned} \tag{6.55}$$

Recalling (6.46), we can obtain

$$v^T(k)v(k) \leq \frac{4\Delta^2\theta}{M^2} u^T(k)u(k) = \frac{4\Delta^2\theta}{M^2} x^T(k)K_m^T K_m x(k). \tag{6.56}$$

It can be derived that

$$\frac{4\Delta^2\theta}{M^2} x^T(k)K_m^T K_m x(k) - v^T(k)v(k) \geq 0. \tag{6.57}$$

According to the conditions (6.55) and (6.57), we have

$$\begin{aligned} &(\mathscr{A}_m(\rho)x(k) + \mathscr{B}_m(\rho)v(k) + D_m(\rho)w(k))^T \sum_{n\in\mathcal{S}} \pi_{mn} P_n \\ &\times (\mathscr{A}_m(\rho)x(k) + \mathscr{B}_m(\rho)v(k) + D_m(\rho)w(k)) \\ &- x^T(k)Ux(k) + x^T(k)L_m^T(\rho)L_m(\rho)x(k) - \gamma^2 w^T(k)w(k) < 0, \end{aligned} \tag{6.58}$$

which means that

$$\begin{aligned} &(\mathscr{A}_m(\rho)x(k) + \mathscr{B}_m(\rho)v(k) + D_m(\rho)w(k))^T \sum_{n\in\mathcal{S}} \pi_{mn} P_n \\ &\times (\mathscr{A}_m(\rho)x(k) + \mathscr{B}_m(\rho)v(k) + D_m(\rho)w(k)) \\ &- x^T(k)P_m x(k) + z^T(k)z(k) - \gamma^2 w^T(k)w(k) < x^T(k)(U - P_m)x(k). \end{aligned} \tag{6.59}$$

Construct the Lyapunov function as:

$$V(k) = x^T(k)P_m x(k). \tag{6.60}$$

Further, we calculate the difference $\Delta V(k)$ along the trajectory of system (6.49) and take expectation

$$\begin{aligned}\mathbb{E}\{\Delta V(k)\} &= x^T(k+1)\sum_{n\in\mathcal{S}}\pi_{mn}P_n x(k+1) - x^T(k)P_m x(k)\\ &= (\mathcal{A}_m(\rho)x(k) + \mathcal{B}_m(\rho)v(k) + D_m(\rho)w(k))^T\sum_{n\in\mathcal{S}}\pi_{mn}P_n\\ &\quad\times(\mathcal{A}_m(\rho)x(k) + \mathcal{B}_m(\rho)v(k) + D_m(\rho)w(k)) - x^T(k)P_m x(k).\end{aligned} \tag{6.61}$$

Combining (6.59) with $w(k) = 0$, there is

$$\mathbb{E}\{\Delta V(k)\} + z^T(k)z(k) < x^T(k)(U - P_m)x(k), \tag{6.62}$$

which means that

$$\begin{aligned}\mathbb{E}\{\Delta V(k)\} &< -\lambda_{\min}(-U + P_m)\, x^T(k)x(k)\\ &< -\varpi x^T(k)x(k)\\ &< 0,\end{aligned} \tag{6.63}$$

where

$$\varpi = \inf\{\lambda_{\min}(-U + P_m),\ m \in \mathcal{S}\} > 0. \tag{6.64}$$

Summing up the inequality of the both sides from 0 to ∞ can obtain that

$$\begin{aligned}\sum_{k=0}^{\infty}\mathbb{E}\left\{x^T(k)x(k)\right\} &< \frac{1}{\varpi}\mathbb{E}\{V(0) - V(\infty)\}\\ &< \frac{1}{\varpi}\mathbb{E}\{V(0)\}\\ &< \infty.\end{aligned} \tag{6.65}$$

Thus, considering Definition 6.1, we can find that system (6.49) is stochastically stable.

Next, from (6.53), inequality (6.59) implies that

$$\begin{aligned}&(\mathcal{A}_m(\rho)x(k) + \mathcal{B}_m(\rho)v(k) + D_m(\rho)w(k))^T\sum_{n\in\mathcal{S}}\pi_{mn}P_n\\ &\times(\mathcal{A}_m(\rho)x(k) + \mathcal{B}_m(\rho)v(k) + D_m(\rho)w(k))\\ &- x^T(k)P_m x(k) + z^T(k)z(k) - \gamma^2 w^T(k)w(k) < 0.\end{aligned} \tag{6.66}$$

Furthermore, we introduce

$$\Omega(k) = \mathbb{E}\left\{z^T(k)z(k) - \gamma^2 w^T(k)w(k)\right\} + \mathbb{E}\{\Delta V(k)\} < 0. \tag{6.67}$$

In order to analyse $\mathcal{H}_\infty$ performance of the system (6.49), we define the cost functional as follows

$$\mathcal{J}(k) \triangleq \mathbb{E}\left\{\sum_{s=0}^{k-1}\left(z^T(s)z(s) - \gamma^2 w^T(s)w(s)\right)\right\}. \tag{6.68}$$

In view of the definition of $\Omega(k)$, we can rewrite (6.68) as

$$\mathcal{J}(k) = \mathbb{E}\left\{\sum_{s=0}^{k-1}\Omega(s)\right\} - \mathbb{E}\{V(k)\} + x^T(0)\mathcal{P}_{r_0}x(0), \tag{6.69}$$

where $\mathcal{P}_{r_0} = \mathcal{P}_s, s \in \mathcal{S}$, is the Lyapunov matrix corresponding to the system model at initial time.

From the condition (6.65) and $w(k) \in \ell_2[0, \infty)$, we can obtain that

$$\lim_{k\to\infty} \mathbb{E}\{V(k)\} = 0. \tag{6.70}$$

Then, consider the zero initial condition and (6.70), the following inequality can be obtained directly

$$\mathcal{J}(\infty) < 0. \tag{6.71}$$

According to Definition 6.2, it can be seen that the inequality (6.51) is guaranteed, i.e., the system (6.49) satisfies the given $\mathcal{H}_\infty$ performance. The proof is completed.

In the rest of this section, we introduce a design scheme for the $\mathcal{H}_\infty$ controller to ensure that system (6.49) is stochastically stable and satisfies the required $\mathcal{H}_\infty$ performance.

Theorem 6.6 *Consider system (6.49). Then, for known parameters M and Δ, the system is stochastically stable with the required $\mathcal{H}_\infty$ performance γ if there exist scalars $\bar{\theta} < 1$, η_m, matrices $Y_m > 0$, $V > 0$, $\bar{K}_m$, and G_m, $m \in \mathcal{S}$ such that*

$$\begin{bmatrix} \Theta_{11}^m & * & * & * & * & * & * & * \\ 0 & -I & * & * & * & * & * & * \\ 0 & 0 & -\gamma^2 I & * & * & * & * & * \\ L_{mi}G_m & 0 & 0 & -I & * & * & * & * \\ \bar{K}_m & 0 & 0 & 0 & -\frac{M^2\bar{\theta}}{4\Delta^2}I & * & * & * \\ \Theta_{61}^{mi} & \pi_\ell B_{mi}\beta_1 & \pi_\ell D_{mi} & 0 & 0 & -\bar{Y}_\ell & * & * \\ 0 & 0 & 0 & 0 & 0 & \eta_m(\pi_\ell B_{mi}\beta_p)^T & -\eta_m I & * \\ \bar{K}_m & I & 0 & 0 & 0 & 0 & 0 & -\eta_m I \end{bmatrix} < 0, \tag{6.72}$$

$$V > Y_m, \tag{6.73}$$

for $i = 1, ..., r$, *where* $\Theta_{11}^m = -G_m - G_m^T + V$, $\Theta_{61}^{mi} = \pi_\ell A_{mi} G_m + \pi_\ell B_{mi} \beta_1 \bar{K}_m$, $\bar{Y}_\ell = diag\{Y_1, Y_2, \ldots, Y_\ell\}$ *and the controller gain matrix is obtained by*

$$K_m = \bar{K}_m G_m^{-1}. \tag{6.74}$$

Proof From Theorem 6.5, we know that system (6.49) is stochastically stable and guarantees the prescribed $\mathscr{H}_\infty$ performance if inequalities (6.52) and (6.53) hold.

Pre- and post-multiplying (6.52) by $diag\{G_m^T, I, I, I, I, I\}$ and its transpose, we can obtain that

$$\begin{bmatrix} -G_m^T U G_m & * & * & * & * & * \\ 0 & -I & * & * & * & * \\ 0 & 0 & -\gamma^2 I & * & * & * \\ L_m(\rho) G_m & 0 & 0 & -I & * & * \\ K_m G_m & 0 & 0 & 0 & -\frac{M^2}{4\Delta^2\theta} I & * \\ \pi_\ell \mathscr{A}_m(\rho) G_m & \pi_\ell \mathscr{B}_m(\rho) & \pi_\ell D_m(\rho) & 0 & 0 & -\bar{P}_\ell \end{bmatrix} < 0. \tag{6.75}$$

From the fact that $-G_m^T U G_m \leq -G_m - G_m^T + U^{-1}$ with $U > 0$, we can derive a sufficient condition for the above inequality as

$$\begin{aligned} &\begin{bmatrix} -G_m - G_m^T + U^{-1} & * & * & * & * & * \\ 0 & -I & * & * & * & * \\ 0 & 0 & -\gamma^2 I & * & * & * \\ L_m(\rho) G_m & 0 & 0 & -I & * & * \\ K_m G_m & 0 & 0 & 0 & -\frac{M^2}{4\Delta^2\theta} I & * \\ \pi_\ell (A_m(\rho) + B_m(\rho)\beta_1 K_m) G_m & \pi_\ell B_m(\rho)\beta_1 & \pi_\ell D_m(\rho) & 0 & 0 & -\bar{P}_\ell \end{bmatrix} \\ &+ \begin{bmatrix} 0 \\ 0 \\ 0 \\ 0 \\ 0 \\ \pi_\ell B_m(\rho)\beta_p \end{bmatrix} \Delta_f \begin{bmatrix} K_m G_m & I & 0 & 0 & 0 & 0 \end{bmatrix} \\ &+ \begin{bmatrix} (K_m G_m)^T \\ I \\ 0 \\ 0 \\ 0 \\ 0 \end{bmatrix} \Delta_f^T \begin{bmatrix} 0 & 0 & 0 & 0 & 0 & (\pi_\ell B_m(\rho)\beta_p)^T \end{bmatrix} < 0. \end{aligned} \tag{6.76}$$

Then a condition for (6.76) can be obtained as follows by using Lemma 1.2,

$$\begin{bmatrix} -G_m - G_m^T + U^{-1} & * & * & * & * & * \\ 0 & -I & * & * & * & * \\ 0 & 0 & -\gamma^2 I & * & * & * \\ L_m(\rho)G_m & 0 & 0 & -I & * & * \\ K_m G_m & 0 & 0 & 0 & -\frac{M^2}{4\Delta^2\theta}I & * \\ \pi_\ell(A_m(\rho) + B_m(\rho)\beta_1 K_m)G_m & \pi_\ell B_m(\rho)\beta_1 & \pi_\ell D_m(\rho) & 0 & 0 & -\bar{P}_\ell \end{bmatrix}$$

$$+ \eta_m^{-1} \begin{bmatrix} 0 \\ 0 \\ 0 \\ 0 \\ 0 \\ \eta_m \pi_\ell B_m(\rho)\beta_p \end{bmatrix} \begin{bmatrix} 0 & 0 & 0 & 0 & 0 & (\eta_m \pi_\ell B_m(\rho)\beta_p)^T \end{bmatrix} \tag{6.77}$$

$$+ \eta_m^{-1} \begin{bmatrix} (K_m G_m)^T \\ I \\ 0 \\ 0 \\ 0 \\ 0 \end{bmatrix} \begin{bmatrix} K_m G_m & I & 0 & 0 & 0 & 0 \end{bmatrix} < 0.$$

Employing Lemma 1.1 can obtain

$$\begin{bmatrix} \bar{\Theta}_{11}^m & * & * & * & * & * & * & * \\ 0 & -I & * & * & * & * & * & * \\ 0 & 0 & -\gamma^2 I & * & * & * & * & * \\ L_m(\rho)G_m & 0 & 0 & -I & * & * & * & * \\ K_m G_m & 0 & 0 & 0 & -\frac{M^2}{4\Delta^2\theta}I & * & * & * \\ \bar{\Theta}_{61}^m(\rho) & \pi_\ell B_m(\rho)\beta_1 & \pi_\ell D_m(\rho) & 0 & 0 & -\bar{P}_\ell & * & * \\ 0 & 0 & 0 & 0 & 0 & \eta_m(\pi_\ell B_m(\rho)\beta_p)^T & -\eta_m I & * \\ K_m G_m & I & 0 & 0 & 0 & 0 & 0 & -\eta_m I \end{bmatrix} < 0, \tag{6.78}$$

where $\bar{\Theta}_{11}^m = -G_m - G_m^T + U^{-1}$ and $\bar{\Theta}_{61}^m(\rho) = \pi_\ell(A_m(\rho) + B_m(\rho)\beta_1 K_m)G_m$.

By defining $\bar{\theta} = \frac{1}{\theta}$, $\bar{K}_m = K_m G_m$, $V = U^{-1}$ and $Y_\ell = P_\ell^{-1}$, the following matrix inequality

$$\begin{bmatrix} \Theta_{11}^m & * & * & * & * & * & * & * \\ 0 & -I & * & * & * & * & * & * \\ 0 & 0 & -\gamma^2 I & * & * & * & * & * \\ L_m(\rho)G_m & 0 & 0 & -I & * & * & * & * \\ \bar{K}_m & 0 & 0 & 0 & -\frac{M^2\bar{\theta}}{4\Delta^2}I & * & * & * \\ \Theta_{61}^m(\rho) & \pi_\ell B_m(\rho)\beta_1 & \pi_\ell D_m(\rho) & 0 & 0 & -\bar{Y}_\ell & * & * \\ 0 & 0 & 0 & 0 & 0 & \eta_m(\pi_\ell B_m(\rho)\beta_p)^T & -\eta_m I & * \\ \bar{K}_m & I & 0 & 0 & 0 & 0 & 0 & -\eta_m I \end{bmatrix} < 0, \tag{6.79}$$

and (6.73) are obtained, where $\Theta_{61}^m(\rho) = \pi_\ell A_m(\rho)G_m + \pi_\ell B_m(\rho)\beta_1 \bar{K}_m$.

From the result in [6] for the fuzzy parameters, condition (6.72) ensures the above inequality holds. And via simple calculation, (6.53) means that (6.73) holds. The proof is completed.

6.2.3 Simulation Example

In this section, the developed result will be demonstrated by the following example for system (6.38) with $\ell = 2$ and $r = 2$.

Plant Rule 1:

$$A_{11} = \begin{bmatrix} 0.2 & 0.8 \\ 0 & -1 \end{bmatrix}, \ A_{21} = \begin{bmatrix} 0.8 & 0.6 \\ 0 & -1 \end{bmatrix},$$
$$B_{11} = \begin{bmatrix} 0.2 \\ 0.2 \end{bmatrix}, \ B_{21} = \begin{bmatrix} 0.2 \\ 0.4 \end{bmatrix},$$
$$D_{11} = \begin{bmatrix} 0.8 \\ 0.7 \end{bmatrix}, \ D_{21} = \begin{bmatrix} 0.6 \\ 0.1 \end{bmatrix},$$
$$L_{11} = \begin{bmatrix} 0.6 & 0.4 \end{bmatrix}, \ L_{21} = \begin{bmatrix} 0.1 & 0.3 \end{bmatrix}.$$

Plant Rule 2:

$$A_{12} = \begin{bmatrix} 1 & 0 \\ 0.3 & 0.8 \end{bmatrix}, \ A_{22} = \begin{bmatrix} 1 & 0 \\ 0.9 & 0.7 \end{bmatrix},$$
$$B_{12} = \begin{bmatrix} 0.5 \\ 0.2 \end{bmatrix}, \ B_{22} = \begin{bmatrix} 0.1 \\ 0.4 \end{bmatrix},$$
$$D_{12} = \begin{bmatrix} 0.4 \\ 0.9 \end{bmatrix}, \ D_{22} = \begin{bmatrix} 0.6 \\ 0.2 \end{bmatrix},$$
$$L_{12} = \begin{bmatrix} 0.3 & 0.5 \end{bmatrix}, \ L_{22} = \begin{bmatrix} 0.4 & 0.2 \end{bmatrix}.$$

The fuzzy weighting function is defined as

$$\rho_1(k) = |\sin(k)|, \ \rho_2(k) = 1 - \rho_1(k). \tag{6.80}$$

The transition rate matrix of the Markov jump is given by

$$\Omega = \begin{bmatrix} 0.3 & 0.7 \\ 0.9 & 0.1 \end{bmatrix}. \tag{6.81}$$

We suppose that the other parameters are $\beta_1 = 0.9$, $\beta_p = 0.1$, $M = 10$ and $\Delta = 0.01$. We employ the MATLAB LMI toolbox to cope with the conditions (6.72) and (6.73) with $\gamma = 2$ in Theorem 6.6. Then the desired controller gain can be given via (6.74) as the follows

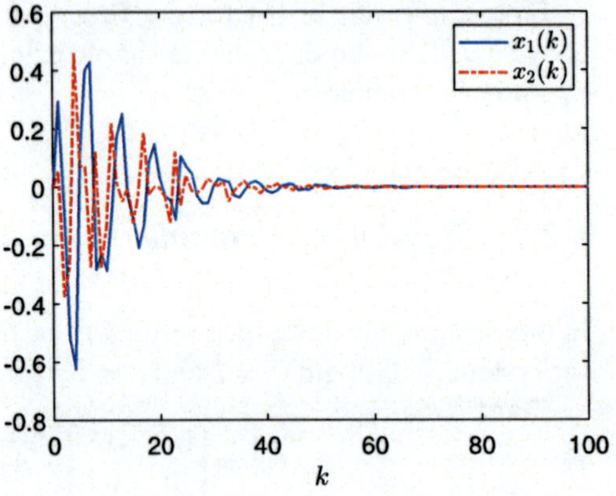

Fig. 6.6 Response of the system state $x(k)$

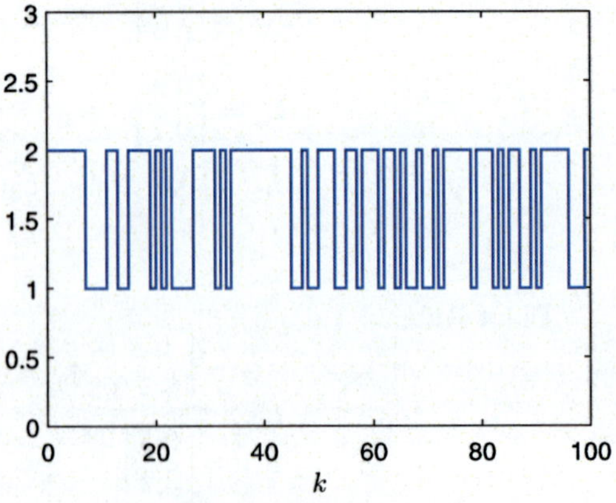

Fig. 6.7 History of the system mode jump

$$K_1 = \begin{bmatrix} -1.5881 & -0.6324 \end{bmatrix}, \; K_2 = \begin{bmatrix} -2.6675 & -0.1496 \end{bmatrix}.$$

Next, we give the simulation results to verify the validity of Theorem 6.6. Adopt $x(0) = \begin{bmatrix} 0 & 0 \end{bmatrix}^T$ and $w(k) = \cos(k)e^{-0.1k}$. Then the simulation results are given in Figs. 6.6, 6.7 and 6.8. Figure 6.6 describes the state trajectories of $x(k)$. Figure 6.7 shows the system mode jumps. Figure 6.8 plots the history of ratio of $\sqrt{\mathbb{E}\left\{\sum_{k=0}^{\infty} z^T(k)z(k) / \sum_{k=0}^{\infty} w^T(k)w(k)\right\}}$. From all the figures, it can be concluded that the fault-tolerant $\mathscr{H}_\infty$ control design approach developed in this section is effective.

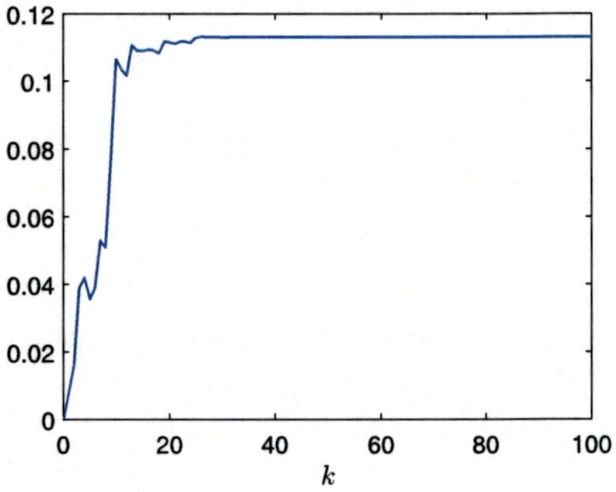

Fig. 6.8 History of $\sqrt{\mathbb{E}\left\{\sum_{k=0}^{\infty} z^T(k)z(k)/\sum_{k=0}^{\infty} w^T(k)w(k)\right\}}$

6.3 Guaranteed Cost Fault-Tolerant Control with Input Quantization

In this section, the quantized guaranteed cost fault-tolerant control problem is studied for the T–S fuzzy system. On the one hand, using the two-step method to design the feedback controller of the quantized closed-loop system can obtain the same guaranteed cost performance as the non-signal quantization system. On the other hand, the dynamic parameters of the quantizer are designed by introducing some auxiliary scalars. The design conditions of dynamic quantizer and controller in LMI form are obtained by decoupling the nonlinear terms with some lemmas.

6.3.1 Problem Formulation

In this section, T–S fuzzy model is introduced as the following form:

Plant Rule i_{th}: **IF** $\varepsilon_1(k)$ is M_{1i}, $\varepsilon_2(k)$ is M_{2i}, and, ..., and $\varepsilon_d(k)$ is M_{di}, **THEN**

$$x(k+1) = A_i x(k) + B_i u(k), \tag{6.82}$$

where $x(k) \in \mathbb{R}^{n_x}$ and $u(k) \in \mathbb{R}^{n_u}$, denote the state vector and control input, respectively. $A_i \in \mathbb{R}^{n_x \times n_x}$ and $B_i \in \mathbb{R}^{n_x \times n_u}$ are system matrices; $\varepsilon_1(k), \varepsilon_2(k), \ldots, \varepsilon_d(k)$ are measurable premise variables; M_{pi}, $i = 1, 2, \ldots, r$, $p = 1, 2, \ldots, d$, are the fuzzy sets, r is the number of fuzzy rules.

Then the T–S fuzzy model (6.82) is inferred as follows:

$$x(k+1) = A(\rho)x(k) + B(\rho)u(k), \tag{6.83}$$

where

$$A(\rho) = \sum_{i=1}^{r} \rho_i(\varepsilon(k))A_i, \; B(\rho) = \sum_{i=1}^{r} \rho_i(\varepsilon(k))B_i.$$

The guaranteed cost controller is considered as state feedback controller, and the controller is given as follows:

$$u(k) = (K + \Delta K)x(k), \tag{6.84}$$

where $K \in \mathbb{R}^{n_u \times n_x}$ is the controller gain to be determined. ΔK is a time-variable matrix and assumed to be of the form

$$\Delta K = F\Delta_k(k)E, \tag{6.85}$$

where F and E are known real constant matrices with appropriate dimensions, and $\Delta_k(k)$ describes the system uncertainty which is assumed to satisfy $\Delta_k^T(k)\Delta_k(k) \leq I$.

Furthermore, the actuator fault is considered in the closed-loop system

$$\bar{u}(k) = Ru(k), \tag{6.86}$$

where $R = \overline{R} + \Delta R \times \underline{R}$ denotes the actuator fault matrix. $\overline{R} = \frac{\overline{\alpha}+\underline{\alpha}}{2}$, $\underline{R} = \frac{\overline{\alpha}-\underline{\alpha}}{2}$, $\overline{\alpha}$ and $\underline{\alpha}$ are known fault factors, $0 < \underline{\alpha} < \overline{\alpha} < 1$. ΔR is a time-variable matrix and assumed to be of the form

$$\Delta R = D\Delta_r(k)N, \tag{6.87}$$

where D and N are known real constant matrices with appropriate dimensions, and $\Delta_r(k)$ describes the system uncertainty which is assumed to satisfy $\Delta_r^T(k)\Delta_r(k) \leq I$.

When the actuator fault is taken into consideration, the T–S fuzzy system can be referred as

$$x(k+1) = A(\rho)x(k) + B(\rho)\bar{u}(k). \tag{6.88}$$

Then, based on the system (6.88) and the controller (6.84), the closed-loop system can be given in the following form:

$$x(k+1) = \mathscr{A}(\rho)x(k), \tag{6.89}$$

with $\mathscr{A}(\rho) = A(\rho) + B(\rho)R\tilde{K}$, $\tilde{K} = K + \Delta K$. The guaranteed cost index is constructed as

$$J = \sum_{k=0}^{\infty} \left\{x^T(k)Q_1x(k) + \bar{u}^T(k)Q_2\bar{u}(k)\right\}, \tag{6.90}$$

where Q_1 and Q_2 are known positive definite matrices.

6.3.2 Fault-Tolerant Guaranteed Cost Control Without Quantization

For the sake of elaborating the guaranteed cost fault-tolerant control problem, it is necessary to introduce the following definition, which will be significant for the subsequent development of our results.

Definition 6.3 [7] When a positive scalar J^* and a control law (6.84) can ensure that the closed-loop system is asymptotically stable with $J \leq J^*$ for the cost function (6.90), then J^* is the upper bound of the guaranteed cost index and $u(k)$ is the guaranteed cost control law for the system (6.83).

Associated with the resulting closed-loop system (6.89), our goal is to find the control law (6.84) such that the following two conditions are satisfied.

(1) The closed-loop system (6.89) is asymptotically stable.
(2) For the closed-loop system (6.89) with the control law (6.84), the cost function (6.90) meets $J \leq J^*$, where J^* is the upper bound of the guaranteed cost index.

In the following, we present a guaranteed cost performance analysis criterion for the system (6.89) with the control law (6.84), i.e., conducting the analysis for (6.89) under the assumption that the controller gain has been given.

Theorem 6.7 *Consider the system (6.88) and the control input (6.84). For the given positive definite matrices Q_1 and Q_2, if there exist matrices $P > 0$ and $G > 0$ such that the following inequality holds:*

$$\mathcal{A}^T(\rho)P\mathcal{A}(\rho) - P + Q_1 + \tilde{K}^T R^T Q_2 R\tilde{K} < -G, \tag{6.91}$$

then the closed-loop system (6.89) is asymptotically stable with the guaranteed cost performance index J satisfying

$$J < x^T(0)Px(0). \tag{6.92}$$

Proof Firstly, we shall to show that if the condition (6.91) holds, then the closed-loop system (6.89) is asymptotically stable. Let us choose a Lyapunov function as

$$V(x(k)) = x^T(k)Px(k), \ P > 0, \tag{6.93}$$

then one has $\Delta V(x(k)) = V(x(k+1)) - V(x(k))$ and

$$\begin{aligned}\Lambda &= \Delta V(x(k)) + x^T(k)Q_1x(k) + \bar{u}^T(k)Q_2\bar{u}(k) \\ &= x^T(k)\mathcal{A}^T(\rho)P\mathcal{A}(\rho)x(k) - x^T(k)Px(k) \\ &\quad + x^T(k)Q_1x(k) + x^T(k)\tilde{K}^T R^T Q_2 R\tilde{K}x(k) \\ &= x^T(k)\Big(\mathcal{A}^T(\rho)P\mathcal{A}(\rho) - P + Q_1 + \tilde{K}^T R^T Q_2 R\tilde{K}\Big)x(k).\end{aligned} \tag{6.94}$$

Obviously, the condition (6.91) guarantees the establishment of the following inequality:

$$\Lambda < x^T(k)(-G)x(k). \tag{6.95}$$

Because of $G > 0$, then from above inequalities, we can obtain

$$\Delta V(x(k)) < -x^T(k)Q_1x(k) - x^T(k)\tilde{K}^T R^T Q_2 R\tilde{K}x(k). \tag{6.96}$$

It is should be noted that $\Delta V(x(k)) < 0$ with $Q_1 > 0$ and $Q_2 > 0$ for any $x(k) \neq 0$. Hence, if the condition (6.91) is satisfied, we can conclude that the closed-loop system (6.89) is asymptotically stable.

In the following, we are in a position to establish the guaranteed cost index. It is noted that, for all $x(k) \neq 0$, integrating the inequality (6.96) yields

$$\sum_{k=0}^{\infty} \left\{\Delta V(x(k)) + x^T(k)Q_1x(k) + \bar{u}^T(k)Q_2\bar{u}(k)\right\} < 0. \tag{6.97}$$

Based on the definition (6.90), (6.96) is equivalent to

$$J < V(x(0)) - V(x(\infty)). \tag{6.98}$$

With $V(x(\infty)) \geq 0$, the above condition implies that

$$J < V(x(0)) = x^T(0)Px(0). \tag{6.99}$$

Accordingly, the guaranteed cost function (6.90) satisfies (6.92). In view of Definition 6.3, (6.84) is a guaranteed cost control law for system (6.88) and $J^* = x^T(0)Px(0)$ is an upper bound of guaranteed cost index for system (6.89). The proof is completed.

In the following, we study the guaranteed cost controller design problem based on Theorem 6.7, that is, we will give design condition to determine the controller gain K such that the closed-loop system (6.89) is asymptotically stable with the upper bound of the guaranteed cost control index given as (6.92).

Theorem 6.8 *Consider the system (6.88) and the control input (6.84). For the given positive definite matrices Q_1 and Q_2, if there exist matrices T, V, and L, scalars $\rho_1 > 0$ and $\rho_2 > 0$ such that the following LMIs hold:*

$$L < Q_1^{-1}, \tag{6.100}$$

$$\begin{bmatrix} \Xi_{1i} & * & * & * \\ \Xi_2 & \Xi_3 & * & * \\ \Xi_{4i} & \Xi_5 & \Xi_6 & * \\ \Xi_{7i} & \Xi_8 & \Xi_9 & \Xi_{10} \end{bmatrix} < 0, \; i = 1, 2, \ldots, r, \tag{6.101}$$

where

$$\Xi_{1i} = \begin{bmatrix} -V & * \\ A_i V + B_i \overline{R} T & -V \end{bmatrix}, \Xi_2 = \begin{bmatrix} \overline{R} T & 0 \\ V & 0 \end{bmatrix}, \Xi_{4i} = \begin{bmatrix} 0 & \rho_1 (B_i \overline{R} F)^T \\ EV & 0 \end{bmatrix},$$

$$\Xi_5 = \begin{bmatrix} \rho_1 (\overline{R} F)^T & 0 \\ 0 & 0 \end{bmatrix}, \Xi_{7i} = \begin{bmatrix} 0 & \rho_2 (B_i D)^T \\ N \underline{R} T & 0 \end{bmatrix}, \Xi_8 = \begin{bmatrix} \rho_2 D^T & 0 \\ 0 & 0 \end{bmatrix}, \Xi_9 = \begin{bmatrix} 0 & 0 \\ \rho_1 N \underline{R} F & 0 \end{bmatrix},$$

$$\Xi_3 = -\, diag\{Q_2^{-1},\ L\}, \Xi_6 = -diag\{\rho_1 I,\ \rho_1 I\}, \Xi_{10} = -diag\{\rho_2 I,\ \rho_2 I\}.$$

Then the closed-loop system (6.89) is asymptotically stable with an upper bound of the guaranteed cost control index in (6.92) with $P = V^{-1}$. The controller gain in (6.84) can be derived as

$$K = T V^{-1}. \tag{6.102}$$

Proof Applying the Lemma 1.1 to (6.91) gives

$$\begin{bmatrix} -P + Q_1 + G & * & * \\ \mathscr{A}(\rho) & -P^{-1} & * \\ R\tilde{K} & 0 & -Q_2^{-1} \end{bmatrix} < 0. \tag{6.103}$$

By defining $V = P^{-1}$, pre- and post-multiplying the matrix inequality (6.103) by $diag\{V,\ I,\ I\}$ and applying the Lemma 1.1 to it, one produces

$$\begin{bmatrix} -V & * & * & * \\ \mathscr{A}(\rho) V & -V & * & * \\ R\tilde{K} V & 0 & -Q_2^{-1} & * \\ V & 0 & 0 & -(Q_1 + G)^{-1} \end{bmatrix} < 0. \tag{6.104}$$

.

Obviously, there exist uncertain items ΔK and ΔR in condition (6.104). In order to design the feedback controller, one should remove these uncertain factors. Firstly, separating uncertain items ΔK from (6.104), one obtains

$$\begin{bmatrix} -V & * & * & * \\ (A(\rho) + B(\rho) R K) V & -V & * & * \\ R K V & 0 & -Q_2^{-1} & * \\ V & 0 & 0 & -(Q_1 + G)^{-1} \end{bmatrix} + He\{\Phi_1^T \Delta_k(k) \Phi_2\} < 0, \tag{6.105}$$

where $\Phi_1 = \begin{bmatrix} 0 & (B(\rho) R F)^T & (RF)^T & 0 \end{bmatrix}$, $\Phi_2 = \begin{bmatrix} EV\ 0\ 0\ 0 \end{bmatrix}$. By using Lemma 1.2, one can produce

$$\begin{bmatrix} -V & * & * & * & * & * \\ (A(\rho)+B(\rho)RK)V & -V & * & * & * & * \\ RKV & 0 & -Q_2^{-1} & * & * & * \\ V & 0 & 0 & -(Q_1+G)^{-1} & * & * \\ 0 & \rho_1(B(\rho)RF)^T & \rho_1(RF)^T & 0 & -\rho_1 I & * \\ EV & 0 & 0 & 0 & 0 & -\rho_1 I \end{bmatrix} < 0. \tag{6.106}$$

Secondly, use the same way to eliminate uncertain items ΔR. By defining new variables $T = KV$ and $L = (Q_1 + G)^{-1}$, and according to the definition of fuzzy basis functions, condition (6.101) can be obtained. It should be noted that $G = L^{-1} - Q_1 > 0$ can be converted to (6.100) by adopting Lemma 1.6. The proof is completed.

Remark 6.4 In Theorem 6.8, the two-step method first eliminates the uncertain term ΔK of the controller (6.84), and then eliminates the time-variable matrix ΔR defined in (6.84), so that the design conditions can be solved by standard LMI technique.

Theorem 6.8 has offered a solution for designing the controller (6.84) to ensure that the system (6.89) is asymptotically stable with the upper bound of guaranteed cost control index as (6.92). We can acutely observe that the upper bound of J is related to the initial state. Therefore, we desire to find an upper bound $J^\star$, which is independent of the initial state. As for the initial state, it is supposed to be arbitrary, but belongs to the set [8]

$$S = \{x(0) \in \mathbb{R}^{n_x} : x(0) = Hv_0, v_0^T v_0 \leq 1\},$$

where H is a given matrix (or a matrix to be determined).

In the light of above assumption, we have

$$x^T(0)Px(0) = v_0^T H^T P H v_0. \tag{6.107}$$

Next, selecting a scalar $\delta > 0$ satisfy

$$H^T P H < \delta I, \tag{6.108}$$

based on the Lemma 1.1, it can be rearranged as

$$\begin{bmatrix} -\delta I & * \\ H & -P^{-1} \end{bmatrix} < 0, \tag{6.109}$$

i.e.,

$$\begin{bmatrix} -\delta I & * \\ H & -V \end{bmatrix} < 0. \tag{6.110}$$

From the above analysis, it can be concluded that

$$x^T(0)Px(0) < \delta. \tag{6.111}$$

Define

$$J^{\star} = \delta. \tag{6.112}$$

Thereby, for (6.92), there is the establishment of $J \leq J^{\star} = \delta$. Then, Theorem 6.8 can be deduced as the following one.

Theorem 6.9 *Consider the system (6.88) and the control input (6.84). For the given positive definite matrices Q_1 and Q_2, if there exist matrices V, L, and T, positive scalars δ, ρ_1, ρ_2 satisfy (6.100), (6.101), and (6.110), then the closed-loop system (6.89) is asymptotically stable and the guaranteed cost index J has the upper bound $J^{\star}$. The controller gain K can be derived as (6.102).*

Remark 6.5 For Theorem 6.9, by applying LMI toolbox in MATLAB to solve (6.100), (6.101), and (6.110), the controller (6.84) with the suitable upper bound of the cost performance (6.112) can be derived. From conditions (6.110) and (6.112), it is easy to obtain that the selection of matrix H will have a great influence on the upper bound of guaranteed cost index. It can be given based on the system performance requirement or determined by solving the controller design problem in Theorem 6.9, and in this section, the former one will be adopted.

Remark 6.6 Based on Theorem 6.9, we can also optimize the upper bound of guaranteed cost performance to formulate an optimal guaranteed cost control law via minimizing $J^{\star}$. Then, the guaranteed cost control issue is transformed into a convex optimization problem as

$$\begin{aligned} &\min J^{\star} = \delta, \\ &\textit{subject to LMIs } (6.100),\ (6.101),\ \textit{and } (6.110), \end{aligned} \tag{6.113}$$

such that the closed-loop system (6.89) is asymptotically stable and the guaranteed cost index J has the optimal upper bound $J^{\star}$. The controller gain K can be derived as (6.102).

6.3.3 Fault-Tolerant Guaranteed Cost Control with Input Quantization

Based on (6.88), the T–S fuzzy system with input quantization can be inferred as follows:

$$x(k+1) = A(\rho)x(k) + B(\rho)\tilde{u}(k), \tag{6.114}$$

where $\tilde{u}(k) = Ru_q(k)$, $u_q(k) \in \mathbb{R}^{n_u}$ is the quantized control input which is defined as

$$u_q(k) = f_\theta(u(k)) = \theta(k) f\left(\frac{u(k)}{\theta(k)}\right), \tag{6.115}$$

where $f_\theta(\cdot)$ is a dynamic quantizer with general form which is a one-parameter family of quantizer given in [1], with $\theta(k)$ denoting the dynamic quantizer parameter. Assume that there exist scalars $M > 0$ and $\Delta > 0$ satisfying the following conditions:

$$\|f_\theta(\iota) - \iota\| \leq \theta\Delta, \ \ if \ \ \|\iota\| \leq \theta M, \tag{6.116}$$

$$\|f_\theta(\iota) - \iota\| > \theta\Delta, \ \ if \ \ \|\iota\| > \theta M. \tag{6.117}$$

When the quantizer is unsaturated, condition (6.116) provides a bound for the quantization error. Condition (6.117) gives a way for detecting the possibility of saturation. $\theta\Delta$ and θM represent the quantization error bound and the quantization range of the quantizer $f_\theta(\cdot)$, respectively.

As a consequence, we can rewrite the system model (6.114) as

$$x(k+1) = A(\rho)x(k) + B(\rho)Ru(k) + B(\rho)Re(k), \tag{6.118}$$

where $e(k) = \theta(k)\left(f\left(\frac{u(k)}{\theta(k)}\right) - \frac{u(k)}{\theta(k)}\right)$. Further, based on (6.118) and (6.84), the quantized closed-loop system can be given as:

$$x(k+1) = \mathscr{A}(\rho)x(k) + B(\rho)Re(k), \tag{6.119}$$

where $\mathscr{A}(\rho)$ has been defined in (6.89).

The guaranteed cost index is introduced for constructing a quantized feedback control law (6.115) as

$$J = \sum_{k=0}^{\infty} \left\{x^T(k)Q_1x(k) + \tilde{u}^T(k)Q_2\tilde{u}(k)\right\}. \tag{6.120}$$

Remark 6.7 It should be pointed out that in the closed-loop system (6.119), there exists additional item $B(\rho)Re(k)$ caused by the effect of quantization. Therefore, the effect of quantization should be removed to obtain the same guaranteed cost performance as the one without quantization.

It is important to note that the performance function (6.120) includes the quantized control input, which will cause great difficulty to design guaranteed cost controller once the input signal is quantized. Next, we shall propose an effective quantization design strategy so that the same guaranteed cost performance as the one without quantization can be still achieved despite the input signal is quantized.

Theorem 6.10 *Suppose that the LMIs (6.100) and (6.101) in Theorem 6.8 are satisfied, and the range M of quantizer is chosen to be large enough so that:*

$$M > \frac{1}{\xi}, \tag{6.121}$$

and the quantized control input (6.115) is designed with on-line adjusting strategy for parameter $\theta(k)$ as:

$$\frac{1}{M}\|u(k)\| \le \theta(k) \le \xi \|u(k)\|, \tag{6.122}$$

where

$$\xi = \frac{\lambda_{\min}(G) - h}{\Delta r_2\left(\beta_1 + \sqrt{\beta_1^2 + \beta_2(\lambda_{\min}(G) - h)}\right)}, \tag{6.123}$$

$$0 < h < \lambda_{\min}(G), \tag{6.124}$$

$$\beta_1 = r_3 + \left\|\sqrt{Q_2}\right\|^2 r_1^2 r_2, \tag{6.125}$$

$$\beta_2 = r_4 + \left\|\sqrt{Q_2}\right\|^2 r_1^2, \tag{6.126}$$

with $G = L^{-1} - Q_1$ and r_1, r_2, r_3, r_4 satisfying

$$\|R\| < r_1, \tag{6.127}$$

$$\left\|\tilde{K}\right\| < r_2, \tag{6.128}$$

$$\left\|(A(\rho) + B(\rho)R\tilde{K})^T P B(\rho)\right\| < r_3, \tag{6.129}$$

$$\left\|B^T(\rho)PB(\rho)\right\| < r_4. \tag{6.130}$$

Then, the controller (6.115) with the adjusting strategy (6.122) can ensure that the closed-loop system (6.119) is asymptotically stable with the guaranteed cost performance index J satisfying

$$J < x^T(0)Px(0). \tag{6.131}$$

Proof According to the same definition in (6.93), we have

$$\begin{aligned}\Lambda_q =&\Delta V(x(k)) + x^T(k)Q_1x(k) + \tilde{u}^T(k)Q_2\tilde{u}(k)\\ =&\Big(\mathcal{A}(\rho)x(k) + B(\rho)Re(k)\Big)^T P\Big(\mathcal{A}(\rho)x(k) + B(\rho)Re(k)\Big)\\ &- x^T(k)Px(k) + x^T(k)Q_1x(k) + \tilde{u}^T(k)Q_2\tilde{u}(k)\\ =&x^T(k)\Big(\mathcal{A}(\rho)^T P\mathcal{A}(\rho) - P + Q_1 + \tilde{K}^T R^T Q_2 R\tilde{K}\Big)x(k)\\ &- \tilde{u}^T(k)Q_2\tilde{u}(k) + 2x^T(k)\mathcal{A}^T(\rho)PB(\rho)e(k)\\ &+ e^T(k)B^T(\rho)PB(\rho)e(k) + \tilde{u}^T(k)Q_2\tilde{u}(k).\end{aligned} \tag{6.132}$$

Under the condition (6.91), thus, from (6.132), we have

$$\begin{aligned}\Lambda_q <&x^T(k)(-G)x(k) + \tilde{u}^T(k)Q_2\tilde{u}(k) - \tilde{u}^T(k)Q_2\tilde{u}(k)\\ &+ 2x^T(k)\mathcal{A}^T(\rho)PB(\rho)e(k) + e^T(k)B^T(\rho)PB(\rho)e(k).\end{aligned} \tag{6.133}$$

Next, we consider the following four factual conclusions:

(i) If the range M and dynamic parameter $\theta(k)$ of the quantizer (6.115) are designed as (6.121) and (6.122), one can be given

$$\frac{1}{M}\|u(k)\| \le \theta(k), \tag{6.134}$$

i.e.,

$$\|u(k)\| \le M\theta(k). \tag{6.135}$$

Based on the property of the quantizer $f_\theta(\cdot)$ given in (6.116), it is effortless to obtain that whenever (6.135) holds, then

$$\left\| f\left(\frac{u(k)}{\theta(k)}\right) - \frac{u(k)}{\theta(k)} \right\| \le \Delta, \tag{6.136}$$

i.e.,

$$\|e(k)\| = \theta(k)\left\| f\left(\frac{u(k)}{\theta(k)}\right) - \frac{u(k)}{\theta(k)} \right\| \le \theta(k)\Delta. \tag{6.137}$$

(ii) By recalling the definition of the Euclidean norm, the triangle inequality and compatible properties of norm, one can be given as

$$
\begin{aligned}
\sqrt{\tilde{u}^T(k)Q_2\tilde{u}(k)} &= \left\|\sqrt{Q_2}\tilde{u}(k)\right\| \\
&= \left\|\sqrt{Q_2}R\theta(k)\left(f\left(\frac{u(k)}{\theta(k)}\right) - \frac{u(k)}{\theta(k)}\right) + \sqrt{Q_2}Ru(k)\right\| \\
&\leq \left\|\sqrt{Q_2}Re(k)\right\| + \left\|\sqrt{Q_2}Ru(k)\right\| \\
&\leq \left\|\sqrt{Q_2}R\right\| \theta(k)\Delta + \sqrt{\bar{u}^T(k)Q_2\bar{u}(k)},
\end{aligned} \tag{6.138}
$$

then we have

$$
\begin{aligned}
&\tilde{u}^T(k)Q_2\tilde{u}(k) - \bar{u}^T(k)Q_2\bar{u}(k) \\
\leq &\left\|\sqrt{Q_2}\right\|^2 \|R\|^2 \theta^2(k)\Delta^2 + 2\left\|\sqrt{Q_2}\right\|^2 \|R\|^2 \theta(k)\Delta \|u(k)\| \\
\leq &\left\|\sqrt{Q_2}\right\|^2 \|R\|^2 \theta^2(k)\Delta^2 + 2\left\|\sqrt{Q_2}\right\|^2 \|R\|^2 \theta(k)\Delta \left\|\tilde{K}\right\| \|x(k)\|.
\end{aligned} \tag{6.139}
$$

(iii) For the positive definite matrix G, the following equation

$$
\lambda_{\min}(G)\|x(k)\|^2 \leq x^T(k)Gx(k) \leq \lambda_{\max}(G)\|x(k)\|^2 \tag{6.140}
$$

holds for any $x(k)$, i.e.,

$$
-\lambda_{\max}(G)\|x(k)\|^2 \leq x^T(k)(-G)x(k) \leq -\lambda_{\min}(G)\|x(k)\|^2. \tag{6.141}
$$

(iv) Combining the definition of the spectral norm, the triangle inequality and compatible properties of norm yields the following inequality:

$$
\begin{aligned}
&2x^T(k)\mathscr{A}^T(\rho)PB(\rho)e(k) + e^T(k)B^T(\rho)PB(\rho)e(k) \\
\leq &2\|x(k)\| r_3 \|e(k)\| + r_4\|e(k)\|^2.
\end{aligned} \tag{6.142}
$$

By assembling the factual conclusions in (i)–(iv), we can know that

$$
\begin{aligned}
\Lambda_q < &-\lambda_{\min}(G)\|x(k)\|^2 + 2\|x(k)\| r_3 \|e(k)\| + r_4\|e(k)\|^2 \\
&+ \left\|\sqrt{Q_2}\right\|^2 \|R\|^2 \theta^2(k)\Delta^2 + 2\left\|\sqrt{Q_2}\right\|^2 \|R\|^2 \theta(k)\Delta \left\|\tilde{K}\right\| \|x(k)\| \\
\leq &-\lambda_{\min}(G)\|x(k)\|^2 + \left(r_4 + \left\|\sqrt{Q_2}\right\|^2 \|R\|^2\right)\theta^2(k)\Delta^2 \\
&+ 2\left(r_3 + \left\|\sqrt{Q_2}\right\|^2 \|R\|^2 \left\|\tilde{K}\right\|\right)\|x(k)\|\theta(k)\Delta.
\end{aligned} \tag{6.143}
$$

Under the definitions (6.125), (6.126) and conditions (6.127), (6.128), (6.129), (6.130), one gives

$$
\begin{aligned}
\Lambda_q <& -\lambda_{\min}(G)\,\|x(k)\|^2 + 2\beta_1\,\|x(k)\|\,\theta(k)\Delta + \beta_2\theta^2(k)\Delta^2 \\
=& -h\,\|x(k)\|^2 - (\lambda_{\min}(G)-h)\Big(\|x(k)\|^2 - \frac{2\beta_1}{\lambda_{\min}(G)-h}\|x(k)\|\,\theta(k)\Delta \\
& - \frac{\beta_2}{\lambda_{\min}(G)-h}\theta^2(k)\Delta^2\Big) \\
=& -h\,\|x(k)\|^2 - (\lambda_{\min}(G)-h)\left(\|x(k)\| - \frac{\beta_1+\sqrt{\beta_1^2+\beta_2(\lambda_{\min}(G)-h)}}{\lambda_{\min}(G)-h}\theta(k)\Delta\right) \\
& \times\left(\|x(k)\| - \frac{\beta_1-\sqrt{\beta_1^2+\beta_2(\lambda_{\min}(G)-h)}}{\lambda_{\min}(G)-h}\theta(k)\Delta\right).
\end{aligned}
\tag{6.144}
$$

Since (6.124) means $\lambda_{\min}(G) - h > 0$ which implies that $\frac{\beta_1-\sqrt{\beta_1^2+\beta_2(\lambda_{\min}(G)-h)}}{\lambda_{\min}(G)-h} < 0$.

On the other hand, based on (6.122) and (6.123), we know that

$$
\theta(k) \le \frac{\lambda_{\min}(G)-h}{\Delta(\beta_1+\sqrt{\beta_1^2+\beta_2(\lambda_{\min}(G)-h)})}\|x(k)\|, \tag{6.145}
$$

which means

$$
\|x(k)\| - \frac{\beta_1+\sqrt{\beta_1^2+\beta_2(\lambda_{\min}(G)-h)}}{\lambda_{\min}(G)-h}\theta(k)\Delta \ge 0, \tag{6.146}
$$

then (6.144) yields

$$
\Lambda_q < -h\,\|x(k)\|^2 < 0. \tag{6.147}
$$

Accordingly, the guaranteed cost function (6.120) satisfies (6.131). In view of Definition 6.3, the quantized control strategy (6.115) is a guaranteed cost control law for the system (6.114) and $J^* = x^T(0)Px(0)$ is an upper bound of guaranteed cost index for the closed-loop system (6.119). The proof is completed.

Remark 6.8 For the system (6.114), the matrices $\mathscr{A}(\rho)$, R and $\tilde{K}$ contain uncertain parameters ΔK and ΔR which means that the determination of scalars r_1, r_2, r_3, and r_4 in the conditions (6.127), (6.128), (6.129), and (6.130) is hard. Thus, further efforts in the following are needed to transform the determination problem into a feasible computation problem.

According to the properties of 2-norm of a matrix A, we know that the condition $A^T A < \mu^2 I$ holds if and only if $\|A\| < \mu$.

Based on the conclusion, the conditions (6.127), (6.128), (6.129), and (6.130) are equivalent to the following inequalities:

$$
(\overline{R} + \Delta R \times \underline{R})^T(\overline{R} + \Delta R \times \underline{R}) < r_1^2 I, \tag{6.148}
$$

$$(K + \Delta K)^T (K + \Delta K) < r_2^2 I, \tag{6.149}$$

$$\left(\mathscr{A}^T(\rho) P B(\rho)\right)^T \left(\mathscr{A}^T(\rho) P B(\rho)\right) < r_3^2 I, \tag{6.150}$$

$$\left(B^T(\rho) P B(\rho)\right)^T \left(B^T(\rho) P B(\rho)\right) < r_4^2 I. \tag{6.151}$$

Then, applying the Lemmas 1.1 and 1.2 to (6.148), (6.149), (6.150), and (6.151) yields, respectively

$$\begin{bmatrix} -r_1^2 I & * & * & * \\ \overline{R} & -I & * & * \\ 0 & \eta_1 D^T & -\eta_1 I & * \\ N\underline{R} & 0 & 0 & -\eta_1 I \end{bmatrix} < 0, \tag{6.152}$$

$$\begin{bmatrix} -r_2^2 I & * & * & * \\ K & -I & * & * \\ 0 & \eta_2 F^T & -\eta_2 I & * \\ E & 0 & 0 & -\eta_2 I \end{bmatrix} < 0, \tag{6.153}$$

$$\begin{bmatrix} -r_3^2 I & * & * & * & * & * \\ (A(\rho) + B(\rho)\overline{R}K)^T P B(\rho) & -I & * & * & * & * \\ F^T \overline{R}^T B^T(\rho) P B(\rho) & 0 & -\eta_3 I & * & * & * \\ 0 & \eta_3 E & 0 & -\eta_3 I & * & * \\ D^T B^T(\rho) P B(\rho) & 0 & 0 & 0 & -\eta_4 I & * \\ 0 & \eta_4 N\underline{R}K & \eta_4 N\underline{R}F & 0 & 0 & -\eta_4 I \end{bmatrix} < 0, \tag{6.154}$$

$$\begin{bmatrix} -r_4^2 I & * \\ B^T(\rho) P B(\rho) & -I \end{bmatrix} < 0. \tag{6.155}$$

Define $l_1 = r_1^2$, $l_2 = r_2^2$, $l_3 = r_3^2$, and $l_4 = r_4^2$, we know that above conditions can be guaranteed by the following LMIs:

$$\begin{bmatrix} -l_1 I & * & * & * \\ \overline{R} & -I & * & * \\ 0 & \eta_1 D^T & -\eta_1 I & * \\ N\underline{R} & 0 & 0 & -\eta_1 I \end{bmatrix} < 0, \tag{6.156}$$

$$\begin{bmatrix} -l_2 I & * & * & * \\ K & -I & * & * \\ 0 & \eta_2 F^T & -\eta_2 I & * \\ E & 0 & 0 & -\eta_2 I \end{bmatrix} < 0, \tag{6.157}$$

$$\begin{bmatrix} -l_3 I & * & * & * & * & * \\ (A_i + B_i \overline{R} K)^T P B_j & -I & * & * & * & * \\ F^T \overline{R}^T B_i^T P B_j & 0 & -\eta_3 I & * & * & * \\ 0 & \eta_3 E & 0 & -\eta_3 I & * & * \\ D^T B_i^T P B_j & 0 & 0 & 0 & -\eta_4 I & * \\ 0 & \eta_4 N \underline{R} K & \eta_4 N \underline{R} F & 0 & 0 & -\eta_4 I \end{bmatrix} < 0,\ i,\ j = 1,\ 2,\ \ldots,\ r, \tag{6.158}$$

$$\begin{bmatrix} -l_4 I & * \\ B_i^T P B_j & -I \end{bmatrix} < 0,\ i,\ j = 1,\ 2,\ \ldots,\ r. \tag{6.159}$$

Now, based on above analysis and Theorem 6.9, the following result of guaranteed cost control for the system (6.114) subject to quantization is proposed, which ensures the asymptotic stability of system (6.114) with the upper bound of guaranteed cost performance $J^{\star}$.

Theorem 6.11 *Suppose that the LMI conditions in Theorem 6.9 can be solved, i.e., there exist matrices V, L, and T, scalar δ satisfying the matrix inequalities (6.100), (6.101), and (6.110) to maintain the guaranteed cost performance $J^{\star}$. The range M of quantizer is chosen to be large enough as (6.121), and the on-line adjusting strategy for quantizer's dynamic parameter $\theta(k)$ is designed as (6.122) with $r_1 = \sqrt{l_1}$, $r_2 = \sqrt{l_2}$, $r_3 = \sqrt{l_3}$, and $r_4 = \sqrt{l_4}$ satisfying (6.156), (6.157), (6.158), and (6.159), respectively. Then, the quantized feedback controller (6.115) with the adjusting strategy (6.122) can ensure that the quantized feedback closed-loop system (6.119) is asymptotically stable with the same guaranteed cost performance index $J^{\star}$ as the one without quantization.*

To facilitate the presentation of design for quantized feedback controller (6.115) with dynamic parameter $\theta(k)$, the following algorithm is given first based on Theorem 6.11.

Algorithm 6.1 ***Step 1.*** Solve (6.100), (6.101), and (6.110) to obtain the upper bound of guaranteed cost index $J^{\star}$;

Step 2. Compute the values of controller gain K with (6.102);

Step 3. Compute the values of l_1, l_2, l_3 and l_4 with (6.156), (6.157), (6.158), and (6.159);

Step 4. Compute the values of r_1, r_2, r_3, and r_4 by $r_1 = \sqrt{l_1}$, $r_2 = \sqrt{l_2}$, $r_3 = \sqrt{l_3}$, and $r_4 = \sqrt{l_4}$;

Step 5. Compute the values β_1 and β_2 with (6.125) and (6.126);

Step 6. Select h according to (6.124);

Step 7. Compute the value of ξ with (6.123).

Then, as in [9], according to the condition (6.122), give the adjusting rule of $\theta(k)$ as follows:

The adjusting rule:

Case a: If $\xi \left\| u(k) \right\| - \frac{1}{M} \left\| u(k) \right\| \geq 1, \theta(k) = p = \mathit{floor}(\xi \left\| u(k) \right\|)$ with $p \in \mathcal{N}^{+}$;

Case b-i: If $\xi \|u(k)\| - \frac{1}{M}\|u(k)\| < 1$ and $\frac{1}{M}\|u(k)\| < 1 \leq \xi \|u(k)\|$, $\theta(k) = p = 1$ with $p \in \mathcal{N}^+$;

Case b-ii: If $\xi \|u(k)\| - \frac{1}{M}\|u(k)\| < 1$ and $0 \leq \xi \|u(k)\| < 1, \theta(k) = p \times 10^{-q} = floor\ (\xi\|u(k)\| \times 10^q) \times 10^{-q}$ with $p, q \in \mathcal{N}^+$;

Case b-iii: If $\xi \|u(k)\| - \frac{1}{M}\|u(k)\| < 1$ and $\frac{1}{M}\|u(k)\| \geq 1, \theta(k) = p \times 10^{-q} = floor\ (\xi\|u(k)\| \times 10^q) \times 10^{-q}$ with $p, q \in \mathcal{N}^+$.

where $q = \min\left\{s \in \mathcal{N}^+ | \left((\xi - \frac{1}{M})\|u(k)\| \times 10^s\right) \geq 1\right\}$ and $floor\ (\Omega)$ represents the maximum integer not exceeding Ω.

6.3.4 Simulation Example

A one-link manipulator with the inclusion of motor dynamics [10] is used to illustrate the effectiveness of the proposed method, and it can be given as the following dynamic equation:

$$\mathcal{D}\ddot{q} + \mathcal{B}\dot{q} + \mathcal{N}\sin(q) = \vartheta,$$
$$\mathcal{M}\dot{\vartheta} + \mathcal{H}\vartheta = u - \mathcal{L}\dot{q},$$

where $\ddot{q}$, $\dot{q}$, and q stand for the link acceleration, velocity, and position, respectively, where $q \in [-\pi, \pi]$. ϑ means the torque caused by the electrical subsystem, u represents the input. $\mathcal{D}$, $\mathcal{B}$, $\mathcal{H}$, $\mathcal{M}$, and $\mathcal{L}$ are the mechanical inertia, coefficient of viscous friction at the joint, armature resistance, armature inductance, and back electromotive force coefficient, respectively. $\mathcal{N}$ is a positive constant related to the mass of the load and the coefficient of gravity, and $\mathcal{N} = 1$.

Define $x_1 = q$, $x_2 = \dot{q}$, and $x_3 = \vartheta$. Then the T–S fuzzy model can be given as follows:

$$x(k+1) = \sum_{i=1}^{2} \rho_i(\varepsilon(k))(A_i x(k) + B_i \tilde{u}(k)), \tag{6.160}$$

where

$$A_1 = \begin{bmatrix} 0 & 1 & 0 \\ -1/\mathcal{D} & -\mathcal{B}/\mathcal{D} & 1/\mathcal{D} \\ 0 & -\mathcal{L}/\mathcal{M} & -\mathcal{H}/\mathcal{M} \end{bmatrix},\ B_1 = \begin{bmatrix} 0 \\ 0 \\ 1/\mathcal{M} \end{bmatrix},$$
$$A_2 = \begin{bmatrix} 0 & 1 & 0 \\ 1/\mathcal{D} & -\mathcal{B}/\mathcal{D} & 1/\mathcal{D} \\ 0 & -\mathcal{L}/\mathcal{M} & -\mathcal{H}/\mathcal{M} \end{bmatrix},\ B_2 = \begin{bmatrix} 0 \\ 0 \\ 1/\mathcal{M} \end{bmatrix}.$$

Let $\mathcal{D} = 1\,\text{kg}\cdot\text{m}^2, \mathcal{B} = 1\,\text{N}\cdot\text{m}\cdot\text{s/rad}, \mathcal{H} = 1\Omega, \mathcal{M} = 0.1\,\text{H}, \mathcal{L} = 0.2\,\text{N}\cdot\text{m/A}$, and sampling period $T = 0.3$ s, assuming the membership function as $\rho_1(\varepsilon(k)) = \sin(x_1(k))/x_1(k)$, $\rho_2(\varepsilon(k)) = 1 - \rho_1(\varepsilon(k))$, then the T–S model can be given as follows.

$$A_1 = \begin{bmatrix} 0.9598 & 0.2512 & 0.0179 \\ -0.2512 & 0.6729 & 0.0727 \\ 0.0357 & -0.1454 & 0.0365 \end{bmatrix},\ B_1 = \begin{bmatrix} 0.0223 \\ 0.1785 \\ 0.9278 \end{bmatrix},$$

$$A_2 = \begin{bmatrix} 1.0408 & 0.2589 & 0.0182 \\ 0.2589 & 0.7454 & 0.0768 \\ -0.0364 & -0.1536 & 0.0361 \end{bmatrix},\ B_2 = \begin{bmatrix} 0.0225 \\ 0.1821 \\ 0.9725 \end{bmatrix}.$$

Selecting $W = diag\{0.5, 0.5, 0.5\}$ for the initial condition $x(0) = \begin{bmatrix}0.3 & 0 & 0\end{bmatrix}^T$, then we solve the LMIs (6.100), (6.101), and (6.110) in Theorem 6.9 with $Q_1 = diag\{0.5, 0.5, 0.5\}$ and $Q_2 = 1$, furthermore, related parameters of matrices ΔK and ΔR are selected as:

$$F = \begin{bmatrix}0.5 & 0 & 0\end{bmatrix},\ E = \begin{bmatrix} 1 & 0 & 0 \\ 0 & 1 & 0 \\ 0 & 0 & 1 \end{bmatrix},\ D = 0.1,\ N = 0.1,$$

$$\Delta_k(k) = 0.6\cos(0.05k),\ \Delta_r(k) = 0.6\cos(0.05k).$$

Then the corresponding upper bound of guaranteed cost index and the controller gain in (6.84) are given as

$$J^{\star} = 9236.1,\ K = \begin{bmatrix}-3.8083 & -2.5857 & -0.2728\end{bmatrix}. \tag{6.161}$$

In the following, we aim to construct a suitable feedback controller (6.84) where the input will be quantized by the dynamic quantizer (6.115) with the proposed adjusting rule for $\theta(k)$, so as to stabilize the quantized closed-loop system with an upper bound of guaranteed cost index as the one without quantization.

Follow Algorithm 6.1, we obtain the same controller gain and the same guaranteed cost performance $J^{\star}$. The auxiliary scalars are given as $r_1 = 6.4365$, $r_2 = 7.3265$, $r_3 = 6.7717$, and $r_4 = 6.6890$. Moreover $\xi = 0.0077$ can be calculated by choosing $h = 0.001$ when assuming $\Delta = 0.001$. Then by choosing $M = 200$, for any $\|u(k)\| \leq \theta(k)M$, we can update the dynamic quantizer's parameter $\theta(k)$ by the above adjusting rule to achieve the same performance as the one without quantization.

To demonstrate the proposed design strategy of this section more intuitively, the simulation results are shown in Figs. 6.9, 6.10, 6.11, 6.12, 6.13, and 6.14, where the dynamic parameter $\theta(k)$ in Fig. 6.12 is consistent with the adjusting rule. From Figs. 6.13 and 6.14, it can be concluded that the proposed method can ensure the same performance as the one without quantization for the quantized closed-loop system (6.160).

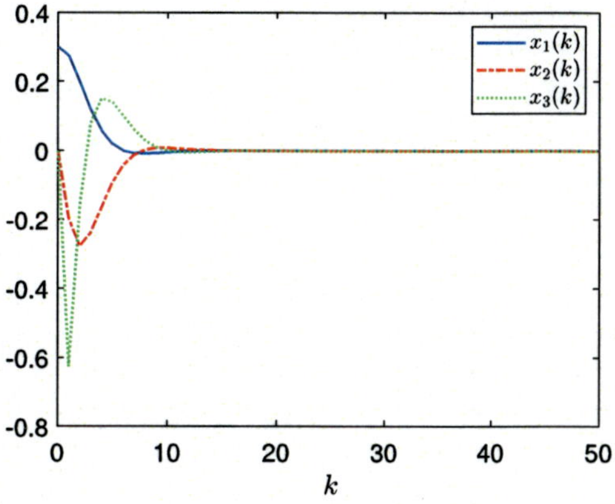

Fig. 6.9 Response of the system state $x(k)$

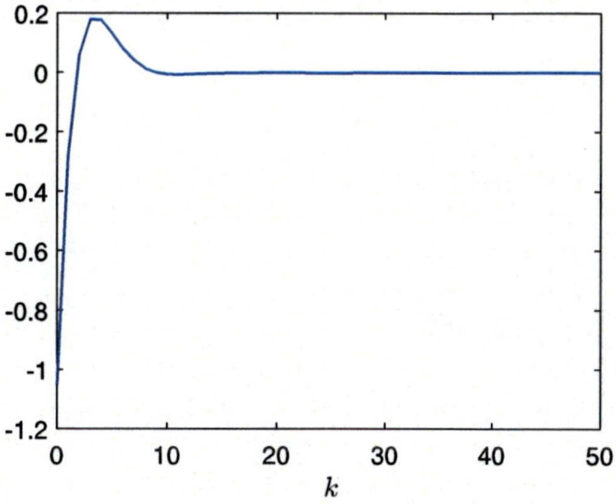

Fig. 6.10 Response of the control input $u(k)$

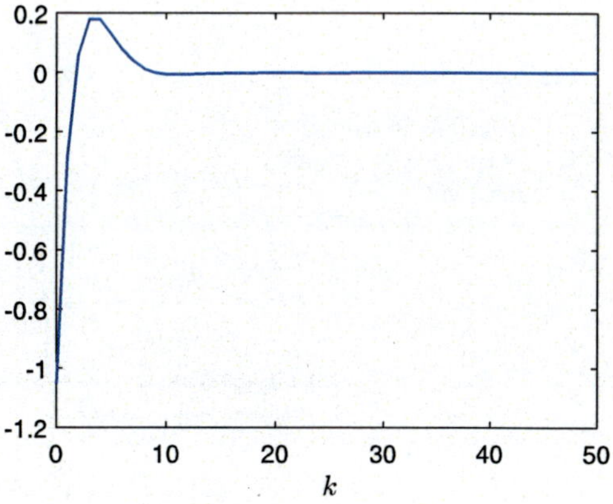

Fig. 6.11 Response of the quantized control input $u_q(k)$

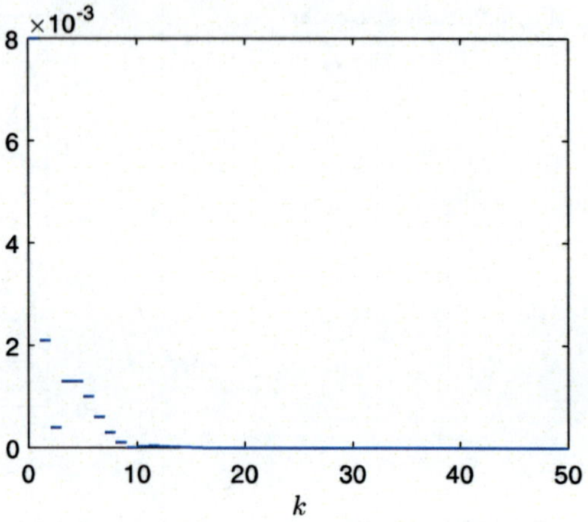

Fig. 6.12 Dynamic quantizer's parameter $\theta(k)$

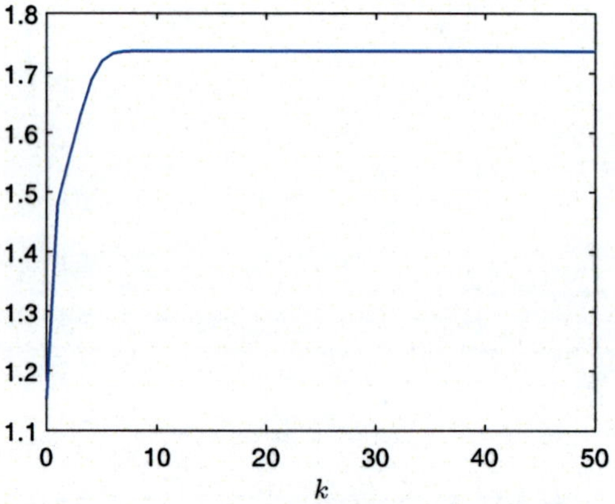

Fig. 6.13 History of $J = \sum_{i=0}^{k} \{x^T(i)Q_1x(i) + \tilde{u}^T(i)Q_2\tilde{u}(i)\}$ for system (6.160) with quantization

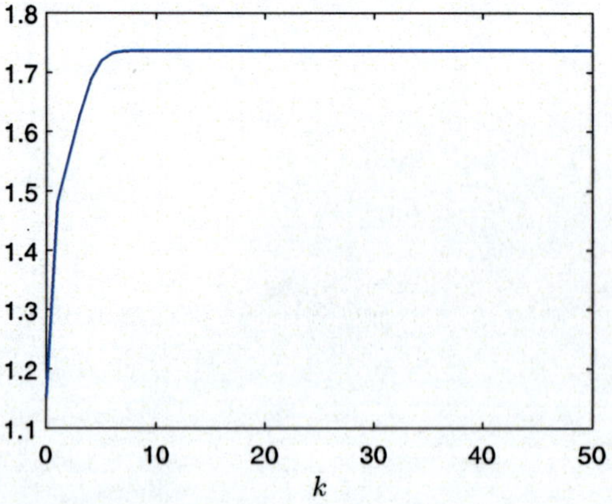

Fig. 6.14 History of $J = \sum_{i=0}^{k} \{x^T(i)Q_1x(i) + \tilde{u}^T(i)Q_2\tilde{u}(i)\}$ for system (6.160) without quantization

6.4 Conclusion

This chapter has investigated fault detection and fault-tolerant problems for T–S fuzzy system with the quantization effects. Firstly, the resilient quantized fault detection problem for T–S fuzzy system has been investigated. To obtain the available residual signal, the fault detection filter has been adopted as a residual generator. And consider the coupling term of the uncertainties from system and the fault detection filter, a two-step design method has been proposed to obtain the fault detection filter design scheme to guarantee the asymptotic stability and desired $\mathscr{H}_\infty$ performance of the residual system. Furthermore, consider the system with Markov jumps, the problem of quantized fault-tolerant control has been studied for T–S fuzzy system with actuator fault. Based on the introduced time-varying coefficient matrix related to the fault signal and the input signal of actuator, the resulting fuzzy Markov jump closed-loop system with actuator fault has been modeled as an uncertain system. And then, according to the robust and mode-dependent stochastic $\mathscr{H}_\infty$ stability analysis approaches, an $\mathscr{H}_\infty$ fault-tolerant controller has been designed such that the closed-loop system with actuator fault is stochastically stable and meets a prescribed $\mathscr{H}_\infty$ performance index. Moreover, the problem of the guaranteed cost fault-tolerant control problem has been studied for the T–S fuzzy system with quantization. The sufficient conditions on the design method of the required controller have been developed such that the closed-loop system is asymptotically stable and achieves a guaranteed cost performance even if there exists the actuator fault in the considered system, and an adjusting rule for the dynamic quantizer has been proposed to ensure the quantization error is within the appropriate range. Further, we have also proposed a computation method for the minimum upper bound of guaranteed cost performance of the closed-loop system suffering from actuator fault. Finally, some simulation examples have been proposed to verify the effectiveness of the algorithms.

References

1. Chang XH, Huang R, Wang H, Liu L (2020) Robust design strategy of quantized feedback control. IEEE Trans Circuits Syst II Exp Briefs 67:730–734
2. Zhao Y, Lam J, Gao H (2009) Fault detection for fuzzy systems with intermittent measurements, IEEE Trans Fuzzy Syst 17:398–410
3. Chen J, Patton RJ (2000) Standard $\mathscr{H}_\infty$ filtering formulation of robust fault detection, IFAC Proceedings Volumes 33:261–266
4. Song X, Men Y, Zhou J, Zhao J, Shen H (2017) Event-triggered $\mathscr{H}_\infty$ control for networked discrete-time Markov jump systems with repeated scalar nonlinearities. Appl Math Comput 298:123–132
5. Mustafa G, Chen T (2011) $\mathscr{H}_\infty$ filtering for nonuniformly sampled systems: A Markovian jump systems approach. Syst Control Lett 60:871–876
6. Chen H, Zheng Q (2021) Nonfragile peak-to-peak filtering for discrete-time Takagi–Sugeno fuzzy systems with dynamic quantizations. IEEE Trans Industr Inf 358:7863–7882
7. Ren J, Zhang Q (2012) Robust normalization and guaranteed cost control for a class of uncertain descriptor systems. Automatica 48:1693–1697

8. Yu L, Gao F (2002) Output feedback guaranteed cost control for uncertain discrete-time systems using linear matrix inequalities. J Optimiz Theory App 113:621–634
9. Niu Y, Ho DWC (2014) Control strategy with adaptive quantizer's parameters under digital communication channels. Automatica 50:2665–2671
10. Niu B, Ahn C K, Li H, Liu M (2017) Adaptive control for stochastic switched nonlower triangular nonlinear systems and its application to a one-link manipulator. IEEE Trans Syst, Man, Cybern: Syst 48:1701–1714